ACS SYMPOSIUM SERIES **664**

Phytoremediation of Soil and Water Contaminants

Ellen L. Kruger, EDITOR
Iowa State University

Todd A. Anderson, EDITOR
Clemson University

Joel R. Coats, EDITOR
Iowa State University

Developed from a symposium sponsored
by the Division of Agrochemicals
and the Division of Environmental Chemistry, Inc.,

American Chemical Society, Washington, DC

Library of Congress Cataloging-in-Publication Data

Phytoremediation of soil and water contaminants / Ellen L. Kruger, editor, Todd A. Anderson, editor, Joel R. Coats, editor.

    p.   cm.—(ACS symposium series, ISSN 0097–6156; 664)

    "Developed from a symposium sponsored by the Division of Agrochemicals and the Division of Environmental Chemistry, Inc., at the 212th National Meeting of the American Chemical Society, Orlando, Florida, August 25–29, 1996."

    Includes bibliographical references and indexes.

    ISBN 0–8412–3503–1

    1. Phytoremediation. 2. Soil remediation. 3. Water-Purification.

    I. Kruger, Ellen L., 1957– . II. Anderson, Todd A., 1963– . III. Coats, Joel R. IV. American Chemical Society. Division of Agrochemicals. V. American Chemical Society. Division of Environmental Chemistry. VI. American Chemical Society. Meeting (212th: 1996: Orlando, Fla.) VII. Series.

TD192.75.P48   1997
628.5—dc21   97–5740
                                                  CIP

This book is printed on acid-free, recycled paper.

PRINTED IN THE UNITED STATES OF AMERICA

Advisory Board

ACS Symposium Series

Foreword

THE ACS SYMPOSIUM SERIES was first published in 1974 to provide a mechanism for publishing symposia quickly in book form. The purpose of this series is to publish comprehensive books developed from symposia, which are usually "snapshots in time" of the current research being done on a topic, plus some review material on the topic. For this reason, it is necessary that the papers be published as quickly as possible.

Before a symposium-based book is put under contract, the proposed table of contents is reviewed for appropriateness to the topic and for comprehensiveness of the collection. Some papers are excluded at this point, and others are added to round out the scope of the volume. In addition, a draft of each paper is peer-reviewed prior to final acceptance or rejection. This anonymous review process is supervised by the organizer(s) of the symposium, who become the editor(s) of the book. The authors then revise their papers according to the recommendations of both the reviewers and the editors, prepare camera-ready copy, and submit the final papers to the editors, who check that all necessary revisions have been made.

As a rule, only original research papers and original review papers are included in the volumes. Verbatim reproductions of previously published papers are not accepted.

ACS BOOKS DEPARTMENT

Contents

PHYTOREMEDIATION OF INDUSTRIAL CHEMICALS

PHYTOREMEDIATION OF METALS

INDEXES

Preface

PHYTOREMEDIATION, or the use of plants to remediate contaminated soils and water environments, has recently become an area of intense study. Ten years ago there was a realization that the root zone of plants was special with respect to its capacity for biotransformation of organic molecules. The rhizosphere, as it was named, has since been studied for its important role in nutrient availability, as well as for the enhanced microbial degradation of pesticides, polynuclear aromatic hydrocarbons, and other synthetic chemicals. Plants impact contaminant reduction principally by providing an optimal environment for microbial proliferation in the root zone. This often leads to enhanced degradation of chemicals in soils that are vegetated, compared to nonvegetated soils. Contamination can also be reduced as a result of plant uptake into the tissue where it can be further degraded to innocuous substances, or removed from the site. In the latter case, plants can be used to extract contaminants from the environment, a process referred to as phytoextraction.

Within the last few years, the vegetation-based bioremediation approach has evolved into an important, viable option for cleanup of contaminated sites. It has now earned the new label "phytoremediation" and is recognized for some especially attractive features, such as its low input and environmentally friendly character. With the development of risk-assessment approaches to environmental contamination, it is apparent that in certain situations a slow, sure, and economical strategy is the most feasible. Phytoremediation demonstrates clear advantages over another newly named process called "natural attenuation".

Traditional methods of remediating contaminated soils and water, such as excavation or combustion, are environmentally disruptive and quite expensive. Phytoremediation may provide a cost-effective remediation technique for some contamination situations. In addition, it may make it possible to treat most contaminants in situ. By eliminating or minimizing the need to move contaminated soils, the risk of causing secondary contamination is also greatly reduced.

The symposium on which this book is based took place at the 212th National Meeting of the American Chemical Society in Orlando, Florida, August 25–29, 1996. This book provides an accumulation of some of the most recent research on phytoremediation. An overview is presented that describes the current understanding and use of plants for stabilization and cleanup of contaminants. An additional overview of rhizosphere ecology is also presented. The remainder of the book discusses three topics: phytoremediation of agrochemicals, industrial organic chemicals, and metals.

We believe that this book provides a good update of the field of phytoremediation since a previous rhizosphere–bioremediation symposium sponsored by the American Chemical Society in 1993. Our book that resulted from that assemblage of experts in the field was entitled *Bioremediation Through Rhizosphere Technology* (ACS Symposium Series No. 563), and it attempted to confirm and define the "rhizosphere effect" and explore ways to exploit it for improving environmental quality. The current book provides information related to remediation of metals and aquatic systems not included in the previous volume and focuses on direct application of the phytoremediation approach for specific contamination problems.

Acknowledgments

We thank the participants for their contributions to this volume. We are grateful to the ACS Division of Agrochemicals and the Division of Environmental Chemistry, Inc., for providing financial assistance and a forum for the symposium upon which this book is based. We also thank the peer reviewers of the chapters published in this volume for their expertise and efforts. We appreciate the efforts of Maureen Matkovich, Laura Manicone, and ACS Books in the culmination of this book. We thank Jennifer Anhalt, Pamela Tank, Brett Nelson, John Ramsey, Chris Schmidt, and Nancy Kite for their assistance in the preparation of this book. We also acknowledge Willis Wheeler for his help in coordinating the symposium and book.

We dedicate this book to our children Adam Kruger; Teagen Anderson; Sarah, Jesse, and Aaron Coats; and Beth and Annie Butin.

ELLEN L. KRUGER
JOEL R. COATS
Pesticide Toxicology Laboratory
Department of Entomology
Iowa State University
Ames, IA 50011–3140

TODD A. ANDERSON
The Institute of Wildlife and Environmental Toxicology
Department of Environmental Toxicology
Clemson University
Pendleton, SC 29670

January 9, 1997

OVERVIEW

Chapter 1

Phytoremediation of Contaminated Water and Soil

S. D. Cunningham[1], J. R. Shann[2], David E. Crowley[3], and Todd A. Anderson[4]

[1]DuPont Central Research and Development, Environmental Biotechnology,
GBC–301, P.O. Box 6101, Newark, DE 19714–6101
[2]Department of Biology, University of Cincinnati, Cincinnati, OH 45221
[3]Department of Soil and Environmental Sciences, University of California,
Riverside, CA 92521
[4]The Institute of Wildlife and Environmental Toxicology, Department
of Environmental Toxicology, Clemson University, Pendleton, SC 29670

Phytoremediation is the use of green plant-based systems to remediate contaminated soils, sediments, and water. Relative to many traditional remediation engineering techniques, phytoremediation is a fledgling technology intended to address a wide variety of surficial contaminants. Phytoremediation targets currently include contaminating metals, metalloids, petroleum hydrocarbons, pesticides, explosives, chlorinated solvents, and industrial by-products. The primary market driver for continued research in this area is the significant cost reduction these systems appear to afford. Phytoremediation, however has inherent limitations in that plants are living organisms with specific oxygen, water, nutrient and pH limits that must be maintained. In addition, significant depth, concentration, and time frame limitations also apply. Despite these limitations, many forms of phytoremediation have emerged from the laboratories and are currently in practice. Commercial phytoremediation systems for clean up of shallow aquifers and water born contaminants are now in place. Field tests for the phytoextraction of metals from contaminated soils are underway as well as advanced stabilization trials. For the most part, the current practices are technically sound, but far from optimized. Field tests have generally been met by good regulatory and public acceptance, yet improvements and extensions can and will be made on many of them. The biological resource for phytoremediation remains largely untapped. Bringing multi-disciplinary teams consisting of biologists, chemists, engineers, as well as lawyers, accountants, and public advocates should continue to yield additional solutions and possibilities for continued application of phytoremediation.

Significant quantities of air, water, and soil have been contaminated as a by-product of the industrial revolution and increased urbanization of the landscape. Increasingly stringent standards for water and air quality have propelled whole industries to re-engineer their fundamental processes and products. Through these types of change, contaminant loading to water and air has generally decreased. In addition to decreased input, the higher rates of mixing and dissipation of contaminants in air and water allow the environmental footprint of contaminants to rapidly fade. This is not the case in soil, a more static media where contaminants with low mobility (e.g. metals and lipophilic compounds) may cast longer shadows. Due to slow kinetics and reduced or absent dissipation processes, contaminated soils may show little inherent improvement over many decades. Unfortunately, contaminated soil is often the primary problem at many of the larger sites in need of remediation. Off site migration of the contaminants at such sites, if left unattended, can continue to further affect groundwater, neighboring areas, and bodies of surface water. Addressing these sites is proving litigious, time-consuming, and expensive.

The art of remediation begins with a site assessment, followed by containment of the identified problem, and theoretically ends with the clean-up of impacted areas. The term "remediation" often has more of a legal connotation than a technical one. Remediation can imply: a) "clean-up" where either the contaminant is removed from the matrix (leaching, bioremediation, etc.) or the entire contaminated matrix is removed from the site (excavation and landfilling at a second site), or b) "stabilization" where the physical or chemical form of the matrix or contaminant is transformed to a more inert condition. The intended remedy must be logistically and technically possible, accomplished within the required time frame, economically feasible, and in compliance with all legal requirements. The public and most site owners often prefer techniques that actually clean-up a site, as it provides the greatest degree of confidence and flexibility for future land use. At many sites, however, "clean-up" is not always possible or practical and "stabilization" techniques are then employed.

Traditional methods of remediating contaminated soils, sediments, and groundwater are often based on civil and chemical engineering technologies that have developed over the last 20 years. These include a wide variety of physical, thermal, and chemical treatments, as well as manipulations to accelerate or reduce mass transport in the contaminated matrix. In certain cases, however, biological (especially microbial) processes have shown some applicability. Recent flexibility in the legal requirements associated with environmental clean-up has increased the acceptability of such "passive" approaches to remediation. In spite of this, a majority of the plans developed for site remediation do not rely on "natural attenuation". The reasons for this are clear. Engineering technologies are often faster, relatively insensitive to heterogeneity in the contaminant matrix, and can function over a wide range of oxygen, pH, pressure, temperature, and osmotic potentials. Biological processes are at a significant disadvantage in most of these areas. The perceived advantage of bioremediation is the often prohibitive cost of effective engineering approaches. If remediation based on traditional technologies were inexpensive, there would appear to be no driving force for the development of alternative strategies based on biological activity.

In our experience, the total costs of remediation (calculated on a m^3 basis) ranges from \$10 to \$100 for remediation methods that can be performed *in-situ* and \$30 to \$300 for *ex-situ* processes. Specialized techniques such as *in-situ* vitrification can easily surpass \$1,000/ m^3. In comparison, the agronomic community can move and process large volumes of soil at a relatively low cost. Techniques for "land-farming" the top meter of soil are orders of magnitude less (e.g. as low as \$0.05/ m^3 per year). It is this cost differential, as well as the reduced capital requirements and lengthened expenditure schedule, that is the source of excitement over agriculturally-based remediation. Agricultural practices, developed for the management of soil and crops, are well understood and tested. As agriculture has as its primary goal the maintenance of soil quality, many of its techniques may have applicability for soil remediation. Most agricultural soil management strategies rely on the physical protection offered by a vegetative cover and on the biological activity of the plants and their associated microbial communities. These same components (stabilization and biological activity) are the basis of plant-based approaches to clean-up of inorganic and organic environmental contaminants.

Low-Input, Agronomic, Stabilization Techniques

The use of agronomic techniques to stabilize contaminated sites in-situ (1) is continuing to be advanced. Under this form of remediation, also called "phytostabilization", soil amendments are applied to contaminated soil to reduce the bioavailability of the contaminants. The site is then planted into vegetation, which reduces off site migration of the stabilized soil matrix and contaminant through water usage and erosion control. Plants are chosen to maximize root uptake of any small amounts of contaminant that would escape from the stabilizing mix and to sequester the material in the root tissue with little translocation to the shoot and potentially help catalyze the formation of insoluble contaminant species such as Pb-pyromorphite.

There are many examples that suggest this strategy is sound. It is known, for example, that Pb in soil and dust from natural sources is often less biologically available than similar concentrations of Pb from anthropomorphic sources. This is likely because many of the natural forms of Pb have low bioavailabilities. These natural forms do not dissolve well in either soil solution or mammalian gastro-intestinal tracts. Researchers involved with advanced stabilization techniques postulate that if anthropomorphic forms of Pb could be converted to more stable forms, it may be safe to leave these materials in place. Conversion of Pb to less toxic forms has been shown to take place in a natural environment with the application of low-cost amendments (2-7). More importantly, these techniques are less likely to harm the soil's potential for sustaining plant growth than most engineering solutions. Similar processes for sequestration of organic contaminants are also beginning to be examined.

Work is ongoing to further develop and validate in-situ stabilization as a viable technology for eliminating the hazard of contaminated soils. The five primary goals of this effort are to: 1) understand the mechanisms by which these technologies work, 2) develop appropriate testing protocols and methodologies that illustrate their utility, 3) improve predictive capabilities, 4) facilitate validation of the effectiveness and persistence of the technique, and 5) prepare guidelines for its implementation.

Phytoremediation

Phytoremediation is a word formed from the Greek prefix "phyto" meaning plant, and the Latin suffix "remedium" meaning to cure or restore. Although the term is a relatively recent invention, the practice is not. The use of plants to improve water quality in municipal and more recently industrial water treatment systems, is well documented (8-9). Vegetation has long been used for the restoration of disturbed areas (1), and tolerant vegetation is often found on or planted into contaminated soils (10). There has also been the opportunity to study the plant-contaminant interactions that have resulted from the application of sewage sludge to land *(11)* and from our 50 years of pesticide use *(12)*. Given our strong agriculturally-based experience with planted soils and the more recent issues of environmental contamination, it is natural to explore the use of plants to remediate contaminated soils, aquifers, and wetlands.

The legal and technical process of soil remediation is complex and continually in flux. The guidelines are clearer when the contaminant is in water. In this media, there exists excellent toxicological data that allows a legal determination of when remediation and pollution prevention techniques are necessary. In soils, however, equivalent concentrations of contaminants often exhibit widely varying toxicity effects. Contaminant speciation, soil pH, other ions in solution, types of clays and oxide surfaces, presence of organic matter, vegetation and rainfall all affect the availability of the contaminant and its potential to cause harm. As a general rule, remediation appears to be necessary when anthropological activities result in an "unacceptable risk". As risk assessment is an imprecise science, and as the tolerance for risk and availability of economic resources vary widely at a community level, it is difficult to determine the exact extent of soils that need remediation. Despite this imprecision, the need for remediation at some sites is apparent even with a cursory overview. Sites that are devoid of vegetation, in proximity to residences, and those sites likely to represent a source of continued off site migration of contaminants are likely candidates for some type of remediation. Other cases are less apparent. A technical basis for decision making is of critical importance to the research and development community involved with phytoremediation.

The market for any remediation technology is entirely dependent on compliance with the law. Plants grown for phytoremediation purposes are not like an agronomic crop where 50 bu is half as good as 100 bu. Phytoremediation systems that do not result in site "closure" (reduction to a legally acceptable level) are failures. In addition, a phytoremediation technique that can reduce contaminant levels twice as low as the legally acceptable limit may have no increased value over a technology that simply meets that level. The marketplace for remediation does not necessarily have value for a "cleaner" clean. Phytoremediation, like all remediation technologies under development is sensitive to the ongoing evolution of regulations. If the regulations for allowable contaminant concentrations in soil or water are set high, the market for a technology is affected in two ways. First, the volume of contaminated soil and water to be remediated decreases dramatically. (There is much more low level contamination than high level contamination). Second, the technology must work at a still higher level and reduce it down to the new "acceptable" level. (Technologies that reduce Pb contamination in soil from 1,000 to 500 ppm have significantly different chemical and physical constraints than those that mitigate 5,000 ppm to a 3,000 ppm guideline).

Phytoremediation of Inorganic Contamination

The elemental composition of normal soils is dependent on the geological and physical processes that occurred during its formation. Soils derived from marine sediments vary from those derived from rock outcroppings abundant in heavy metals. In addition to this inherent variability, anthropomorphic activities have increased soil heterogeneity. The most commonly cited sources of anthropogenic inorganic contamination are the mining and smelting of metalliferous ore, fossil fuel handling and use, industrial manufacturing, and the application of fertilizers and municipal sludges to land.

Decontamination of heavy metals in soils and sediments is among the most technically awkward clean-ups. Many inorganic contaminants bind tightly to the soil matrix. At older contaminated sites this is nearly always true, since more mobile contaminants have already migrated off site in areas of even moderate rainfall. Few remediation options exists for the contaminants that remain. Some advanced treatments include soil washing or electroosmosis, but for the most part many site clean-ups involve excavation and removal to a secure landfill area. Alternatively, some of these areas have also been simply covered in place in a process known as "capping". For still larger areas, particularly in the mining and smelting regions, site stabilization with some soil modification is done and revegetation occurs where possible.

Phytoextraction. Phytoextraction involves the use of plants to extract contaminants from the environment. The term was originally applied almost exclusively to heavy metals in soils but has since come to apply to many other materials in other media as well. The phytoextraction R & D community involved with inorganics has gradually evolved into two groups. The first group uses phytoextraction for remediation purposes (primarily targeting Pb and radionuclides with some efforts on Cr, As, and Hg). The second group targets inorganics with intrinsic economic value (primarily Ni, some Cu and a few with precious metals). This latter technical area (also known as "bio" or "phyto"- mining) is still in its infancy, but significant progress is being made with Ni *(13)*.

Water. The initial research on phytoextraction of inorganic contaminants in the environmental field began with the use of wetlands (constructed, natural, and reed beds) for water purification *(8)*. In these systems the inorganic contaminants often precipitated out of the water into the sediments making them difficult to recover. Floating plant systems followed *(14)* in which the contaminants could be removed in the harvested biomass. Unfortunately, these systems are not particularly efficient or economical, especially in temperate zones. More recently, greenhouse based hydroponic designs using terrestrial plants *(15)* have been field tested for heavy metals and radionuclides. These latter systems have been developed with plants selected for high contaminant root uptake/affinity and poor translocation to the shoots. These terrestrial plant (and more recently seedling) based applications were developed to take the place of synthetic resin chelates for water purification. Again the contaminants are removed from the system by harvesting root biomass. Rhizofiltration fits well with many of the biological limitations inherent in phytoextraction (e.g. poor translocation of many contaminants from root to shoot). These systems appear technically promising and require less R & D innovation than the soil remediation systems. Unfortunately, there currently exists much competition from other waste water purification technologies. Many excellent, non-plant based technologies exist to reduce water borne contaminant to meet regulatory guideline levels (e.g., alkali precipitation and cation exchange). For sites that are currently in compliance with local regulations the impetus for adopting new technology may be low. This is particularly true when the site has capital investments in water purification equipment already in place. Barring a change in water quality standards, an additional capital outlay for another technique (even if it were proven technically superior at a lower unit cost) is unlikely. New sites, additional polishing steps or large scale uses would appear to be the best target for "rhizofiltration"

Soils. In the soil environment, phytoextraction has fewer competing technologies for the clean-up of inorganic contaminants. The few that exist have significant technical and economic disadvantages. Stabilization technologies, however, may at times be in direct competition with this technology.

At the beginning of this decade, most researchers believed the success of phytoextraction of inorganic contaminants from soils centered on four requirements: 1) the bioavailability of the contaminant in the environmental matrix, 2) root uptake, 3) translocation internal to the plant, and 4) plant tolerance *(16)*. This set of minimum requirements has since changed, as will be addressed later in this review, although the original set of requirement remains a long-term goal of many active research programs.

The fundamental paradigm of phytoextraction is that most inorganic contaminants in soils are difficult to extract with engineering technologies such as thermal, chemical, and physical techniques. Soil is a complex matrix consisting primarily of Si, Al, O and Fe based materials that are difficult to separate from contaminating inorganics. The premise behind phytoextraction is that plants can be used to extract inorganic contaminants from the soil matrix and transfer them into a primarily carbon-based matrix, the plant material. Engineering techniques that were ineffective on the soil matrix are more readily employed on the plant material. Chemical, thermal, microbial and leaching processes separate many of these inorganics from the plant matrix. The most cited example is the smelter analogy where an energy source and an ore are used to produce a product. In soil, the contaminating materials are often at too low a level and the energy requirements in smelting soil are too large to be effective. When plants are used to transfer and concentrate soil-borne contaminants, however, the process becomes feasible. Plant material can have both significant stored energy potential and metal concentrations and can act simultaneously as both energy source and ore for a smelting process.

In certain cases, inorganic contaminants have volatile forms and remediation strategies that end with the volatilization of the contaminant have been proposed and field tested. Selenium phytoremediation as phytoextraction *(17)* and "phytovolatilization" *(18)* have both been explored. More recently, transgenic plants have been shown to reduce Hg from the more hazardous ionic and methylated forms to Hg(0), which is then volatilized *(19)*. In most remediation strategies, however, phytoextraction involves biomass removal and processing.

Exploiting Botanical Variation

All plants accumulate a wide variety of mineral elements but, for the most part, plants are quite adept at excluding those elements that are non-essential for their growth and survival. This general rule has notable exceptions. The first exception is that certain classes of plants take up large amounts of some non-essential, and relatively non-toxic elements. The clearest example of this is the element Si, which is taken up by many plants (particularly grasses) to levels that may exceed 1% of their dry weight. From this researchers have learned that if an element does not interfere with normal cellular metabolic processes and is appropriately sequestered, plants can tolerate relatively large loadings of inorganics. The second exception is that to a small degree all plants reflect the environment in which they grow. Many geologists have recognized this fact and have used plant-tissue analysis to locate buried ore bodies. The use of plants in this manner is referred "geobotanical prospecting" and is used to locate ore bodies containing elements such as uranium. The third exception to the text book rule is perhaps the most remarkable. There exists a small group of plants known as hyperaccumulators that can take up, translocate, and tolerate levels of certain heavy metals that would be toxic to any other known organism *(20)*.

Hyperaccumulators. Despite widely varying soil concentrations of most elements, with rare exceptions almost all plants exist within a narrow spectrum of relative concentration of elements *(21)*. Hyperaccumulating plants, on the other hand can take up, translocate and tolerate shoot concentrations of heavy metals in excess of 0.1% Ni, Co, Cu, Cr, Pb or 1% Zn on a dry weight basis *(22)*. These plants, which often evolved on metalliferous outcroppings, are remarkable not only for their high levels of accumulation and tolerance, but also for their nearly insatiable desire to concentrate these elements from even "normal" soil. Although taxonomically widespread, this hyperaccumulating trait is relatively rare, indicating a rather late appearance in the evolution of modern species *(22)*. Although the study of these plants still represents a focus of many labs, practical phytoremediation with these plants has remained elusive. Many of these species seem to hyperaccumulate only one metal while most sites have mixed metal contaminants. In addition, like any "weed" species little is known about their management. Many of these species are slow growing and, although they have high metal concentrations, produce low biomass. This is particularly problematic as the measure of phytoextraction success is often based on the amount of contaminant removed/hectare -y. High metal concentrations alone are insufficient. Slow-growing, low-biomass hyperaccumulating plants, no matter how fascinating biologically, are insufficient to form the backbone of a technology useful for remediation or biomining.

Hyperaccumulators, however, remain a significant motivational factor to the agronomic and molecular biology communities. These plants prove that biological systems can be developed with plants maintaining up to 4% metal in their tissues without significant yield decreases. Additional effort will be required to engineer, breed or adapt plants to obtain this goal, but some efforts are now under way. Progress is being made on a number of fronts and the area has recently been reviewed *(23)*.

Induced Hyperaccumulation. Not long after hyperaccumulators showed the inherent botanical potential for phytoextraction *(24)*, many additional laboratories began active research programs with these plants. It soon became apparent that the existing populations of hyperaccumulators were not ideal for commercial level phytoextraction. In 1991, screening projects began in a number of labs to identify higher biomass crop or weed species which might also accomplish the same feat. Various species were proposed by different groups ranging from Indian mustard *(25)*, through ragweed *(16)*. All of these plants showed improvement in biomass yet reduced metal uptake compared to hyperaccumulators. These experiments have continued with cultivar / ecotype screening in both Brassica juncea and Arabidopsis as they afford two ways to directly obtain or to engineer future high biomass hyperaccumulators. It is well known that there is much heterogeneity in many germplasm sources and screening these for metal uptake, translocation, and tolerance has proven no exception *(26)*.

Paralleling these efforts, however, the fundamental limiting factors of phytoextraction are being explored. Much of this work has been done with Pb, the most common heavy metal contaminant in the environment. In the case of this element, three limitations were uncovered. The first was that Pb soil solution levels were low even when total Pb concentrations were high. The second was that although most plant roots took up the Pb well, translocation from roots to shoots was poor. Lastly, tolerance to Pb in plant tissue at high rates severely reduced biomass *(27)*. In retrospect, this troubling lack of translocation from root to shoot is not surprising given what is known about the chemistry of Pb under cellular conditions of near neutral pH and in the presence of cytoplasmic concentrations of phosphate, proteins and carbonate anions.

The lack of availability of the contaminant and poor translocation and tolerance in plants were disappointing to many and led to the testing of potential techniques to circumvent these limitations.

Much is known about the soil chemical processes that influence the solubility and availability of many inorganic contaminants. Many researchers working in the phytoextraction of inorganics sought to alter such factors as soil pH, organic matter, phosphate level, etc. in an attempt to increase metal concentrations in plant tissue. To some degree this worked, and a doubling of Pb concentration in tissue was reported. This 200 to 1000 ppm level, however, was far short of the hyperaccumulator levels reported in some plants as well as the phytoextraction goal of 1%, calculated from engineering and economic parameters. Other remediation hybrid technologies were used, including electrokinetics, with only marginal increases in plant tissue concentrations of heavy metals. It was only when phytoextraction was combined with techniques learned from soil washing experiments that tissue concentrations began to approach target levels.

Chelating solutions have been used in soil washing experiments for some time. They increase the solution concentration of many heavy metals in soils and have also been used in agronomic and horticultural environments to deliver micronutrients to plants (most notably Fe-EDTA). Initial trials with chelates in many labs were too tentative. Many researchers were afraid of injuring the plants with high levels of chelates or chelate-Pb complexes. It was only when a "non-botano-centric" rethinking of the problem occurred that high enough doses of chelates were used to obtain the desired Pb level in plants *(27)*. In this case, chelates are applied to the soil and, as expected, Pb soil solution levels are increased. A remarkable, and serendipitous, benefit occurs, however, when the chelate Pb complex enters the plant and prevents Pb from precipitating in the root. The Pb-chelate complex continues up into the plant shoot through the xylem *(28-29)*. In this scenario, plants are grown in soil, chelates and other amendments are added, and Pb complexes are taken into the plant under the plant transpiration gradient. In this case plant tolerance is more or less irrelevant as the plant lives most of its life without much tissue Pb. The amendments are applied to the soil and within one to two weeks the plants are harvested. The plant sustains significant damage; however, the Pb-laden plant can be readily harvested regardless of its physiological state. Chelates thus act to eliminate limitations in Pb solubility in the soil, root to shoot translocation and tolerance.

The use of chelates is not, however, without a significant risk management and cost penalty. The technology is currently in its first field testing season with commercial customers slated for 1997. In conjunction with regulators and with local and state oversight, these techniques are being field tested by Phytotech, Inc. a New Jersey based firm. Chelates, albeit even ones used in foods like EDTA, raise safety concerns. Chelates increase the mobility of Pb in the soil. Downward movement must be monitored. Pb and EDTA have approximately the same molecular weights, so that if a 1-1 ratio exists, removing a ton of Pb may require a ton of chelate. Mechanisms to reduce this requirement are being explored along with chelates with significantly different properties. Given the value inherent in clean (vs. contaminated) soil, ton quantities of chelates and chelated-assisted phytoextraction may be a viable remediation technology, but most view this as only a first phase in its development. Chelate assisted phytoextraction may not be applicable for all areas, but with appropriate irrigation management techniques and site management, it may eventually become a viable remediation alternative. It is not, however, the best or probably even the ultimate answer for Pb phytoextraction.

Future of Phytoextraction

To date, no inorganic contaminated site has been remediated using phytoextraction. To these authors' knowledge perhaps a dozen field tests of phytoextraction as a remediation technique have been conducted with an additional four geared toward phyto-mining. The largest economic opportunity identified to date is in Pb phytoextraction. Unfortunately, the chemical and biological constraints are among the most difficult for this contaminant. Zinc or Ni would have been preferable targets from a biological and chemical perspective. It is clear that at least for Pb, where most work has been done at a field/practical level, some mechanism of increasing Pb solubility around the root zone will be required. Plant or plant derived chelating complexes and pH shifts have been proposed but these hypotheses remain untested. Internal chelating mechanisms will also be needed to prevent Pb sequestration in the root tissue and to allow it to move from root to shoot tissue. Lastly, the ultimate Pb-phytoextraction technique, unlike the current chelate-assisted one, will probably also require Pb tolerant plants.

This review represents only a brief snapshot in time of a field arguably only a decade old. Significant breakthroughs in many laboratories are continuing with reports, papers, and patent activity dramatically increasing for the phytoextraction of a wide range of elements and contaminated environments. To date, relatively few plants, people, and years have been spent on the development of phytoextraction as either a mining or remediation technique. Given that no plant in the history of the world was ever selected for maximum metal yield, the biological potential remains largely unplumbed. Futurists remark on a time when we may need to recycle most of our inorganics from our waste streams. Biomining and plant phytoextraction may provide one mechanism to do so.

Phytoremediation of Organic Contaminants

Unlike inorganic pollutants which are immutable at an elemental level, most organic pollutants can be altered by biological systems. This increases the inherent economic advantage of such techniques by potentially eliminating the needs for harvesting plant material.

Direct Plant Effects- Uptake of Organics. Plant uptake of organic compounds is an important component which must be considered in the evaluation of phytoremediation. The majority of the data on plant uptake of nonnutritive substances comes from studies of agricultural chemicals, primarily herbicides. The effectiveness of many herbicides depends on their ability to enter the target plant. Since these compounds were commercially designed with this goal in mind, the principles regarding their uptake provide a strong basis for understanding that of other chemicals *(30-31)*.

Root uptake of organic compounds from soil is affected by three factors: (1) physicochemical properties of the compound, (2) environmental conditions, and (3) plant characteristics *(32-33)*. Plant characteristics such as root surface area can substantially alter absorption. Surface area may be increased in plants with large root morphologies, or in those with a high number of fine root hairs. As water mediates the transfer of solutes to the root, a plant characteristic which affects evapotranspiration could also influence the potential for uptake of organic contaminants. A systematic approach to selecting plant species and varieties for maximizing traits such as these has rarely been attempted, but would be warranted in the development of phytoremediation technology. The bioavailability of organic contaminants for plant uptake is primarily under the control of environmental soil factors such as organic matter content, pH, and moisture. Even in this aspect, however, species and families of plants with traits that allow them to modify the environment (such as pH) surrounding their roots, could be investigated.

Assuming constant plant and environmental characteristics, root uptake has been shown to be directly proportional to the n-octanol/water partition coefficient (Kow) for the chemical. More lipophilic compounds can better partition into roots and this structure-activity relationship has been used to develop empirical models of the uptake of different classes of organic compounds (primarily pesticides) by plants. Unlike root absorption, translocation to aboveground plant tissues via the transpiration stream (measured as the transpiration concentration factor or TSCF) is most efficient for compounds with intermediate polarity (e.g. log Kow = 1.8) *(34)*. Briggs and co-workers found that this relationship (maximum TSCF = log Kow of 1.8) held true for literature data despite a variety of plant species, compounds, and experimental techniques. Plant uptake of organic contaminants from soil has been primarily investigated using hybrid poplar trees by a number of groups *(35-36)*. These authors now report extending the model proposed by Briggs et al. *(34)* for organics ranging from pesticides to volatile organic compounds (VOCs) *(37)*. Results of these studies indicate that VOCs with log Kow from 1.0 to 3.0 are taken up and translocated by rooted poplar cuttings in hydroponic solutions. It has been noted, however, that the modeling parameters developed for plants in hydroponic systems can be compromised in soil where fairly strong sorption to soil occurs. In soil system, it seems that the log Kow for TSCF maximum may be shifted down by two units *(38)*.

Many of the field tests currently underway with the phytoremediation of organics involve plant uptake of the contaminant from a water phase. These water borne contaminants include surface applied water, wetland areas, ponds and impoundments as well as shallow aquifers *(35)*.

Direct Plant Effects- Fate of Contaminants in the Plant.   Much is known about the general fate of xenobiotics in plants. Unfortunately most of the literature base has evolved around increased understanding of the fate of pesticides with specific information on many of the priority pollutants lacking. In general, however, plants have a wide array of metabolic capacities which can effect the fate of a chemical once it enters the plant. General plant metabolism of many xenobiotics often appears remarkably similar to metabolic detoxification processes that occur in mammalian livers *(39)*. In addition to herbicides, much metabolic work in plants has also been done with PCB's and more recently TCE. In many cases much of the contaminant entering plants get incorporated into cell biomass that is chemically difficult to extract and characterize. The rate and extent of these metabolic processes, relative to total contaminant uptake rate by the plant remain poorly characterized. It now appears, that transpiration rate, xylem mobility of the contaminant, and volatility may play a large fate in determining contaminant fate/partitioning. Evidence appears to be gathering that our understanding of the fate of contaminants may be greatly influenced by how these experiments are carried out. Sealed jars for mass balance studies may accentuate metabolism, studies under high transpiration streams found in a laboratory hood may accelerate volatilization of the contaminant through the plant transpiration stream. It is for this reason that carefully controlled field studies are needed.

Preliminary results from one such study are reported in this volume *(36)*. In this report, laboratory and field tests on phytoremediation of TCE-contaminated groundwater are reported. Initial studies in the laboratory indicated that a hybrid poplar clone, H11-11, was able to absorb trichloroethylene (TCE) from groundwater and that cell cultures of H11-11 could metabolize ^{14}C-TCE as well as incorporate ^{14}C into the cells.

There is also increasing evidence that plants can directly affect xenobiotic concentrations outside the plant itself. Root enzymes, such as those involved with oxidative coupling. have long been suspected of influencing concentrations of xenobiotics both internal and external to the plant upon release during normal or accelerated plant turn over *(40)*. This process has been extended into sediment remediation by plant produced enzymes as well *(35)*. The nature and extent of this process is under active investigation by a number of groups.

Indirect Plant Effects: Plant-Microbe Interactions. The previous sections described ways in which the plant may act directly on organic contaminants through plant based uptake and/or transformation. Although the plant may often metabolize or sequester environmental toxins, plants are at a significant disadvantage in two ways. Plants are primarily autotrophic and derive their living primarily by construction of cell materials from CO_2, light, water, and minerals. As such, unlike microbial systems, they have not had to evolve with the necessity to degrade chemically intransigent materials. The result of this is that plants metabolize a more restrictive set of chemical structures than do their microbial counterparts. Secondly, plants often detoxify xenobiotics by chemically altering them (e.g., by hydroxylation and glycosilation) to more water soluble forms in the plant tissue. In many cases the parent material, or chemically altered form, is then sequestered in the cell wall matrix or cytoplasm. In contrast, microbial metabolism often ends with the compound being reduced to CO2, water, and cellular biomass. In many ways, combining the plant structural functions (water transpiration, root surfaces and soil penetration) with microbial degradative processes is technically, economically and regulatorily attractive.

In soils contaminated with lipophilic organic compounds, the plant may play a less direct role - perhaps only that of a support system for degradative microorganisms and microbial communities *(41)*. Since the late 1970s, numerous studies have shown that plants (or planting) may enhance the degradation of selected compounds, including organo-phosphates *(42)*, parathion *(43)* polyaromatic hydrocarbons *(44)*, and chlorinated organics *(45-46)*. To date, much of the research has been descriptive. Observed increases in degradation have been generally attributed to microbial activity in the rhizosphere, however, a mechanistic understanding of the process, is lacking. Successful application of plant-microbial systems for bioremediation of a wide range of contaminants will require that we understand how microorganisms that bring about chemical transformations are influenced by plant roots. The plant-contaminant-soil interactions are already exceedingly complex. Adding to this a microbial community component which varies in time and space along the length of a root and in the bulk soil adds another technical dimension (and required skill base) and dramatically increases complexity of the system.

Plants might influence microorganisms that degrade organic contaminants by providing substrates for microbial growth or cometabolism, by allowing the assemblage of unique communities (analogous to biofilms) on root surfaces, and by alteration of soil chemical and physical conditions such as redox, pH and inorganic nutrient availability. These plant-microbial interactions are not necessarily distinct from the more direct activities of the plants themselves. For example, contaminants which are taken up by the roots may be transformed by the plant and subsequently redeposited into the rhizosphere in an altered chemical form.

One of the first steps in dissecting how plants might accelerate biodegradation of organic soil contaminants is to consider the factors that influence the survival and activity of bacteria along a root. During growth, plants release significant quantities of carbon into the rhizosphere as amino acids, organic acids, sugars, and numerous other structurally diverse materials, which can comprise from 15 to 40% of the carbon assimilated through photosynthesis. Most of these materials are released in the zone of elongation, just behind the root tip, and support a diverse microbial community. In older root zones, carbon becomes limiting and the rhizosphere community includes nematode and protozoa predators that graze on bacteria associated with the roots. This bacterial turnover leads to selection for microorganisms that are adapted to coexistence in a crowded, oligotrophic environment. Subsets of the oligotrophic community undergo further succession as new carbon substrates become available when soil animals feed on the roots, or at sites where lateral roots emerge and rupture through the cortex tissue. Mycorrhizae formation with symbiotic root-colonizing fungi in the older root zones further changes the quantity and composition of root exudates to the rhizosphere microbiota. Eventually, as plant roots die and decompose, all of the compounds produced by plants and their associated microbiota, are released into the soil where they are degraded by yet another microbial community that includes specialized degrader organisms capable of growth on cellulose, chitin, and lignin.

The above concerns ways in which plant roots and their depositions determine the basic structure and makeup of the rhizosphere microbial community. The plant may also foster the development of degradative activity in the community by bringing together dense populations of diverse microorganisms that contain catabolic plasmids or genetic material. Following recombination, this proximity may lead to new pathways for degradation. Since bacterial mating and genetic recombination is dependent on the active growth of microorganisms, plants may accelerate the evolution of new catabolic pathways for degradation of xenobiotics by providing a mating surface. In addition to plasmids carrying catabolic genes, other plasmids may confer the genes for chemoattractant sensors, or antibiotic or heavy metal resistance that are beneficial to growth and survival of introduced degrader organisms. Much of the latter remains speculative, but it is well known that genes for catabolism of a variety of substrates are plasmid borne, and recent work has shown that plasmid stability for degradation of a normally recalcitrant substrate, 2,5 dichlorobenzoate, is increased in the rhizosphere *(47)*. In this manner, bioaugmentation of indigenous microbial communities using bacterial vectors that contain specific plasmids may be more easily accomplished in planted soils.

Although the effect of plants on the physicochemical environment is a large and separate topic, one particular note should be made in regard to biodegradation of highly chlorinated compounds. Plant roots and their associated microflora alter soil redox through respiration. This could increase rates of reductive dehalogenation as electrons are generated during metabolism of root exudates. Plants can deplete soil oxygen while also removing the nitrate which may serve as an alternate microbial electron acceptor in lieu of oxygen. Depending on water management of the soil, or the presence of plants with aerenchyma to transport oxygen into the rhizosphere, it may be possible to generate conditions favorable to both aerobic and anaerobic microorganisms that could transform chlorinated compounds through sequential anaerobic and aerobic metabolism. The microbial consortium would be analogous to a biofilm through which the contaminant is delivered during bulk flow of water. This idea is supported by reports of dechlorinating microorganisms being found in the rhizosphere. Many primary root colonizers (Pseudomonas, Arthrobacter, Achromobacter) are known to degrade various chlorinated hydrocarbons (see review- 48). In addition to bacteria, a rhizosphere fungal species, Aspergillus niger, degrades PCBs (49) and 2,4-D, as well as carboxin fungicides (50). Still other rhizosphere fungi have been investigated for their ability to degrade chloroaniline-based pesticides (51).

Given the dynamic nature of the rhizosphere, it is an enormous challenge to figure out the influence of the plant on the microbial community. Methods for examining microbial activity in the rhizosphere may involve the use of reporter genes in which the promoter for selected degradative genes is coupled to a reporter system such as bioluminescence (47), or may employ PCR to amplify genes lifted directly from soil or root sites (52). Other more general approaches have involved the plating of microorganisms associated with plant roots to determine their degradative abilities, or by simply sampling rhizosphere soil which contains microorganisms that have been previously enriched in soil associated with plant roots. Regardless, the growing number of techniques available are now allowing the exploration of the rhizosphere and providing the information needed to assess the potential for phytoremediation of organic contamination. This research area is receiving much current attention (53).

Future

This overview has only touched on our current understanding and use of plants for stabilization or clean-up of inorganic and organic contamination. Increasingly phytoremediation literature and reviews are appearing in many venues (54). All indications to date suggest that multiple challenges and opportunities remain in the development and application of this as a viable technology. Perhaps the greatest hurdle is the elucidation of the mechanisms involved in all forms of phytoextraction and phytoremediation. Without a better understanding of many of these processes, it is difficult to exploit the selection and engineering of plant and microbial populations for process optimization. The research presented in the remainder of this volume is aimed at closing the gap between our current position and where we need to be; often by identifying key areas which require further investigation to advance this new technology.

While the above suggests that this is still largely in the research and development phase, many scientific, economic, and societal factors support its development. The pace of research in this area is quickening, as evidenced by this volume. In a field that was of little interest 10 years ago, research is now on-going in dozens of labs. Because of the potential for significant economic return on these endeavors, groups conducting and funding research extend beyond the traditional academic and environmental sectors. Small entrepreneurial companies and spin-off technologies have recently emerged with names like: Phytotech, Phytokinetics, PhytoWorks, Ecolotree and Treemediation.

Even as scientific progress is made, the societal view of environmental contamination and remediation is changing. The number of sites and volume of materials being listed in the U.S. alone under CERCLA (ie. superfund) and RCRA is staggering and unmanageable given the technologies and resources now available. While as a society we identify cleanup of environmental contamination as a national priority, there is a growing awareness of the economic realities associated with addressing this problem. This awareness is reflected in changes in legal standards and methods and measurements used to assess and prioritize remediation. In the past, cleanup was based on the total loading of a contaminant - regardless of its biological and chemical availability or the intended future use of the site. At all levels the regulatory community is now leaning towards a more risk-based decision making process, where corrective action at a given site would be based on qualitative and quantitative assessment of the hazardous compounds available for interaction with living systems as well as future land use. This "bioavailable" pool may be only a small percentage of the total amount of the compounds present. One of the pressing needs in environmental risk assessment, therefore, is the establishment of methods and approaches to evaluate the status of contaminants at a site. With these methods, environmentally acceptable (treatment) endpoints could be established as targets for traditional and alternate remediation technologies. With treatment guidelines based on biologically available quantities of contaminants, and a treatment technology which, almost by definition, remediates the biologically available fraction phytoremediation would have increased and renewed interest. These changes in technology, societal views on remediation, as well as increasing flexibility in the legal system would appear to bode well for the development of phytoremediation as a viable remediation technology of the future.

References

1. Bradshaw, A.D.; Chadwick, M.J. The Restoration of Land: The Geology and Reclamation of Derelict and Degraded Land. Univ. of CA Press. Berkeley. CA. **1980.**
2. Rabinowitz, M.B. *Bull. Environ. Contami. Toxicol.* 1993, 51, 438-444.
3. Mench, M.J.; Didier, V.L.; Loffler, M.; Gomez, A; Masson, P. *J. Environ. Qual.* **1994**, 23, 58-63.
4. Ruby, M.V.; Davis, A; Nicholson, A. *Environ. Sci. Technol.* **1994**, 28, 646-654.
5. Ma, Q.Y.; Logan, T.J.; Traina, S.J. *Environ. Sci. Technol.* **1995**, 29, 1118-1126.
6. Berti, W.R.; Cunningham; S.D. *Environ. Sci. Technol.* **1997**, (in press)
7. Vangronsveld, J.;.Van Assche, F.; Clijsters, H. *Environ. Pollu.* **1995**, 87, 51-59.
8. Kadlee, R.H.; Knight, R.L. Treatment Wetlands. CRC Lewis Publishers, NY, NY. **1996.**
9. Moshiri, G.A. Constructed Wetlands for Water Quality Improvement. Lewis Publishers. NY, NY. **1993.**
10. Baker, A.J.M., Proctor, J.; Reeves, R.D. The vegetation of Ultramafic (Serpentine) Soil. Proceedings of the first international conference on serpentine ecology. Intercept. Andover Hampshire. UK. **1992.**
11. Clap, C.E.; Larson, W.E.; Dowdy, R.H.; Sewage Sludge Land Utilization and the Environment. SSSA Miscellaneous publication. Soil Science Society of America. Madison WI. **1994.**
12. Mansour, M. Fate and Prediction of Environmental Chemicals in Soil, Plants and Aquatic Systems. Lewis Publishers. Boca Raton, FL. **1993.**
13. Nicks, L.J.; Chambers, M.F. Mining environmental management. Sept 1995. Edenbridge, Kent UK. **1995.** pp 15-18.
14. Dierberg, F.E.; DeBusk, T.A.; Goulet, N.A. Jr. In Aquatic Plants for Water Treatment and Resource Recovery. Reddy, K.B.; W.H. Smith eds. Magnolia Publishing Inc. FL.**1987.** p 497-507.

15. Salt, D.E., Blaylock, M.; Kumar, N.P.B.A.; Dushenkov, V.; Ensley, B.D.; Chet, I.; Raskin, I. *Biotechnol.* **1995**, 13, 468-474.

16. Cunningham, S.D.; Berti, W.R. *In Vitro Cell. Dev. Biol.* **1993**, 29P, 207-212.

17. Banuelos, G.S.; Cardon, G.E; Phene, C.J.; Wu, L.; Akohoue, S.; Zambrzuski, S. *Plant and Soil.* **1993**,148, 253-263.

18. Terry, N.; Zayed, A.M. In Selenium in the Environment. Frakenberger, W.T.; Benson, S. (eds). Marcel Dekker, Inc. NY, NY. **1993**. pp 343-367.

19. Rugh, C.L.; Wilde, H.D.; Stack, N.M.; Thompson, D.M.; Summers, A.P.; Meagher. R.B. *PNAS* **1996**, 93, 3182-3187.

20. Baker, A.J.M.; Brooks, R.R. *Biorecovery.* **1989**,1, 81-126.

21. Adriano, D.C. Trace Elements in the Terrestrial Environment. Springer-Verlag, NY, NY. **1987**.

22. Baker, A.J.M.; Walker, P.L. In Heavy Metal Tolerance in Plants: Evolutionary Aspects. Shaw, A.J. ed. CRC Press. Boca Raton. FL. **1990**. pp 155-177.

23. Cunningham, S.D.; Ow, D. W. *Plant Physiol.* **1996**, 110, 715-719.

24. Baker, A.J.M.; Reeves, R.D.; McGrath, S.P. In In-Situ Bioreclamation. Applications and Investigations for Hydrocarbon and Contaminated Site Remediation; Hinchee, R.E.; Olfenbuttel, R.F. (eds). Butterworth-Heinemann. Boston, MA. **1991**. p 600-605.

25. Kumar, NPBA, Dushenkov, V.; Motto, H; Raskin, I. *Environ. Sci. Technol.* **1995**, 29, 1232-1238.

26. Chen, J.; Cunningham, S.D. **1997,** (this volume).

27. Huang, J.W.; Cunningham, S.D. *New Phytol.* **1996**, 134, 75-84.

28. Huang, J.W.; Chen, J.; Berti, W.R.; Cunningham, S.D. *Environ. Sci. Technol.* **1997,** (in press).

29. Blaylock, M.J; Salt, D.E.; Dushenkov, S.; Zakharova, O; Gussman, C; Kapulnik, Y.; Ensley, B.D; Raskin, I. *Environ. Sci. Technol.* **1997,** (in press).

30. Shone, M. G. T.; Wood, A. V. *J. Exp. Botany* **1974**, 25, 390-400.

31. Shone, M. G. T.; Barlett, B. B.; Wood, A.V. *J. Exp. Botany.* **1974**, 25, 401-409.

32. Ryan, J. A.; Bell, R. M.; Davidson, J. M.; O'Connor, G. A. *Chemosphere.* **1988**, 17, 2299-2323.

33. Paterson, S.; Mackay, D; Tam, D.; Shiu, W. Y. *Chemosphere.* **1990**, 21, 297-331.

34. Briggs, G. G.; Bromilow, R. H.; Evans, A. A. *Pest. Sci.* **1982**,13, 495-504.

35. Schnoor, J.L., Licht, L.A.; McCutcheon, S.M.; Wolfe, N.L.; Carreira, L.H. *Environ. Sci. Technol.* 1995, 29(7), 318-323.

36. Gordon et al. **1997,** (this volume).

37. Schnoor et al. **1997,** (this volume).

38. Hsu, F. C., Marxmiller, R. L.; Yang, A. Y. S. *Plant Physiol.* **1990**, 93, 1573-1578.

39. Sandermann, H. Jr. *Trends Biochem. Sci.* **1992**, 17, 82-84.

40. Bollag, J-M, Meyers, C.; Pal, S.; Huang, P.M. In Environmental Impacts of Soil Component Interactions. Huang, P.M., Bollag, J-M.; McGill; W.B. Page, AL. ed. LewisPublishers, Chelsea MI. **1995**. pg. 297-308.

41. Shimp, J.F. , Tracy, J.C.; Davis, L.C.; Lee, E.; Huang, W.; Erickson, L.E.; Schnoor, J.C. *Crit. Rev. Environ. Sci. Technol.* **1993**, 23, 41-77.

42. Hsu, S.; Bartha, R. *Appl. Environ. Microbiol.* **1979**, 37, 36-41.
43. Reddy, B. R.; Sethunathan, N. *Appl. Environ. Microbiol.* **1983**, 45, 826-829.
44. Aprill, W. and Sims, R.C. *Chemosphere* **1990**, 20, 253-265.
45. Walton, B.T., Anderson,T.A. *Appl. Environ. Microbiol.* **1990**, 56, 1012-1016.
46. Boyle, J.J.: Shann; J.R. *J. Environ. Qual.* **1995**, 24(4), 782-785.
47. Crowley, D.E; Brennerova, M.V.; Irwin, C.I.; Brenner, V.; Focht; D.D. *FEMS Microbiol. Ecol.* **1996**, 20, 79-89.
48. Chaudry, G.R.; Chapalamadugu, S. *Microbiol. Rev.* 1991, 55(1), 59-79.
49. Dmochewitz, S.; Ballschmiter, J. M. *Chemosphere* 1988, 17, 111-121.
50. Agnihotri V. P. *Indian Phytopath* **1986,** 39, 418-422.
51. Kaufman, D.D.; Blake, J. *Soil Biol. Biochem.* **1973**, 5, 297-308.
52. Koh, S.C.; Marschner, P.; Crowley, D.E.; Focht, D.D. *FEMS Micro Ecol.* **1997**, (in press).
53. Cunningham, S. D.; Anderson, T. A. ; Schwab, A. P., Hsu, F. C. *Adv. Agron.* **1996**, 56, 55-114.
54. Crowley, D.E., Alvey, S.: Gilbert, E.S. **1997**, (this volume).

RHIZOSPHERE ECOLOGY

Chapter 2

Rhizosphere Ecology of Xenobiotic-Degrading Microorganisms

David E. Crowley, Sam Alvey, and Eric S. Gilbert

Department of Soil and Environmental Sciences, University of California, Riverside, CA 92521

The rhizosphere fortuitously enhances the population numbers and activity of certain microorganisms that degrade xenobiotic soil contaminants. This review examines the ecology of degrader microorganisms in the rhizosphere, and summarizes prior research that has examined the influence of plants on biodegradation of chlorobenzoates, chlordane, polychlorinated biphenyls (PCBs), and the herbicide atrazine. Degradation rates of most xenobiotics examined to date are not significantly influenced by the presence of a rhizosphere. However, a major benefit of the rhizosphere may be to harbor certain degrader organisms at higher cell numbers, thereby shortening the acclimation period. Another benefit may be enhanced transfer of degradative plasmids, or in some instances, enhanced cometabolism of compounds which can not be directly utilized as substrates for microbial growth.

The use of plant-microbial systems for bioremediation is a rapidly developing technology that may be particularly beneficial for in situ treatment of contaminated soils (1, 2). Since the late 1970s, sporadic reports have appeared in the literature demonstrating that plants may enhance the degradation of several different compounds, including organophosphates (3), parathion (4) polyaromatic hydrocarbons (5), pentachlorophenol (6) and organic solvents such as trichloroethylene (7). However, to date there is very little knowledge of the specific mechanisms by which plants influence degrader organisms or the ecology of microbial communities that promote xenobiotic degradation. Successful application of plant-microbial systems against a wide range of contaminants that occur in the field will require that we have a better understanding of how soil microorganisms are influenced by the presence of plant roots, and the benefits and limitations of using plants to increase the activity of introduced or indigenous xenobiotic degraders.

The most commonly proposed explanation for the influence of plants on xenobiotic biodegradation is the general increase in microbial cell numbers and microbial activity that occur in the rhizosphere as a result of growth on carbon substrates provided by rhizodeposition. However, there are also a number of other effects plants might have on the soil microbial community as shown in Figure 1. In the rhizosphere, diverse species of heterotrophic microorganisms are brought together at high population densities, which may enhance stepwise transformation of xenobiotics by consortia, or provide an environment that is conducive to genetic exchange and gene rearrangements.

In some instances, structural analogs of various xenobiotics contained in root exudates, cell wall components and lysates, as well as secondary products of decomposition of these materials may fortuitously select for microbes that metabolize or cometabolize xenobiotics. Evapotranspiration of water by plants also may influence the transport of water soluble compounds by increasing their mass flow to the root surface where they can be acted upon by the rhizosphere microflora. In other instances, it is possible the rhizosphere may have little or no effect on degradation of xenobiotics; for example, if a degrader population is absent from the soil or is noncompetitive in the rhizosphere, or if the degrader organisms can grow independently on the substrate such that the presence of the rhizosphere is superfluous. Prediction of the rhizosphere effect for a specific compound requires knowledge of its chemical characteristics, as well as its likely degradation pathway, and an understanding of the influence of plant root growth on the microorganisms that are involved in the degradation process.

Ecology of rhizosphere degrader organisms

Although details of plant effects on xenobiotic degradation are still generally lacking for most compounds, several reports have given an indication of the various processes that may occur in the presence of roots and root exudates. In one of the earliest studies on the "rhizosphere effect", Hsu and Bartha (*3*) reported that the presence of plants or irrigation of soil with root exudates enhanced the rate of parathion mineralization relative to nonplanted soil. Because viable counts of microorganisms on soil extract agar plates were not significantly changed in planted or root-exudate-treated soils, population selection and possibly enhanced cometabolic activity were concluded to be the most likely mechanisms for the increased degradation. However, population estimates of the number of parathion degraders and cometabolizers were not conducted.

In another early study, Sandmann and Loos (*8*) found very high R/S (rhizosphere/bulk soil) ratios of 2,4-D-degrading organisms in sugarcane rhizosphere soil, and much lower R/S ratios for African clover rhizosphere soil. The higher R/S ratios of 2,4-D-degrading organisms in some soils were too large to be explained by the general microbial population increase in the rhizosphere. Thus, the 2,4-D degraders appeared to have been selectively enriched in the sugarcane rhizosphere soils. Kunc (*9*) subsequently studied the effect of continuously-supplied, synthetic root exudate on changes in the number and proportion of organisms capable of

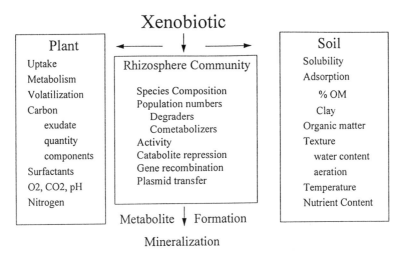

Figure 1. Conceptual model of plant and soil influences on bioavailability and biodegradation of xenobiotic soil contaminants by microbial communities in the rhizosphere.

degrading 2,4-D. After 2 wk, the proportion of 2,4-D degraders in the control soil had increased from 9.4% to 77.0 % of the total microbial population, while the proportion in the simulated rhizosphere soil increased slightly more from 9.4% to 93.7%. Since 2,4-D is readily used as a carbon and energy source for growth, presumably, the major benefit of elevated degrader populations in the rhizosphere for enhancing 2,4-D degradation would be to shorten the acclimation period prior to rapid growth linked disappearance of the herbicide. However, a particular mechanism explaining how the biodegradation was enhanced was not identified.

In comparison to studies involving growth-linked degradation of soil contaminants, very few studies have yet examined the influence of the rhizosphere on xenobiotic-cometabolizing microorganisms. Fournier (*10*) quantified 2,4-D-cometabolizing and metabolizing microbes in soil, which were estimated using a most probable number technique employing two types of 2,4-D-containing media. The first was a mineral salts medium used to estimate numbers of microbes capable of growth on 2,4-D as a sole carbon source (metabolizers). The second was a soil extract medium (0.2 g L^{-1} organic carbon) that was used to estimate microbes capable of degrading 2,4-D in the presence of additional carbon sources (cometabolizers). The results showed that in soils not previously exposed to 2,4-D, there were much higher populations of cometabolizers than microorganisms that were able to use 2,4-D as a sole carbon source. In soil previously exposed to 2,4-D, however, an increase in the number of metabolizers but not cometabolizers was observed.

These data suggest that in certain instances, plants may fortuitously enhance the populations of cometabolizer organisms. This phenomenon may be of importance when degraders that are capable of using the xenobiotic as a carbon source are absent or have an overly long acclimation period prior to reaching a population size necessary for rapid disappearance of the compound. Rates of cometabolism would be dependent on the availability of substrates provided by rhizodeposition in different root zones, as well induction of the necessary enzymes required for degradation of the xenobiotic. On the other hand, in soils that are contaminated with a xenobiotic that can be utilized as a growth substrate, and which is present at a concentration that is high enough to support active growth of the degrader populations, plants may do little to improve the overall rate of xenobiotic degradation in a soil that is continuously exposed to the contaminant.

Spatial Heterogeneity in Microbial Activity

In many studies on the rhizosphere effect, a common practice is to compartmentalize the soil into rhizosphere and bulk soil fractions, which unfortunately may lead to speculative conclusions regarding the effects of plants on biodegradation of soil contaminants. For example, rhizosphere soil may be separated from bulk soil by pulling plants and shaking the soil that adheres to the roots into a beaker, after which it is sieved and placed into flasks for further study. Samples from these fractions typically are characterized in terms of viable cell counts on rich agar media, microbial activity (e.g., respiration or dehydrogenase activity), and for the ability of microorganisms in these fractions to degrade xenobiotic contaminants in experimental systems that no longer include plants. The data are then compared to those obtained for bulk soil samples. In some instances, comparisons have been made with soils from field locations that apparently could not support plant growth, or with soil that did not adhere to the roots, but which had almost certainly been subjected to the previous influence of a plant rhizosphere (11, 12, 13). While these simple experimental methods are useful for characterizing the short term biodegradation potential of microbial communities associated with these crude soil fractions, the structural and dynamic nature of the rhizosphere has been destroyed, and consequently, the relevance of these data to in situ plant-microbial interactions is limited.

The rhizosphere is commonly perceived as a site of high microbial activity and bacterial population numbers. Microbial population densities in the rhizosphere are typically measured at 10^9 cells per gram of soil or root (14), a number as great as that obtained in nutrient rich media used to culture cells in the laboratory and 100 to 1000-fold greater than occurs in bulk soil. However, high numbers do not necessarily translate to high activity for all microorganisms associated with the rhizosphere. Rhizosphere ecologists have long recognized that much of the rhizosphere is oligotrophic, containing stationary phase or very slow dividing bacteria that are growth limited by lack of carbon substrates (15). In actuality, the only sites of relatively high microbial activity are confined to the root tips or sites of lateral root emergence where microorganisms have received a temporary pulse of carbon provided by root exudates or root cell lysates. These spatial variations in nutrient availability, as well as differences in the composition of plant derived compounds in different zones may impact both growth-linked and cometabolic degradation of xenobiotics.

Different root zones can also influence the heterogeneity of the rhizosphere community, which reflects microbial succession after the primary colonization of new roots. Early studies of the rhizosphere microbial community suggested that pseudomonads were the predominant microorganisms associated with plant roots, comprising anywhere from 30 to 90% of the culturable organisms isolated on agar media (16). More recent studies using media that select for oligotrophic microorganisms or DNA probing methods show that there is tremendous diversity in the community of microorganisms associated with roots, and that these communities vary for different plant species, or even for the same plant species in different soils (17). Microbial communities associated with individual plants may even differ from one root to the next, depending on stochastic events that lead to the successful colonization of a particular root by different microorganisms. Redox and gas diffusion gradients, as well as the structure and transition of microbial communities in a biofilm that extends radially from the root surface further delimit environmental conditions that affect microbial transformations of xenobiotics. Lastly, protozoa and nematode grazing of bacteria associated with the root surface, results in turnover of microorganisms, release of nitrogen mineralized from the microbial biomass, and changes in the types and quantities of substrates that are available to competing microorganisms in the rhizosphere over time.

An illustration of the differences in microbial activity that occur along plant roots is shown in Figure 2, in which roots of bean plants were inoculated with a root-colonizing pseudomonad containing a bioluminescence marker. In this marker, light production is regulated by a ribosomal promoter, such that light is produced only when the bacterium is actively growing. This organism also carried a plasmid which conferred the ability to degrade the recalcitrant xenobiotic, 2,5-dichlorobenzoate (*18*). The soil used for this study had been uniformly inoculated with the degrader, and contaminated with the xenobiotic prior to transplanting the plants into the microcosms. After three days, the plants were removed and exposed to x-ray film for autophotography to locate actively growing cells. As shown in the Figure 2, growth-linked bioluminescence of the pseudomonad was apparent only at the root tips and at sites of lateral root emergence. Several root tips did not show any bioluminescence and were colonized only at very low population densities by the degrader organism. Other locations were found to contain high population densities, but the cells were not bioluminescent, indicating that the cells were in stationary phase.

To further investigate the physiological status of this microorganism in different locations of the rhizosphere, we have since developed novel methods to quantify the relationship between the starvation state of this bacterium and expression of its bioluminescence marker (*19*). Starvation-related changes in cell physiology include a decrease in physiological activity (*20*), as well as sharp decreases in ATP, protein, DNA, and RNA per cell (*21*). When starved cells are subsequently provided with nutrients, one of the first events in their recovery is induction of rRNA required for production of new proteins. This leads to an longer lag phase in starved cells, which increases in proportion to the degree of prior starvation (*22, 23*). In *Pseudomonas fluorescens* 2-79RL, stationary phase cells are nonbioluminescent, whereas during recovery from starvation the first step prior to cell division is to synthesize new ribosome. Since the bioluminescence genes are coupled to a ribosomal promoter, the onset of light production marks the end of lag phase. In principle, this methodology is similar to other luminescence-based marker and reporter systems that have been used to assay the activity of pseudomonads in the rhizosphere (*24, 25*) and in soil (*26*).

Typical data showing the physiological status (lag phase duration) and cell numbers of *P. fluorescens* 2-79 are shown in Figure 3. Short lag phases (line graph data) are observed primarily at locations behind plant root tips or at sites of lateral root emergence, reflecting the relative availability of carbon in these root zones. In soil and on older root parts, the lag phase ranges from 3 to 16 h depending on the degree of starvation and concomitant reduction in physiological status. Population densities (bar graph data) are highest in the zone of elongation 3 cm behind the root tip, and decline on the older root parts as the bacterium is displaced by other indigenous bacteria that are presumably better adapted to low carbon availability. Pulse-labeling by providing the plants with $^{14}CO_2$ allows us to quantify the relative amounts of carbon in different roots zones using a modified filter-paper technique.

These studies show the heterogeneity of the plant rhizosphere in terms of physiological status and microbial distribution. Our data also reveal that there may be tremendous differences in the physiological status of microorganisms in the

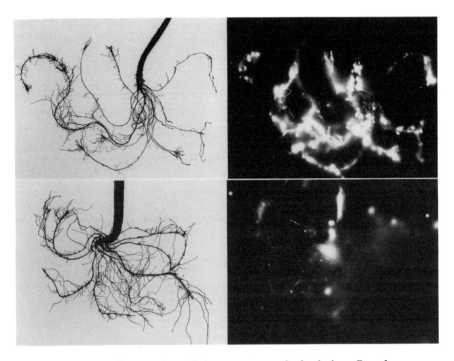

Figure 2. Autophotographs of bean roots colonized by *Pseudomonas fluorescens* 2-79 RLD containing a bioluminescence marker. Plants were cultured for 3 days and 10 days in soil microcosms containing 10 mg kg^{-1} 2,5-dichlorobenzoate. Soil was inoculated with 10^6 cfu g soil. Film exposure times were 3 and 36 h for top and bottom autophotographs, respectively. Top: light production by actively growing cells at 3 d is associated with most of the root surface, but is particularly intense at root tips and sites of lateral root emergence. Bottom: After 10 d, light production by *P. fluorescens* is greatly diminished, except at the root tips and at base of the plant stem.
Copyright 1996 with kind permission from Elsevier Science - NL.

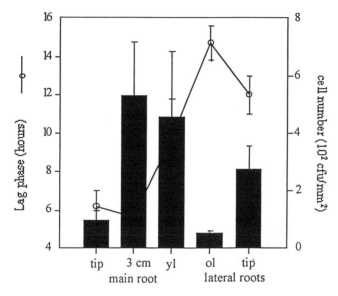

Figure 3. Length of lag phase (line graph) and population density (bar graph) of *Pseudomonas fluorescens* 2-79RL in different root zones. Main root: tip (seminal root apex); 3 cm (site of highest root exudation distal to root apex); yl (young lateral roots, including sites where main root cortex tissue has been ruptured by lateral root emergence). Lateral roots: ol (older suberized lateral roots); tip (root apex).

rhizosphere than under laboratory conditions in rich media where xenobiotic degraders are frequently characterized. Although further study is necessary to extend these observations, the rapid decline to a status reflecting extreme starvation shows that the rhizosphere is a crowded, highly competitive, oligotrophic environment.

Rhizosphere Effects on Biodegradation of Selected Xenobiotics

There are a number of compounds that have been examined in studies with plants (Table 1). Most of these studies show an increase in the degradation of the xenobiotic in planted soil over nonplanted soil, and make a strong case that phytoremediation may be useful for cleanup of a variety of soil contaminants. However, in addition to demonstrating that there are statistical differences, it is also important to consider whether the measured differences have practical significance. For example, a difference between 2 and 4 days for disappearance of relatively nonrecalcitrant compounds, such as 2,4-D or certain chlorobenzoates, in soil with and without plants suggests that compound is not likely a problem in the environment provided that degrader organisms are present (18). At the other extreme, it may be possible to show slight increases in rates of disappearance, but the practical value of these slight differences may be negligible if excessively long periods are required to phytoremediate the soil.

Recently, there are data from a number of studies that suggest plants may have little effect on the actual degradation rates of many xenobiotics, but may still be important for enhancing the interim survival of degrader organisms in soils between pulses of contamination, and for shortening the duration of the acclimation period prior to appearance of significant degradation activity after a soil has been contaminated. Pertinent findings of these studies are summarized in the sections below, while comparative data on plant effects on xenobiotic disappearance and effects of plants on cell numbers of culturable degrader organisms or consortia are presented in Table 2.

Chlorobenzoates. In initial research conducted by our laboratory, chlorobenzoates were chosen as model compounds for studies on the effects of rhizodeposition and selected organic substrates on microbial populations that metabolized the xenobiotic for growth in the absence of supplemental carbon, or that degraded chlorobenzoate by growth-linked metabolism or cometabolism requiring supplemental nutrients. Halogenated benzoic acids have previously been used as model compounds to study the degradation and metabolic pathways of other halogenated aromatic compounds (33,34). Chlorobenzoates can either be cometabolized (35-38) or used as a sole source of carbon and energy (39-42). Monosubstituted chlorobenzoates are only moderately recalcitrant, allowing degradation studies to be conducted in a period of several weeks. Along with being model compounds, chlorobenzoates are themselves environmental contaminants. They occur as a result of the use of chlorobenzoate-based pesticides (40, 43), as breakdown products of other classes of pesticides (44) and as metabolites of polychlorinated biphenyl degradation (45).

Table 1. Rhizosphere Studies on Xenobiotic Degradation

Reference	Compound	Rhizosphere Effect
Reddy, 1983 (*4*) Hsu, 1979 (*3*)	parathion	Rate of mineralization was increased in both studies; the former a 4 fold increase with a mineralization of 23% in 15 days.
Walton, 1990 (*7*)	TCE, trichloroethylene	Increased mineralization in rhizosphere soil, but overall rates for both rhizosphere and bulk soil were slow.
Haby, 1996 (*12*)	3-chlorobenzoate	Increased numbers of degraders, and shorter lag phase until disappearance of the compound.
Brazil, 1995 (*27*)	biphenyl	inserted *bph* genes into a rhizosphere pseudomonad
Krueger, 1991 (*28*)	dicamba (3, 6-dichloro-2-methoxybenxoic acid)	Faster disappearance in planted soil, plants protected by inoculation with degraders.
Aprill, 1990 (*5*)	polyaromatic hydrocarbons	Higher rate of disappearance in planted soil.
Hsu, 1979 (*3*)	diazanon	5% mineralization in bulk soil, 13% in planted soil.
Sandmann, 1984 (*8*)	2,4-D	Higher R/S ratios of 2,4-D degraders in sugarcane.
Ferro, 1994 (*6*)	phenanthrene	No change in mineralization in planted verses nonplanted soil
Ferro, 1994 (*6*)	pentachlorophenol	24% mineralization in planted soil verses 6% in nonplanted.
Singh, 1991 (*29*)	chlordane, heptachlor aldrin	Similar persistence in cropped verses noncropped fields.
Siebert, 1981 (*30*) Mandelbaum 1993 (*13*) Anderson, 1995 (*31*)	atrazine	Formation of hydroxyatrazine in the corn rhizosphere. Enrichment cultures from roots mineralized atrazine more slowly. Anderson, demonstrated mineralization in rhizosphere enrichment cultures, but rates for nonrhizosphere soil were not reported.

Table 2. **Influence of Plants on Degradation Rates of Selected Xenobiotics and Final Population Numbers of Culturable Degrader Organisms**

Chemical	% Disappearance		Degrader Cell Numbers**		Plant
	No plants	Plants	No Plants	Plants	
3-CB*	100	100	5×10^4	1×10^7	ryegrass (12)
2,5-DCB	63	100	5×10^2	8×10^4	bean (18)
chlordane	0	0	0	0	Bermuda grass
4,4'-DCDB	<1	<1	9×10^6	1.8×10^7	ryegrass
atrazine	84	71	270	8100	corn (32)

*Abbreviations: 3-CB 3-chlorobenzoate; 2,5-DCB 2,5-dichlorobenzoate; 4,4'-DCDB 4, 4'-dichlorobiphenyl (PCB). **Microcosms were bioaugmented with *P. fluorescens* 2-79RLD for experiments lasting 14 to 21 d with 2,5-DCB, and with *Arthrobacter* sp B1B in 50 d studies with 4, 4'-DCDB. Atrazine disappearance was measured as mineralization ($^{14}CO_2$). Atrazine degrader cell numbers are based on most probable number assays of an atrazine-degrading bacterial consortium after 4 wk. 3-CB degradation was studied for indigenous microorganisms after 6 months in soil continuously exposed or nonexposed to the xenobiotic; chlordane biodegradation experiments were conducted with soil containing an indigenous microflora that had been exposed to ca 100 ppm chlordane for several years.

3-Chlorobenzoate. Experiments were conducted with soil containing indigenous microorganisms capable of degrading the model compound 3-chlorobenzoate (3-CB) (*12*). The results showed degradation of 3-CB in 1:1 soil:water slurries from ryegrass rhizosphere soil that had not been previously exposed to 3-CB had a faster initial rate of 3-CB degradation than nonrhizosphere soil. This rhizosphere effect could be simulated in nonrhizosphere soil using glucose, mannitol, or benzoate. A long term experiment further showed that the population size of microorganisms that degraded 3-CB in the absence of supplemental carbon (catabolizers), and those that degraded 3-CB in the presence of supplemental carbon, e.g., via growth-linked metabolism or cometabolism, were enriched by 40- and 250-fold, respectively, in rhizosphere soil. In contrast, planted and nonplanted microcosms that were repeatedly exposed to 3-CB over 24 wk were similarly enriched for 3-CB degraders, which had a MPN of 4 x 10^8 g^{-1} soil. Our results suggested that under conditions where there is not constant exposure of the microbial population to 3-CB as a carbon source, carbon provided by rhizodeposition may enhance the population numbers of microorganisms that degrade 3-CB by cometabolic or growth-linked metabolism, and thereby promote rapid depletion of 3-CB from soil during short term or low level exposure to the xenobiotic.

2,5-Dichlorobenzoate. 2,5-DCB is normally a highly recalcitrant substrate and was not significantly degraded in soil unless the soil was bioaugmented with a degrader bacterium. Genes for catabolism of this compound are carried on a catabolic plasmid pPB111, which was originally isolated from *Pseudomonas putida* P111, and transferred by conjugation to *P. fluorescens* 2-79RL (*18*). The major benefit of the plant rhizosphere was to promote a greater rate of degradation and to improve long-term stability of the plasmid in *P. fluorescens* 2-79 RLD. Part of the difference in the rate of 2,5-DCB disappearance appeared to be due to plant uptake of the compound, but the major effect appeared to involve enhancement of degradation activity. In the presence of plants, the xenobiotic was completely mineralized after 4 days, whereas, in nonplanted soil 2,5-DCB disappeared after 7 days to 14 days. One possibility for the enhanced disappearance rate may have been that the plasmid was more stably maintained in the rhizosphere, which is supported by the fact that most of the introduced bacteria continued to carry the plasmid even after the 2,5-DCB substrate was gone. Another benefit may have been enhanced transfer of the degradative plasmid to the indigenous microflora (*18*).

4,4'-Dichlorobiphenyl. A microcosm study was conducted to investigate the influence of the ryegrass rhizosphere on polychlorinated biphenyl (PCB) biodegradation. Each microcosm contained 100 g of 4,4'-dichlorobiphenyl contaminated soil (100 mg kg^{-1}), selected as a model PCB congener (*46*), and received a single inoculation with *Arthrobacter* sp. strain B1B, a natural soil bacterium isolated after enrichment on biphenyl and known to cometabolize PCBs (*47*). The inoculated populations declined approximately 10-fold over 50 days with respect to the initial cell densities in both the planted and unplanted treatments; the cell densities at the conclusion of the experiment were not statistically different (p<0.32; Table 2). Trace amounts (<1 ppm) of 4-chlorobenzoate (4-CB), a metabolic intermediate of aerobic cometabolism of 4,4'-dichlorobiphenyl, were detected in both microcosms. The presence of 4-CB indicated that a biologically-mediated transformation occurred; however, there were no differences between the planted and unplanted treatments. Although a well-developed rhizosphere was established in the planted microcosms, the results of our study found that this was not sufficient to promote cometabolism of the model PCB. These results support the concept that phytoremediation of recalcitrant compounds via bacterial cometabolism will only be effective if the plants provide a substrate which increases production of the requisite enzymes involved in catabolism of the target compound.

Atrazine. Among the various xenobiotics that may have enhanced degradation in the rhizosphere, atrazine is a particularly good candidate since, like most hydrophobic compounds, it accumulates on clay and organic matter in the upper soil horizon occupied by plant roots. At normal application rates, it is present at relatively low concentrations that may preclude the growth and maintenance of an effective degrader population size (*48*). This latter problem is particularly problematic for atrazine, which has a highly oxidized triazine ring that provides no energy for microbial growth (*49*) Thus atrazine-degrader organisms can only acquire energy by utilizing carbon in the side chains attached to the ring or by growth on other carbon compounds contained in soil organic matter or root exudates. Atrazine has also been shown to serve as a nitrogen source for microbial growth (50-52), which may be enhanced by plant uptake of nitrogen and simultaneous deposition of carbon, thereby providing a selective advantage for microorganisms that are able to derive nitrogen from the triazine ring.

Recent studies in our laboratory have shown that despite optimistic scenarios for enhanced atrazine biodegradation in the rhizosphere, that rates of atrazine biodegradation were not significantly improved by the presence of plants when atrazine degraders were added to soil at a high population density (*32*). Bioaugmentation of soil contaminated with 6 ppm atrazine resulted in mineralization ca. 75-80% of that applied in 4 wk in the presence or absence of plants. The only benefit of the plants was to slow the decline in population size of the consortium over that which occurred in bulk soil without plants. Thus if atrazine degraders were ever utilized for removing atrazine residues at the end of a growing season, there might be some benefit to utilizing plants in conjunction with bioaugmentation. This might also be employed for bioaugmentation of riparian strips to reduce atrazine runoff into surface waters. However, these data again point to the consistent observation that the primary benefit of the rhizosphere is to harbor the degrader organisms at a larger population size that may shorten the acclimation period rather than enhancing the actual rate of degradation by the degrader organisms.

Data from our experiments showed that corn plants themselves may have been responsible for transformation of the parent compound to nonphytotoxic hydroxyatrazine, since high levels of hydroxyatrazine accumulated in the rhizosphere even under axenic conditions. Differences in the level of hydroxyatrazine formation between greenhouse and growth chamber experiments further revealed the importance of environmental conditions on the rate of hydroxyatrazine transformation mediated by plants (*32*). Further incubation of the planted soils under fallow conditions demonstrated that hydroxyatrazine was a relatively transient molecule with subsequent formation of hydroxylated N-dealkyl metabolites (*32*). These data highlight some of the complexity that must be considered in evaluating disappearance of xenobiotics from soil or in characterizing rhizosphere effects on biodegradation by soil microorganisms.

In soils where atrazine is present at concentrations that will support direct growth of a degrader population it has been shown that bioaugmentation results in high levels of degradation. Mandelbaum and coworkers inoculated soils from a hazard waste site (*51*) and found ca. 65% disappearance of the atrazine upon inoculation with the atrazine mineralizing *Pseudomonas sp.* in 3 wk (51). Other studies have demonstrated similar rates of mineralization of atrazine in soil, Alvey and Crowley found 60 % mineralization of 1000 ppm of atrazine in soil after inoculation with a consortium and Kontchou and Gschwind found 60 % mineralization of 10 ppm in soil (*32, 53*). These data suggest that bioaugmentation alone might be sufficient for remediation of atrazine contaminated soils.

Future Directions

The successful application of plant-microbial systems to bioremediation problems requires an understanding of rhizosphere ecology to predict the ways in which plants influence microbial biodegradation of specific xenobiotics. There are numerous scenarios by which plants might affect the biodegradation process. As reviewed by Alexander (54) there are several aspects of biodegradation which can be considered independently; including acclimation, microbial population sizes and growth kinetics, bioavailability, and environmental factors, all of which contribute to the rate of degradation. In some instances, plants might directly influence biodegradation by promoting growth-linked degradation or by enhancing cometabolism via rhizodeposition of cosubstrates that induce the enzymes for catabolism of soil contaminants. Plants may also have an indirect influence on biodegradation processes; for example, by creating microenvironments that are conducive to increased genetic exchange of degradative plasmids. Finally, plants may physically alter the soil environment for xenobiotic degradation by altering pH and redox conditions in the rhizosphere, or by influencing the spatial arrangement of microbial communities that develop in the vicinity of plant roots. To optimize the processes that control the rates of degradation, it is important to understand the specific ways in which plants might influence microbial ecology.

Although there is increasing evidence that plants may under certain conditions enhance the biodegradation of xenobiotics, it should be kept in mind that there are also circumstances where plants may have no effect on biodegradation, or even inhibit biodegradation. For example, when a compound can be used as a growth substrate, there are probably very few instances where plants will benefit the degradation rate if the indigenous microflora has already acclimated to the substrate. This was observed in our studies with chlorobenzoates and atrazine. Even in experiments with 2,5-DCB, the primary factor causing disappearance of this recalcitrant compound was bioaugmentation with a degrader organism, not the effects of plants on the activity of this degrader.

The importance of plants for harboring degrader organisms in the rhizosphere, and thereby shortening the acclimation phase, also has to be considered in relation to the bioremediation problem. For instance, this effect of the rhizosphere is probably not relevant for phytoremediation of waste sites in which the indigenous microflora have already been subject to long term exposure to the contaminant. On the other hand, if the degradation process is dependent on specific plant exudate components or other rhizodeposition products for cometabolism, the rhizosphere may exert a real effect on the disappearance of the contaminant, provided that the cometabolizers have first access to this substrate. In some cases, there may also be a priming effect in which the rhizosphere fortuitously enhances the growth of a degrader organism to a population size that subsequently becomes effective for metabolizing a contaminant otherwise present at a level too low to support growth of the degrader.

By knowing the rate limiting factors that control degradation, it is also possible to speculate on situations in which bioremediation might be inhibited by plants, or influenced by the nutritional management of the plant microbial system. For example, in situ bioremediation of petroleum hydrocarbons under nitrogen-limiting conditions is a case where plants might increase the length of time necessary for cleanup. Because the target compound is composed primarily of carbon, achieving the most rapid degradation occurs by providing the limiting nutrients that permit the carbon to be metabolized (primarily nitrogen and phosphorus), as was done following the Exxon *Valdez* oil spill (*55*). Because the rhizosphere is relatively rich in carbon-containing root exudates, enhanced root exudation could have the effect of suppressing metabolism of the target compounds by providing a superior source of carbon. On the other hand, if the system is not nitrogen limited, the development of an active biofilm of microorganisms on the root surface, in combination with bulk flow of contaminants through evapotranspiration, might promote more rapid degradation.

Conclusions

While it is possible to demonstrate significant rhizosphere effects in the laboratory using various model systems, it is also possible to create experimental systems that unrealistically demonstrate positive effects of plants on biodegradation. When a compound is being considered for cleanup with phytoremediation, it is important to consider its bioavailability and the potential pathway by which it will be broken down, in order to decide whether plants will be of benefit to the process, and the best ways in which the plant-microbial system can be manipulated to achieve this purpose. This will likely vary for every class of chemical compounds that are targeted for remediation, but will provide a rational approach that considers the ecology of soil microorganisms and the influence of the plant rhizosphere on their growth and activity.

Literature Cited

1. Shimp, J. F.; Tracy, J. C.; Davis, L. C.; Lee E. *Crit. Rev. Environ. Sci. Technol.* **1993**, *23*, 41-77.
2. Anderson, T. A.; Coats, J. R. *Env. Sci. Health.* **1995**, *30*, 473-484.
3. Hsu, T. S.; Bartha R. *Appl. Environ. Microbiol.* **1979**, *37*, 36-41.
4. Reddy, B. R.; Sethunathan N. *Appl. Environ. Microbiol.* **1983**, *45*, 826-829.
5. Aprill, W.; Sims, R. C. *Chemosphere.* **1990**, 20, 253-266.
6. Ferro, A. M.; Sims, R. C.; Bugbee, B. *J. Environ. Qual.* **1994**, *23*, 272-279.
7. Walton, B. T.; Anderson, T.A. *Curr. Opin. Biotechnol.* **1992**, *3*, 267-270.
8. Sandman, E. R.; Loos, M. A. *Chemosphere.* **1984**, *13*, 1073-1084.
9. Kunc, F. In: *Interrelationships between Microorganisms and Plants in Soil.* Elsevier. New York, NY, 1989; pp 329-334
10. Fournier, J. C. *Chemosphere.* **1980**, *9*, 169-174.
11. Walton, B. T.; Anderson, T. A. *Appl. Environ. Microbiol.* **1990**, *56*, 1012-1016.
12. Haby, P.; Crowley, D. E. *J. Environ. Qual.* **1995**, *25*, 304-310.
13. Mandelbaum, R. T.; Wackett, L. P.; Allan, D. L. *Appl. Environ. Microbiol.* **1993**, *59*, 1695-1701.
14. Curl E. A.; Truelove, B. *The Rhizosphere.* 1986. Springer-Verlag, NY.
15. Bowen, G. D.; Rovira, A. D. *Ann. Rev. Phytopath.* **1976**, *14*, 121-144.
16. Vancura, V. *Folia Microbiol.* **1980**, *25*, 168-173
17. Laguerre, G.; Allard, F.; Lemanceau, P. *Appl. Environ. Microbiol.* **1996**, *62*, 2449-2456.
18. Crowley, D. E.; Brennerova, M. V.; Irwin, C. I.; Brenner, V.; Focht, D. D. *FEMS Microbiol. Ecol.* **1996**, *20*, 79-89.
19. Marschner, P.; Crowley, D. E. *Soil Biol. Biochem.* **1996**, in press.
20. Kurath, G.; Morita, R. Y. *App. Environ. Microbiol.* **1983**, *45*, 1206-1211.
21. Morita, R. Y. In: *Starvation in Bacteria.* 1993. New York, Plenum Press.
22. Amy, P. S.; Pauling, C; Morita, R.Y. *Appl Environ. Microbiol.* **1983**, *45*, 1685-1690.
23. Vandenhove, H.; Merckx, R.; Wilmots, H.; Vlassak, K. *Soil Biol. Biochem.* **1991**, *23*, 233-263.
24. Meikle, A.; Glover, A.; Killham, K.; Prosser J. *Soil Biol. Biochem.* **1994**, *26*, 747-755.
25. de Weger, L. A.; Dunbar, P.; Mahafee, W. F.; Lugtenberg, B. J. J.; Sayler G. S. *Appl Environ. Microbiol.* **1991**, *57*, 3641-3644.
26. Meikle, A.; Kilham, K.; Posser, J. I.; Glover, L. A. *FEMS Microbiol. Lett.* **1992**, *99*, 217-220.
27. Brazil, G. M.; Kenefick, L.; Callanan, M.; Haro, A.; Lorenzo, V.; Dowling, D. N.; O'Gar, F. *Appl. Environ. Microbiol.* **1994**, *61*, 1946-1952.
28. Krueger, J. P.; Butz, R. G.; Cork, D. J. *J. Agric. Food Chem.* **1991**, *39*, 1000-1003.

29. Singh, G.; Kathpal, T. S.; Spencer, W. F.; Dhankar, J. S. *Env. Pollution.* **1991**, *66*, 253-262.
30. Siebert, K.; Fuehr, F.; Cheng, H. H.. In: *Theory and Practical Use of Soil Herbicides Symposium;* European Weed Resource Society: Paris, 1981, pp. 137-146.
31. Anderson, T. A.; Coats, J. R.. *J. Environ. Sci. and Health.* **1995**, *30*, 473-484.
32. Alvey, S.; Crowley, D. E.. *Env. Sci. Technol.* **1996**, *30*, 1596-1603.
33. Marks, T. S.; Smith, A. R. W.; Quirk, A. V. *Appl. Environ. Microbiol.* **1984**, *48*, 1020-1025.
34. Focht, D. D;. Shelton, D. *Appl. Environ. Microbiol.* **1987**, *53*, 1846-1849.
35. Horvath, R. S.; Alexander, M. *Appl. Microbiol.* **1970**, *20*, 254-258.
36. Walker, N.; Harris, D. *Soil Biol. Biochem.* **1970**, *2*, 27-32.
37. Horvath, R. S. *Bacteriol. Rev.* **1972**, *36*, 146-155.
38. Horvath, R. S. *Appl. Microbiol.* **1973**, *25*, 961-963.
39. Briggs, G. G.; Alexander, M.. *Soil Biol. Biochem.* **1972**, *4*, 187-190.
40. Dorn, E.; Hellwig, M.; Reineke, W.; Knackmuss, H. J. *Arch. Microbiol.* **1974**, *99*, 61-70.
41. Ferrer, M. R.; Ruiz-Berraquero, F.; Ramos-Cormenzana, A. *Agrochimica* **1986**, *30*, 458-463.
42. Focht, D. D.; Shelton, D. *Appl. Environ. Microbiol.* **1987**, *53*, 1846-1849.
43. Hartmann, J.; Reineke W.; Knackmuss H. J. *Appl. Environ. Microbiol.* **1979**, *37*, 421-428.
44. Rouchaud, J.; Moons, C.; Benoit, F.; Ceustermans, N.; Maraite, H. *Weed Sci.* **1987**, *35*, 469-475.
45. Ahmed A.; Focht, D. D. *Can. J. Microbiol.* **1973**, *19*, 47-52.
46. Adriaens, P.; Kohler H.-P.E.; Kohler-Staub; D.; Focht, D.D. *Appl. Environ. Microbiol.* **1989**, *55*, 887-892.
47. Kohler, H. P. E.; Kohler-Staub D.; Focht. D. D. *Appl. Environ. Microbiol.* **1988**, *54*, 1940-1945.
48. Pahm, M. A.; Alexander, M. *Microbial Ecol.* **1993**, *3*, 275.
49. Cook, A. M. *FEMS Microbiol. Rev.* **1987**, *46*, 93.
50. Alvey, S.; Crowley, D. E. *J. Environ. Qual.* **1995**, *24*, 1156-1162.
51. Mandelbaum, R. T.; Wackett, L. P.; Allan, D. L. *Appl. Environ. Microbiol.* **1995**, *61*, 1451-1457.
52. Mandelbaum, R. T.; Wackett, L. P.; Allan, D. L. *Environ. Sci. Technol.* **1993**, *27*, 1943-1946.
53. Yanze-Kontchou, C.; Gschwind, N. *J. Agric. Food Chem.* **1995**, *43*, 2291-2294.
54. Alexander, M. *Biodegradation and Bioremediation.* 1994, Academic Press. San Diego, CA
55. Pritchard, P.H.; Costa, C. F. *Env. Sci. Technol.* **1991**, *25*, 372-379

PHYTOREMEDIATION OF AGROCHEMICALS

Chapter 3

Aromatic Nitroreduction of Acifluorfen in Soils, Rhizospheres, and Pure Cultures of Rhizobacteria

Robert M. Zablotowicz, Martin A. Locke, and Robert E. Hoagland

Southern Weed Science Laboratory, Agricultural Research Service, U.S. Department of Agriculture, P.O. Box 350, Stoneville, MS 38776

Reduction of nitroaromatic compounds to their corresponding amino derivatives is one of several pathways in the degradation of nitroxenobiotics. Our studies with the nitrodiphenyl ether herbicide acifluorfen showed rapid metabolism to aminoacifluorfen followed by incorporation into unextractable soil components in both soil and rhizosphere suspensions. Aminoacifluorfen was formed more rapidly in rhizospheres compared to soil, which can be attributed to higher microbial populations, especially of Gram-negative bacteria. We identified several strains of *Pseudomonas fluorescens* that possess nitroreductase activity capable of converting acifluorfen to aminoacifluorfen. Factors affecting acifluorfen nitroreductase activity in pure cultures and cell-free extracts, and other catabolic transformations of acifluorfen, ether bond cleavage, are discussed. Plant rhizospheres should be conducive for aromatic nitroreduction. Nitroreduction by rhizobacteria is an important catabolic pathway for the initial degradation of various nitro-herbicides and other nitroaromatic compounds in soils under phytoremediation management.

Nitroaromatic compounds are of environmental concern because they represent a class of chemicals that are widely used as herbicides, explosives, pharmaceuticals and industrial chemicals. Understanding their biotransformations can lead to more effective management techniques to minimize their persistence and subsequent environmental contamination. Phytoremediation is an environmentally desirable approach for remediating contaminated matrices (1). Enhanced degradation of xenobiotics under phytoremediation strategies can be due to both uptake and transformation by the plant as well as by the general rhizosphere effect that enhances a microbial community capable of biodegrading the contaminant (2, 3).

Several mechanisms are possible in bacterial transformations of nitroaromatic compounds: oxidative pathways, reductive pathways and partial reductive pathways.

Nitrophenols can undergo an oxidative liberation of nitrite and replacement by a hydroxyl group prior to ring cleavage and complete mineralization (4, 5). Reductive pathways are mediated by the enzyme nitroreductase which transforms the nitro moiety to the corresponding amino derivative. Aromatic nitroreductases have been widely studied in gastrointestinal microflora, especially in anaerobic bacteria such as *Bacteroides* (6), *Clostridium* and *Eubacterium* (7) and in certain aerobic and facultative anaerobic bacteria, i.e., *Nocardia* and *Pseudomonas fluorescens* (8) *Escherichia coli* (9), *Enterobacter cloacae* (10) *Mycobacterium* sp. (11) and *Salmonella typhimurium* (12). However, little attention has been given to the contribution of this enzyme to biotransformations of nitroherbicides by terrestrial bacteria.

Acifluorfen, 5-[2-chloro-4-(trifluoromethyl)phenoxy]-2-nitrobenzoic acid, is a widely used herbicide in soybean, peanut and rice production to control many broadleaf and certain grass weeds (13). The major degradation metabolite of the nitrodiphenyl ethers observed in soil, especially under low redox conditions, is the amino derivative (14-17). Limited studies on the degradation of acifluorfen in soil are available in the literature. The half-life of acifluorfen can range from 23 to 122 d in soil (18). Recently, the ability of mixed bacterial cultures to degrade acifluorfen under anaerobic (19) and aerobic conditions (20) has been reported with aminoacifluorfen as the major accumulating metabolite under both conditions.

Enhanced degradation of the nitroaromatic herbicide trifluralin has been reported in rhizospheres of tolerant plants colonizing pesticide-contaminated sites (21). Little is known about the fate of other nitroaromatic herbicides in the rhizosphere. In this report we examined the potential for metabolism of acifluorfen in soils, rhizospheres, and pure bacterial cultures. This information may be useful in exploiting the potential of enhanced degradation of nitroaromatic herbicides in the rhizosphere for phytoremediation.

Materials and Methods

Chemicals. Acifluorfen, (free acid, technical grade) was obtained from Chem Service (West Chester, PA). Radiolabeled acifluorfen, ^{14}C-(UL)-4-trifluoromethyl-2-chlorophenol-ring label (99% purity, 667 MBq mmol^{-1} specific activity), aminoacifluorfen, decarboxyacifluorfen, and 4-trifluoromethyl-2-chlorophenol (TFMCP) were obtained from BASF (Berlin, Germany and Research Triangle, NC). NADH, NADPH, FMN, and 2-nitrobenzoic acid (2-NBA) were from Sigma Chemicals (St. Louis, MO).

Degradation of acifluorfen in soil and rhizosphere suspensions. Soil (Dundee silt loam) with a long-term history of acifluorfen application (10 yr) and soybean (*Glycine max L.* Merr.) roots were collected from a field near Stoneville, MS. Loosely adhering soil was removed upon excavation and approximately 3 kg of roots and rhizosphere soil were collected. Soil, to a depth of 15 cm, was collected from areas free of vegetation. Outer rhizosphere soil was removed by shaking and gently brushing the roots with a camel hair brush. Root-free and outer rhizosphere soil was sieved (2-mm sieve) to remove coarse material and nodules were removed with forceps. Inner rhizosphere soil was removed from the processed roots by vigorous shaking in 2.5 L potassium phosphate buffer (0.125 M, pH 7.0) for 2 h followed by mild sonication in a sonicator bath. The inner rhizosphere soil suspensions were concentrated by centrifugation (13 k x g, 20 min.) and resuspended in a final volume of 225 ml phosphate buffer that contained 6.23 mg of soil (oven dry weight ml^{-1}). Five g (oven dry weight equivalent) of root-free and outer rhizosphere soil were added to sterile 25 ml centrifuge tubes, and phosphate buffer was added to a final volume of 3.0 ml water. Eight ml (500 mg of soil) of inner rhizosphere soil suspension were added to the tubes. Eight ml of 30 μM acifluorfen (free acid) solution prepared in phosphate buffer containing 3250 Bq ^{14}C- labeled acifluorfen was added to each tube. Eighteen tubes were prepared for each soil sample with 2 replicates sampled at time of treatment and four replicates each at 24, 48, 96 and 144 h after treatment.

Aqueous soluble ^{14}C-material was recovered by centrifugation (9.75 k x g, 10 min). Acifluorfen and metabolites were further recovered from soil by a 24-h extraction with 10 ml methanol and a second 24-h extraction with 10 ml methanol:NaOH (0.5 N) 95:5 (v:v). Each fraction was weighed and recovery of ^{14}C was determined from duplicate aliquots by liquid scintillation spectroscopy using Ecolume cocktail. Each corresponding extract was diluted with water, acidified with HCl (pH 3.0), and then passed though a C-18 column. Acifluorfen and metabolites were eluted from the C-18 column with 4.0 ml methanol. Methanol extracts were concentrated as needed under N$_2$ and analyzed by RAD-HPLC and RAD-TLC.

Bacterial populations were determined in soil suspensions prepared as above, except non-radioactive acifluorfen solution was used. Total aerobic heterotrophic bacteria, Gram-negative bacteria and fluorescent pseudomonads were enumerated by serial dilution and spiral plating on appropriate media as described elsewhere (22). Three replicates of root-free and outer rhizosphere suspensions, and four replicates of inner rhizosphere suspensions were evaluated at 0, 24, 48, 96 and 144 h after treatment.

RAD-HPLC and RAD-TLC Analytical Methods. A Waters (Waters, Milford, MA) HPLC system (Model 510 pump, 712 WISP autosampler, 490E UV detector, System Interface Module) with an Altima C-18 reversed phase column (5 μM, 250 mm) (Deerfield, IL) was used. Mobile phase conditions were 1.0 ml min^{-1}, 60:40 isocratic acetonitrile:H$_2$O (pH 3.2 adjusted with H$_3$PO$_4$). Acifluorfen and metabolites were monitored using UV detection (230 nm) and ^{14}C was monitored using a Beta-Ram detector (INUS Systems Inc., Tampa, FL). HPLC retention times were: acifluorfen = 10.1 min, aminoacifluorfen = 13.0 min, and TFMCP = 8.5 min.

TLC was conducted on silica gel plates (250 μM thickness) with fluorescent indicator (Whatman, Clifton, NJ). Two solvent systems were used for acifluorfen and metabolites : A = ethyl acetate:toluene:acetic acid:water (50:50:1:0.5) and B = ethyl acetate: hexane:acetic acid (95:5:1). Acifluorfen had R$_f$ values of 0.20 (A) and 0.35 (B); aminoacifluorfen had R$_f$ values of 0.35 (A) and 0.65 (B), TFMCP R$_f$ = 0.75 (A) and 0.90 (B); and decarboxyacifluorfen R$_f$ = 0.95 (A) and 0.98 (B). Radioactivity in chromatograms was analyzed with a Bioscan System 200 imaging Scanner (Bioscan, Washington, DC).

Acifluorfen Metabolism by Pure Cultures of Bacteria. Acifluorfen metabolism by pure bacterial cultures were evaluated in several *Pseudomonas fluorescens* strains BD4-13, UA5-40, (23), and RA-2 (22) that have been shown to possess other herbicide catabolic pathways. Additionally, *Enterobacter cloacae* strain ATCC 43560 previously reported to have aromatic nitroreductase activity (10) was also evaluated. In all studies, bacterial strains were grown on tryptic soy broth (TSB) for 48 h, cells were harvested by centrifugation (9.75 k x g, 10 min.), and washed twice in potassium phosphate buffer (KPi, 0.05 M, pH 7.6). Washed cells were resuspended in KPi to cell concentrations of optical density (o.d.) 10.0 to 16 determined at 660 nm [ca. 4.2 to 5 mg cells dry weight ml-1]. In some studies frozen cell suspensions (-5° C) were used for preparation of cell-free extracts. Cultures of ATCC 43560 were also grown in the presence or absence of 50 μM 2-nitrobenzoic acid (2-NBA) to assess possible induction of acifluorfen nitroreductase.

Initial studies assessed acifluorfen metabolism in cell suspensions by BD4-13, RA-2, UA5-40 and ATCC 43560. Cell suspensions (500 μλ, o.d. = 10, about 2 mg cells) were placed in 2.0 ml polypropylene microcentrifuge tubes. Then a stock solution of technical grade acifluorfen and ^{14}C- labelled acifluorfen were added to achieve a concentration of 8 or 80 μM and 4,000 Bq ml-1 in a final volume of 750 μλ. KPi with acifluorfen was included as a control. The tubes were capped, contents were vigorously mixed, and incubated statically at 30 °C. Aliquots (150 μλ) were removed periodically, transferred to microfuge tubes, then 300 μλ of methanol and 15 μl sonication for 15 min. The suspension was clarified by centrifugation (22 k x g, 10 min), and the supernatant analyzed by either RAD-TLC and/or RAD-HPLC as previously described.

The effects of aeration on acifluorfen metabolism by BD4-13 and RA-2 were evaluated in cell suspensions and TSB cultures. Cell suspensions (4.0 ml) of BD4-13 (o.d. = 16, 17 mg ml-1]. and RA-2 (o.d. = 12, 5 mg ml-1]. were placed in sterile centrifuge tubes (25 ml) and 4.0 ml of sterile unlabeled acifluorfen (320 mM in KPi) was added, and the tubes loosely capped. Duplicate tubes were incubated statically at 28 °C, and two replicate tubes were incubated on a rotary shaker at 150 rpm at 28 oC. Erlenmeyer flasks (250 ml) containing 20 ml of 50% strength TSB and 160 mM acifluorfen were inoculated with 1.0 ml of the same cell suspensions of BD4-13 and RA-2. Duplicate flasks of each strain were incubated at both aeration regimes. Following a 24 h incubation, 2.0 ml aliquots were removed, acidified with 200 μλ of 1.0 N HCl, and phase-extracted twice with 8 ml of methylene chloride. The combined methylene chloride extracts were concentrated to 4 ml under N$_2$ and analyzed by HPLC.

The effects of preconditioning cells under microaerophilic conditions and exposure to 2-nitrobenzoic acid as an inducer of acifluorfen metabolism were investigated in strain RA-2. Cell suspensions (10 ml, od = 10, 4.2 mg ml-1]. were placed in sterile screw cap Erlenmeyer flasks (25 ml). Then 2.0 ml of 200 mM 2-NBA in KPi or 2.0 ml of KPi was added, and flasks were loosely capped. The flasks were incubated on a rotary shaker (150 rpm, 28 °C) or incubated statically at 28 oC, three replicates per treatment. Following a 24 h incubation, acifluorfen metabolism was evaluated in whole cells or cell-free extracts (cfe). Whole cell assays were conducted in 4 ml conical vials with a screwcap with septum containing 700 μl of cells and acifluorfen in a final volume of 1.0 ml and a final concentration of 75 μM acifluorfen, with four replicates per treatment. KPi controls were included for both assays. Following 2.5 h incubation the reaction was terminated by addition of 250 μl of 1.0 N HCl and 3 ml of methanol. The suspensions were extracted as described above and analyzed by HPLC. Cfe's were prepared by sonication as described elsewhere (23). Acifluorfen nitroreductase assays were conducted in 4.0 ml conical vials containing 700 μl of cfe (approximately 1.0 mg protein), 75 nmol acifluorfen, 320 nmol NADH, and 10 nmol FMN in a final volume of 1.0 ml under a N_2 atmosphere for 2.5 h. These reductant concentrations (320 μM NADH and 10 μM FMN) are similar to those used by others (10, 11). Assays were terminated with 250 μl of 1.0 N HCl and 3 μM of methanol and analyzed like the whole cell suspensions.

Further studies on acifluorfen nitroreductase activity were conducted on cfe's of *P. fluorescens* strains RA-2, UA5-40, and EC43560 to ascertain the role of reductant co-factors and oxygen. Assay mixtures contained approximately 125 μl cfe (0. 3 to 0.38 mg protein), 14.22 nmol of acifluorfen (4000 Bq) in a final volume of 208 μl. Cfe's treated with reductant included 64 nmol NADH (320 μM) and 2.08 nmol of FMN (10 μM) in a final volume of 208 μl. Three replicates of each treatment per strain were incubated under a N_2 atmosphere or air for 2 h at 28 oC. Assays were terminated with a similar ratio of HCl and methanol as described above, and extracted by sonication prior to analysis by TLC.

Kinetics parameters of aerobic acifluorfen nitroreductase activity were determined in cfe's of RA-2 and UA5-40. Microfuge tubes contained 125 μl cfe (RA-2 = 0.32, UA5-40 = 0.38 mg protein), and 0.2 to 3.02 nmol of acifluorfen (2000 Bq) in a final volume of 200 μl of KPi, with three replicates per concentration. The reaction was initiated by substrate addition, mixed vigorously, and incubated at 28 oC under static conditions for 2 h. The reaction was terminated by addition of 20 μl 1.0 N HCl and 400 μl methanol, and extracted as described above. Production of aminoacifluorfen and TFMCP was determined by RAD-TLC. Kinetic parameters (K_m and V_{max}) were determined from the Michaelis-Menten equation using a nonlinear regression model.

Results and Discussion

Acifluorfen Metabolism in Soil and Rhizosphere Suspensions. Acifluorfen (I; see Fig. 5 for structures of parent and metabolites) was rapidly degraded in rhizosphere and root-free soils (Fig. 1 and 2). After a 6 d incubation no acifluorfen was recovered from inner or outer rhizosphere soils, and 32 % recovered from root-free soil. Because different amounts of soil were used in the inner rhizosphere studies (500 mg) compared to 5.0 g for root-free and outer rhizosphere soils, only limited comparisons can be made. Half-lives for acifluorfen were 1.3 d in inner rhizosphere soil, 2.5 d in outer rhizosphere soil, and 4.5 d in root-free soil. Metabolism of acifluorfen by soil and rhizosphere suspensions produced one major metabolite (Fig. 1 and 2). This compound has been identified as aminoacifluorfen, based upon co-chromatography with an authentic pure standard in HPLC and TLC analyses in two solvent systems. Based on production of aminoacifluorfen, reduction of the nitro moiety proceeded more rapidly in inner rhizosphere > outer rhizosphere > root-free soil (Fig. 1 and 2). Rapid incorporation of ^{14}C into methanol and alkaline methanol-unextractable residues in the soil was concurrent with aminoacifluorfen formation. Throughout the study, significantly

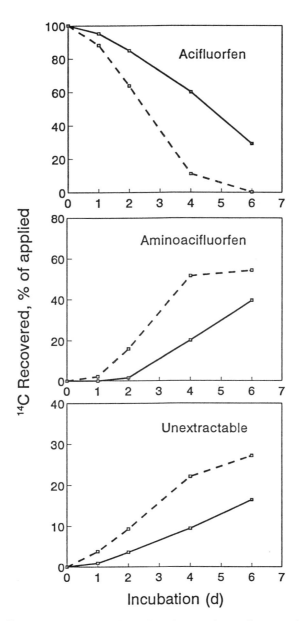

Figure 1. ^{14}C- Acifluorfen transformations in root-free soil suspensions (solid lines) and outer rhizosphere soil suspensions (dashed lines) over a 6 d incubation.

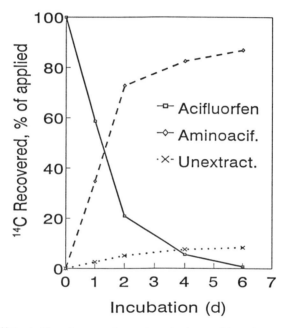

Figure 2. [14]C- Acifluorfen transformations in inner rhizosphere suspensions. Acifluorfen (solid line), aminoacifluorfen (dashed line), unextractable (dotted line) over a 6 d incubation.

greater amounts of the applied ^{14}C were incorporated into unextractable material in outer rhizosphere soils than in root-free soils. During the course of the incubation, two other metabolites (TFMCP and decarboxyacifluorfen) were found in methanol extracts (each less than 2% of total ^{14}C applied). Decarboxyacifluorfen was also found in sterile acifluorfen solution during the incubation and may be a product of photolysis (24). Since we used only ^{14}C label in one of the two acifluorfen rings, we are only able to account for the fate of the trifluoromethyl, 2-chlorophenol ring. The basic degradative pathway for transformation of acifluorfen in soils and rhizospheres appears to be initial nitroreduction as reported for other nitrodiphenyl ether herbicides (14-17), followed by ether bond cleavage and incorporation of these products into unextractable humic components. Oxidative coupling reactions for incorporation of phenols (25) and anilines (26) into humic acids have been described for many pesticide metabolites.

Table I. Aerobic, Heterotrophic Bacterial Populations Associated with Root-free Soil, and Outer and Inner Rhizosphere Soil Suspensions During a 144 h Incubation

Time (h)	Fluorescent pseudomonads*	Gram-negative bacteria*	Total bacteria*
	----------------------- log (10) cfu g[-1] soil -------------------------		
Root-free soil			
0	4.99 d	6.17 d	7.43 c
24	5.21 c	6.30 d	7.82 b
48	5.46 b	6.85 b	8.09 b
96	5.81 a	6.85 b	8.10 b
144	5.96 a	7.13 a	8.17 a
Outer rhizosphere			
0	6.83 d	7.32 d	8.24 d
24	7.00 c	7.65 c	8.67 c
48	7.41 a	8.10 b	9.10 a
96	7.40 a	8.25 a	9.09 a
144	7.16 b	8.21 a	8.79 b
Inner rhizosphere			
0	8.07 d	9.00 d	9.51 c
24	8.32 bc	9.16 b	9.63 a
48	8.38 a	9.24 a	9.57 ab
96	8.34 ab	9.08 c	9.41 d
144	8.26 c	9.08 c	9.36 d

* Mean of three replicates of root-free and outer rhizosphere soil and four replicates of inner rhizosphere soil. Means within a column for a particular soil followed by the same letter do not differ at the 95% confidence level.

The kinetics of acifluorfen metabolism is dependent on initial microbial populations associated with the soil and rhizosphere samples and subsequent population changes observed during the incubation (Table I). Numbers of all three classes of bacteria were greatest in inner rhizosphere, with outer rhizosphere populations greater than in root-free soil. Gram-negative bacteria became more dominant during the incubation; i.e., in root-free soil, they accounted for 5 % of the total bacteria initially and 9 % after 6 d. In outer rhizosphere soil, Gram-negative bacteria comprised 12 % of the total bacteria initially and 27 % after 6 d, while in inner rhizosphere soils 31 % initially and 53 % after 6 d. Fluorescent *Pseudomonas* populations exhibited a similar enhancement as observed in the Gram-negative bacteria during the study.

Acifluorfen Metabolism by Pure Cultures of Rhizobacteria. Several strains of rhizosphere competent *P. fluorescens* strains were capable of transforming acifluorfen to aminoacifluorfen in resting cell suspensions under static (microaerophilic) conditions (Fig. 3). Reduction of the nitro moiety proceeded more rapidly in the three *P. fluorescens* strains compared to the *Enterobacter cloacae* strain. In the three *Pseudomonas* strains incubated in 80 mM acifluorfen, rapid conversion to aminoacifluorfen was observed initially during the incubation, followed by a period of lower activity. Similar patterns of aminoacifluorfen reduction were noted in enrichment cultures (19), which was attributed to either insufficient reductant or possible toxicity of aminoacifluorfen. No metabolites other than aminoacifluorfen were observed in any of these strains at this concentration. A more complete reduction of acifluorfen was observed when either strain RA-2 or UA5-40 was incubated in 8 mM acifluorfen (Fig. 3). Only strain UA5-40 produced TFMCP, at about 3 % of the initial ^{14}C after 20 h at 8 μM acifluorfen (data not shown). Isolates obtained from enrichment cultures capable of acifluorfen nitroreduction (27) have been characterized as spore-forming bacteria (*Bacillus thuringiensis, Clostridium perfringes* and *C. sphenoides*). Our work indicates that this transformation can also be mediated by obligate aerobic organisms; i.e., *P. fluorescens*, and that perhaps this transformation may be mediated by a wide variety of bacteria as has been suggested for other nitrodiphenyl ether herbicides (15). *E. cloacae* ATCC 43560, has aromatic nitroreductase activity on a wide range of substrates including trinitrotoluene and nitrobenzoic acid (10), in addition to acifluorfen as demonstrated in our study. Thus, acifluorfen may be a substrate for nitroreductases from many sources. The nitroreductase of ATCC 43560 is considered inducible (10). However in our studies, exposure to 2-NBA, prior to acifluorfen exposure had no effect on acifluorfen reduction (data not shown). Similar low rates of aminoacifluorfen reduction (12 % of acifluorfen transformed to aminoacifluorfen in 24 h) were observed under both conditions.

The effects of aeration on formation of aminoacifluorfen by strains BD4-13 and RA-2 were evaluated in cell suspensions and TSB cultures (Table II). Under well-aerated conditions, minimal production of aminoacifluorfen was observed by both strains in either cell suspensions or TSB cultures. Although lower amounts of aminoacifluorfen were formed by static broth cultures, the cell density was 10% of that of cell suspensions. Studies by others (19) indicated that acifluorfen reduction was much more rapid under anaerobic compared to aerobic conditions. Further studies on acifluorfen metabolism with strain

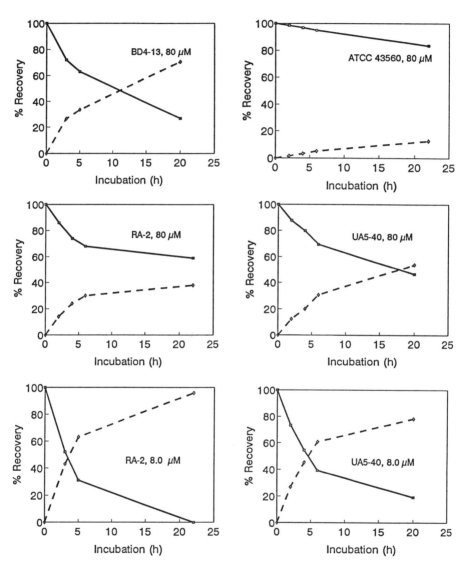

Figure 3. Acifluorfen transformations by static bacterial cell suspensions over a 20 to 22 h incubation. Acifluorfen (solid line), aminoacifluorfen (dashed line).

RA-2 evaluated the effects of preconditioning the cells under microaerophilic conditions (static incubation) or by exposure to an alternative substrate 2-NBA (Table III). Even when cell suspensions of RA-2 were well-aerated, they expressed 75% of the nitroreductase activity of statically incubated RA-2 cell suspensions during a 2.5 h microaerophilic assay. Prior exposure of RA-2 cells to 2-NBA under either incubation condition had no significant effect on aminoacifluorfen formation by whole cells. However, cfe's obtained from these cells had significantly different activities under a N_2 atmosphere. Crude extracts from microaerophilic cultures exhibited 51 to 139% greater total activity ml^{-1} compared to cells maintained under well-aerated conditions. Cells previously maintained under well-aerated conditions in the presence of 2-NBA had significantly greater acifluorfen nitroreductase activity compared to well-aerated cells not exposed to 2-NBA only in total activity (not specific activity). Cells exposed to 2-NBA under microaerophilic conditions were not different than cells incubated under microaerophilic conditions in the absence of 2-NBA. These studies indicate that a certain level of aminoacifluorfen nitroreductase activity in these strains is constitutive, however this activity can be stimulated under low oxygen tensions.

Table II. Production of Aminoacifluorfen in Cell Suspensions and TSB Cultures of Strains BD4-13 and RA-2 Incubated Statically or Under Well-aerated Conditions

Incubation condition	Aminoacifluorfen formed nmole ml^{-1} cells in 22 h	
	BD4-13	RA-2
Cell suspensions, well-aerated	1.0 *	1.3
Cell suspensions, static	52.3	17.6
TSB culture, well-aerated	0.3	0.4
TSB culture, static	17.6	6.8

* Average of duplicate samples

Table III. Effects of Preconditioning Aeration and 2-Nitrobenzoic acid (2-NBA) on Acifluorfen Nitroreductase Activity of Strain RA-2, in Whole Cell Suspensions and Cell-free Extracts

Precondition regime	Whole Cells nmol aminoacifluorfen formed ml^{-1} cells h^{-1}	Cell-free extracts nmol aminoacifluorfen formed	
		ml^{-1} cfe h^{-1}	mg protein^{-1} h^{-1}
Microaerophilic, 2-NBA	6.96 a*	25.5 a	6.04 a
Microaerophilic, none	6.67 a	27.0 a	6.37 a
Aerated, 2-NBA	5.23 b	16.8 b	4.27 b
Aerated, none	5.17 b	11.3 c	4.73 b

* Mean of three replicates, means within a column followed by the same letter do not differ at the 95% level.

Aromatic nitroreductases of various bacteria have been shown to require various reductants, i.e., NADH and /or NADPH (6, 9-12). Flavin mononucleotide is also an important cofactor in many aromatic nitroreductases (6, 9-12). We investigated acifluorfen nitroreductase activity in cfe's as affected by oxygen and exogenous reductant (NADH or NADPH, and FMN) of three bacterial strains ATCC 43560, RA-2 and UA5-40 (Table IV). No acifluorfen nitroreductase activity was observed in cfe's of *Enterobacter cloacae* strain ATCC 43560 under any of the conditions studied, even when assay conditions were extended to 5 h (data not shown). No nonenzymatic nitroreduction observed in acifluorfen controls regardless of reductant. In the presence of exogenous reductant (NADH and FMN), acifluorfen nitroreductase activity of RA-2 and UA5-40 under air was approximately 70% of that under a nitrogen environment, indicating minimal inhibition by oxygen. Addition of NADH and FMN significantly increased aminoacifluorfen formation in strain RA-2 under both atmospheres. In strain UA5-40, increased nitroreductase activity due to NADH and FMN was observed only under nitrogen. Addition of NADPH had no effect on RA-2 activity but significantly reduced UA5-40 acifluorfen nitroreductase activity. TFMCP was produced under most of these assay conditions by both strains. In most cases much more TFMCP was produced in cfe's of UA5-40 compared to RA-2, except when exogenous NADH or NADPH were added under aerobic conditions. In both strains the highest amount of TFMCP was produced when only NADH was added. A general mechanism of resistance to the diphenyl ether herbicides in plants is metabolic inactivation by a glutathione- or homoglutathione-mediated cleavage of the ether bond (28). This has also been observed for the nitrodiphenyl ether herbicide fluorodifen by various rhizosphere pseudomonads (23). Aminoacifluorfen is more electrophilic than acifluorfen and may be more susceptible to glutathione-mediated cleavage of the ether bond compared to acifluorfen. Studies on metabolism of the dinitro herbicide 3,5-dinitro-*o*-cresol by a *Pseudomonas* sp. demonstrated that one nitro group is initially reduced via nitroreductase, followed by hydroxylation, before the second nitro group is reduced (29).

Table IV. Formation of Aminoacifluorfen and 4-Trifluormethyl-2-chlorophenol (TFMCP) by Cell-free Extracts of Strains RA-2 and UA5-40 as Affected by Atmosphere and Reductant

Assay conditions Atmosphere:Reductant	Aminoacifluorfen formed nmol mg protein h^{-1}		TFMCP formed nmol mg protein h^{-1}	
	RA2	UA5-40	RA-2	UA5-40
N_2 : NADH and FMN	15.58 a	3.63 a	<0.15 c	1.25 b
N_2 : none	6.97 c	2.59 b	0.57 b	1.04 b
Air : NADH and FMN	11.22 b	2.67 b	0.18 c	1.07 b
Air : NADH	4.80 d	2.31 b	2.19 a	1.94 a
Air : NADPH	3.52 d	1.60 c	1.40 a	1.61 ab
Air : FMN	3.94 d	2.55 b	<0.15 c	1.06 b
Air : none	3.88 d	2.27 b	<0.15 c	1.20b*

* Mean of three replicates, means for a strain within a column followed by the same letter do not differ at the 95% level.

Aerobic aminoacifluorfen formation in crude enzyme preparations (cfe's) of RA-2 and UA5-40 exhibited acifluorfen concentration dependent kinetics (Fig. 4) that can be described by the Michaelis-Menten equation. Apparent kinetic parameters for acifluorfen nitroreductase in strain RA-2 and UA5-40 were: K_m = 44.2; Vmax = 6.07, and K_m = 60.7 ìM; V_{max} = 6.07 nmol mg protein^{-1} h^{-1}, respectively. The apparent Km for acifluorfen in these *P. fluorescens* crude extracts is similar to that reported for 2,4,6- trinitrotoluene, but ten-fold lower than that for 2,4-dinitrotoluene and nitrofurazone in a purified nitroreductase preparation from *Enterobacter cloacae* 43560 (10). TFMCP was produced at all acifluorfen concentrations in cfe's from UA5-40. In RA-2 preparations, TFMCP was consistently observed in all replicates at acifluorfen concentrations up to 50 μM. But at higher concentrations, TFMCP formation was inconsistent and/or below levels of detection (<2.2 % of total radioactivity) in RA-2. In assays with UA5-40, TFMCP production was 24 to 43% of the aminoacifluorfen formed, with the relative amount decreasing with increased acifluorfen concentration. Based on this and the previous study, it appears that there is competition for reductant between nitroreduction and ether bond cleavage. This is evident from studies with cell suspensions, where TFMCP was rarely observed except for UA5-40 exposed to 8 μM acifluorfen. Coordinate expression of both pathways would be highly desirable for remediating matrices contaminated with nitrodiphenyl ether herbicides. Data from pure culture and cell-free bacterial metabolism and from soil/rhizosphere studies suggest a multi-stepped degradation pathway for acifluorfen (Fig. 5). An understanding of the dynamics of reductant (NADH) and GSH pools of rhizosphere bacteria in relation to the activity of these enzymes is needed. The relationship between expression of nitroreduction and ether bond cleavage is being pursued in further studies.

Implications of Aromatic Nitroreduction in Phytoremediation. Our studies indicate that the potential for aromatic nitroreduction of acifluorfen is enhanced in rhizosphere soil compared to root-free soil. We did not control redox potential or measure oxygen concentration in our experimental system. However, microbial activity, growth and respiration in these soil slurries could have reduced O_2 tensions (especially in a closed system) that favored the transformation of acifluorfen to aminoacifluorfen. Studies by others (14-17) indicated that formation of aminoderivatives of nitrodiphenyl ether herbicides occurs primarily under low redox conditions. Plant rhizospheres have several unique factors that would perhaps make expression of nitroreduction more prevalent in rhizospheres compared to soil. It has been demonstrated here and in many other studies that microbial populations (especially Gram-negative bacteria) are more numerous in the rhizosphere (3, 22). Root exudation of labile carbonaceous materials enhances microbial activity. Higher microbial respiration combined with root respiration would result in lower O_2 and higher CO_2 concentrations. Other processes related to reduced O_2 tension, i.e., denitrification, are more active in rhizospheres compared to soil (30, 31). Our findings that several Gram-negative bacteria (*P. fluorescens* and *Enterobacter cloacae)* capable of nitroreductase activity on acifluorfen indicates this activity is perhaps widely distributed in soil, and especially in rhizosphere bacteria. Bacteria with acifluorfen nitroreductase activity identified in this study also have activity on other nitroaromatic substrates such as the herbicide trifluralin, nitrophenol, and nitrobenzoic acid (unpublished data). It may be feasible to select specific

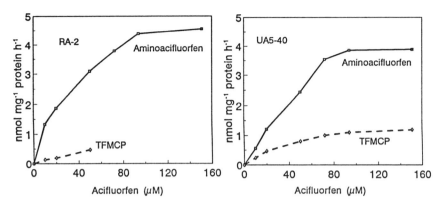

Figure 4. Concentration dependent aerobic catabolism of acifluorfen in cell-free extracts of *Pseudomonas fluorescens* strains RA-2 and UA5-40, without exogenous reductant. Aminoacifluorfen (solid line), TFMCP (dashed line).

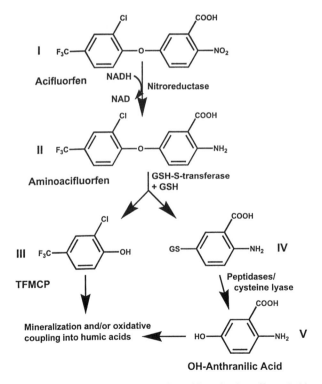

Fgure 5. Proposed degradation pathway for acifluorfen in soils and rhizospheres. **I** = acifluorfen, **II** = aminoacifluorfen, **III** = 4-trifluoromethyl -2-chlorophenol (TFMCP), **IV** = glutathione conjugate of anthranilic acid, **V** = hydroxy anthranilic acid.

organisms with more unique transformations of certain nitroaromatic compounds; i.e., oxidative release of nitrite, or nitroreduction and subsequent ring hydroxylation (29), but this has not been observed in enrichment cultures with acifluorfen (18, 19). Strains that have acifluorfen nitroreductase activity also express other catabolic pathways for detoxification of other herbicides, i.e., glutathione conjugation for UA-5-40 and BD4-13 (23), and/or aryl acylamidase, RA-2 (22). Certain algae and higher plants possess similar nitroreductases, that may have promise in removal of trinitrotoluene and other nitroaromatic contaminants from soil and water (32). It may be possible to exploit both microbes and plants in bioremediation approaches. There is intrinsic potential for degradation of nitroaromatic compounds in the rhizosphere. Understanding factors regulating aromatic nitroreductase activity and maximizing expression of this enzyme(s) by rhizosphere microorganisms may lead to practical methods for the removal of nitroaromatic contaminants using phytoremediation.

Acknowledgments

We are grateful to BASF Corporation for generously providing the ^{14}C-labeled acifluorfen and metabolites and Andy Goetz (BASF) for sharing TLC methodologies. We appreciate the technical assistance of Liz Smyly, Earl Gordon, and Rachael Voth. We also wish to thank F. Dayan, R.H. Kloth, H. Lee and E.M. Thurman for their reviews of the manuscript.

Literature Cited

1. Cunningham, S.D.; Ow, D.W. *Plant Physiol.* **1966,** *110,* 715-719.
2. Anderson, T.A.; Guthrie, E.A.; Walton, B.T. *Environ. Sci. Technol.* **1993,** *27,* 2630-2635.
3. Walton, B.T.; Guthrie, E.A.; Hoylman, B.T. In *Bioremediation Through Rhizosphere Technology;* T.A. Anderson, J.R. Coats, Eds.; ACS Books, Washington DC, **1994,** 11-26.
4. Zeyer, J.; Kearney, P.C. *J. Agric. Food Chem.* **1984,** *32,* 238-242.
5. Raymond, G.G.M.; Alexander, M. *Pestic. Biochem. Physiol.* **1971,** *1,* 123-130.
6. Kinouchi, T.; Ohnishi, Y. *Appl. Environ. Microbiol.* **1983,** *46,* 596-604.
7. Raffi, F.; Franklin, W.; Heflich, R.H.; Ceriglia, C.E. *Appl. Environ. Microbiol.* **1991,** *57,* 962-968.
8. Cartwright, N.J.; Cain, R.B. *Biochem. J.* **1959,** *73,* 305-314.
9. Peterson, F.J.; Mason, R.P.; Hovsepain, J.; Holtzman, J.L. *J. Biol. Chem.* **1979,** *254,* 4009-4014.
10. Bryant, C.; DeLuca, M. *J. Biol. Chem.* **1991,** *266,* 4119-4125.
11. Raffi, F.; Selby, A.L.; Newton, R.K.; Ceriglia, C.E. *Appl. Environ. Microbiol.* **1994,** *60,* 4263-4267.
12. Bryant, D.W.; McCalla, D.R.; Leeksma, M.; Laneville, P. *Can. J. Microbiol.* **1981,** *27,* 81-86.
13. Weed Science Society of America. In *Herbicide Handbook, 7th ed.;* W.H. Ahrens, Ed.; Weed Science Society of America, Champaign Il. **1997,** 7-9.

14. Leather, G.R.; Foy, C.L. *Pestic. Biochem. Physiol.* **1977**, *7*, 437-442.
15. Niki, Y.; Kuwatsuka, S. *Soil Sci. Plant. Nutri.* **1976**, *22*, 223-232.
16. Oyamanda, M.; Kuwatsuka, S. *J. Pestic. Sci.* **1988**, *13*, 99-105.
17. Oyamanda, M.; Kuwatsuka, S. *J. Pestic. Sci.* **1989**, *14*, 321-327.
18. Gennari, M.; Négre, M. In *Study and Prediction of Pesticide Behaviour in Soil, Plants, and Aquatic Systems.* **1990**, *3*, 221-236.
19. Gennari, M.; Négre, M.; Ambosoli, R.; Andreoni, V.; Vincenti, M.; Acquati, A. *J. Agric. Food Chem.* **1994**, *42*, 1232-1236.
20. Andreone, V.; Colombo, M.; Gennari, M.; Négre, M.; Ambosoli, R. *J. Environ. Sci. Health,* **1994**, *B29*, 963-987.
21. Anderson, T.A.; Krueger, E.L.; Coats, J.R. *Chemosphere* **1994**, *28*, 1551-57.
22. Hoagland, R.E.; Zablotowicz, R.M.; Locke, M.A. In *Bioremediation Through Rhizosphere Technology;* T.A. Anderson, J.R. Coats, Eds.; ACS Books, Washington DC, **1994**, 160-183.
23. Zablotowicz, R.M.; Hoagland, R.E.; Locke, M.A. In *Bioremediation Through Rhizosphere Technology;* T.A. Anderson, J.R. Coats, Eds.; ACS Books, Washington DC, **1994**, 184-98.
24. Pusino, A.; Gessa, C. *Pestic. Sci.* **1991**, *32*, 1-5.
25. Bollag, J.-M.; Loll, M.J. *Experientia* **1983**, *39*, 1221-1231.
26. Tatsumi, K; Freyer, A.; Minard, R.D.; Bollag, J.M-. *Environ. Sci Technol.* **1994**, *28*, 210-215.
27. Fortina, M.G.; Acquati, A.; Ambrosoli, R. *Institut Pasteur Res. Microbiol.* **1966**, *147*, 193-199.
28. Frear, D.S.; Swanson, H.R.; Manasager, E.R. *Pestic. Biochem. Physiol.* **1983**, *20*, 299-310.
29. Tewfik, M.S.; Evans, W.C. *Biochem. J.* **1966**, *99*, 31-32.
30. Brar, S.S. *Plant Soil* **1972**, *36*, 713-722.
31 Bailey, L.D.; Beauchamp, E.C. *Can. J. Soil Sci.* **1973**, 219-225.
32. Schnoor, J.L.; Licht, L.A.; McCutcheon, S.C.; Wolfe, N.L. Carreira, L.H. . *Environ. Sci. Technol.* **1995**, *29*, 318-323.

Chapter 4

Atrazine Degradation in Pesticide-Contaminated Soils: Phytoremediation Potential

Ellen L. Kruger[1], Jennifer C. Anhalt[1], Diana Sorenson[1], Brett Nelson[1], Ana L. Chouhy[2], Todd A. Anderson[3], and Joel R. Coats[1]

[1]Pesticide Toxicology Laboratory, Department of Entomology, Iowa State University, Ames, IA 50011–3140
[2]Ministerio de Ganaderia Agricultura y Pesca, Direccion de Sanidad Vegetal, Servicios de Laboratorios, Montevideo, Uruguay
[3]The Institute of Wildlife and Environmental Toxicology, Department of Environmental Toxicology, Clemson University, Pendleton, SC 29670

Studies were conducted in the laboratory to determine the fate of atrazine in pesticide-contaminated soils from agrochemical dealer sites. No significant differences in atrazine concentrations occurred in soils treated with atrazine individually or combinations with metolachlor and trifluralin. In a screening study carried out in soils from four agrochemical dealer sites, rapid mineralization of atrazine occurred in three out of eight soils tested, with the greatest amount occurring in Bravo rhizosphere soil (35% of the applied atrazine after 9 weeks). Suppression of atrazine mineralization in the Bravo rhizosphere soil did not occur with the addition of high concentrations of herbicide mixtures, but instead was increased. Plants had a positive impact on dissipation of aged atrazine in soil, with significantly less atrazine extractable from *Kochia*-vegetated soils than from nonvegetated soils.

High concentrations of pesticides in soils is a serious problem at agrochemical dealer sites where contamination has resulted from inadvertent spillage during mixing and loading of chemicals for application to crops. Expensive remediation technologies may not be economically feasible for such dealerships. In addition, biological approaches (bioremediation) may be inhibited by the presence of mixtures of contaminants at high concentrations. The primary contaminants at many of these sites are herbicides *(1);* thus, using a phytoremediation strategy as an *in situ* solution for cleanup of these soils is challenging. Studies of pesticide-contaminated soils from several agrochemical dealer sites have shown encouraging results, however, in that certain plants can survive moderate contamination *(2)*.

It is important to understand how the fate of individual pesticides in soil can be influenced by the presence of other pesticides. Pesticide fate studies in soil, however, have typically addressed individual compounds only. An additional problem with soils that have been contaminated over the course of many years is the decreased bioavailability of the compounds for further degradation by biological systems. Compounds that remain in soil for

54

an extended time are referred to as aged pesticide residues, with the word *residue* referring to the residual nature of the compound *(3)*. In conducting laboratory studies to better understand the fate of pesticides in contamination scenarios, it is important to consider the influence of aging the compounds in soil.

A phenomenon that can occur in soil that has had long-term exposure to chemicals is enhanced degradation. When the chemical is applied to soil that has previously been exposed to that chemical, accelerated decomposition can occur. This enhanced degradation is induced by the prior treatment to the soil, but does not develop in all soils nor with all pesticides. Microorganisms able to survive or benefit from the imposed selective pressure will dominate and be poised to more rapidly degrade the compound upon the next application *(4)*. In a pest-control situation, this is not ideal because failure to control unwanted plants or insects can result in financial loss to the farmer. In a contaminated-soil scenario, however, this could be a desirable characteristic in that subsequent spills would be more rapidly degraded by the selected microbial population. Also, soils with the capability of rapidly degrading a particular pesticide are primary candidates for isolation of degrading populations of microorganisms, which could be cultured and used in bioremediation approaches, such as bioreactors or on-site inoculations.

Phytoremediation of pesticide-contaminated soils is of great interest as an inexpensive, on-site alternative to agrochemical dealers. Plants can have a positive influence on removal of organic wastes by taking up contaminants into the plant tissue, with the possibility of further metabolism into innocuous compounds. Plants also influence the degradation of contaminants as a result of the increased microbial activity associated with the roots *(5)*. A study conducted in this laboratory indicated that degradation of a mixture of herbicides was greater in soil taken from the rhizosphere of a herbicide-tolerant plant than in soil from a nonvegetated area of an agrochemical dealer site *(6)*.

Studies were conducted in this laboratory on contaminated soils taken from agrochemical dealerships to determine the fate of a herbicide when applied individually or in combination with other herbicides, to determine the degradative capabilities of the soils on freshly applied herbicide wastes, to determine if the degradative ability could be suppressed by high concentrations of pesticide mixtures or if this ability could be transferred from one soil to another, and to assess whether plants could have a positive influence on the degradation of these wastes. Summaries of results from several experiments are presented in this chapter to give an overview of the approach we have taken to solve the problem of remediation of pesticide-contaminated sites.

Degradation of Atrazine Applied in Herbicide Mixtures to Soil From a Pesticide-Contaminated Site

Surface soil samples from an agrochemical dealership in Iowa were obtained by using shovels to remove the top 15 cm of soil. Samples were placed into large 30-gallon metal drums and transported back to the laboratory where they were stored at ambient temperature. Soils were sieved (2.4 mm) and mixed well. A subsample of soil was sent to Midwest Laboratories of Omaha, NE, for determination of physicochemical properties. Background herbicide concentrations were determined by extracting subsamples of soil three times with ethyl acetate (2:1 solvent:soil ratio, v/v). Samples were extracted by mechanical agitation for 20 minutes, followed by vacuum filtration. Combined extracts were concentrated by rotary evaporation and then rediluted to 10 mL. Concentrations of pesticides in the extracts were determined by using gas chromatography. A Shimadzu GC9A gas chromatograph equipped with a nitrogen-phosphorus detector was used under the following conditions: packed glass column, OV17 (1.8 m); injector temperature, 250 °C; detector temperature, 250 °C; column temperature, 235 °C; carrier gas, He (40 mL/min). External standards of known pesticide concentrations were used in determining pesticide concentrations in soil extracts.

Soils were treated at 50 µg/g for each herbicide to obtain the following four treatments (three replicates each): (1) atrazine; (2) atrazine and metolachlor; (3) atrazine and pendimethalin; and (4) atrazine, metolachlor, and pendimethalin. Treated soils were mixed well before aliquots were transferred to French square bottles (250 mL volume) and randomly assigned to one of four incubation lengths (0, 21, 63, and 160 days.) Soil moistures were adjusted to the gravimetric soil moisture content at -33 kPa and maintained at this level throughout the incubation period by adding ultrapure water to maintain the initial weight. The jars were opened to the air once a week to maintain an aerobic headspace. All treatments were incubated at 25°C in the dark. At the end of each incubation period, a 20-g aliquot of soil from each incubation jar was extracted and analyzed as described previously.

The soil used in this experiment had the following physicochemical properties: loam texture; organic matter content, 3.9%; sand content, 50%; silt content, 34%; clay content, 16%; total nitrogen content, 0.26%; pH, 7.0; and cation exchange capacity, 12.7. Background concentrations of atrazine and trifluralin were 1 µg/g or less, whereas concentrations of metolachlor and pendimethalin were 4 to 11 µg/g. Atrazine was very degradable in these soils, and no significant differences in atrazine concentrations were seen among the four treatments (ANOVA; $p = 0.85$) (Figure 1). After 21 days, 6% of the applied atrazine was extractable from soil (mean of all treatments), and 2% of the applied atrazine was extractable after 160 days. The half-life of atrazine, based on extractable atrazine residues, was calculated to be 45 days. The half-life was not significantly effected by the application of multiple herbicides.

Assessing the Degradative Capabilities of Pesticide- Contaminated Soils from Agrochemical Dealer Sites

To determine if rapid atrazine degradation was widespread in pesticide-contaminated soils from agrochemical dealer sites, soils from four locations in Iowa were used for mineralization studies. Surface soils (0 to 15 cm) were taken from nonvegetated areas of dealer sites by using hand trowels. Soils were placed in whirlpak™ bags and transported to the laboratory. From two of the soil-source sites (Alpha and Bravo), *Kochia scoparia* plants were transplanted to small pots and transported to the laboratory. Rhizosphere soil was obtained by shaking the plants and collecting soil that had been in close contact with the roots. Soils were mixed well and sieved (2.4 mm), and soil moistures were determined. Twenty-gram aliquots were extracted and analyzed for background herbicide concentrations, as described earlier (Table 1). Soils with the greatest quantities of herbicide contamination included Echo B, with over 2600 µg/g trifluralin and Foxtrot B with 9 µg/g of atrazine. Soil from Echo A had 1 to 3 µg/g of alachlor and metolachlor, respectively. A treating solution of analytical grade atrazine (ChemService; West Chester, PA) and uniformly ring-labeled ^{14}C-atrazine (Ciba Crop Protection; Greensboro, NC) was made in certified acetone. Aliquots of soil (10 or 20-g) were placed in 250-ml incubation jars (three replications) and treated with the

Table 1. Background pesticide concentrations of soils used in this study

Site	Soil type	Location	Background Herbicide Concentrations (µg/g)				
			Atrazine	Alachlor	Metolachlor	Pendimethalin	Trifluralin
Alpha A	nonvegetated[1]	Iowa	Trace	--	0.3	--	--
Alpha B	rhizosphere[2]	Iowa	Trace	--	0.3	--	0.1
Bravo A	nonvegetated	Iowa	0.3	0.2	0.2	0.3	0.3
Bravo B	rhizosphere	Iowa	Trace	--	0.6	0.4	0.9
Echo A	nonvegetated	Illinois	0.1	1.4	3.4	0.4	0.2
Echo B	nonvegetated	Illinois	--	--	--	--	2638
Foxtrot A	nonvegetated	Nebraska	--	2.7	--	1.4	--
Foxtrot B	nonvegetated	Nebraska	9.0	--	--	--	--

[1] Soil taken from a nonvegetated area at an agrochemical dealer site
[2] Soil taken from the rhizosphere of *Kochia scoparia* at an agrochemical dealer site

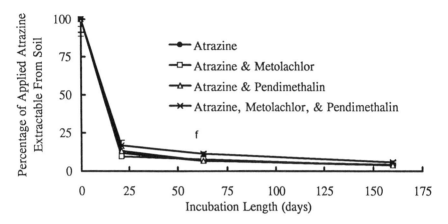

Figure 1. Atrazine degradation in pesticide-contaminated soils applied with mixtures of herbicides (50 μg/g per herbicide). Bars represent standard deviations of the mean (n=2).

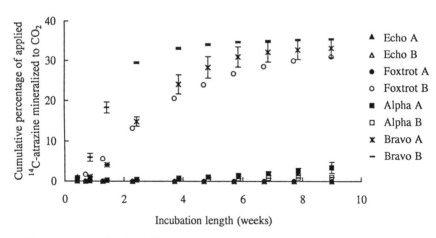

Figure 2. Mineralization of ^{14}C-atrazine applied to pesticide-contaminated soils from agrochemical dealer sites at a concentration of 50 μg/g. Bars represent standard deviations of the mean (n=3).

radiolabeled treating solution to give a final atrazine concentration of 50 μg/g soil. Acetone was evaporated immediately. Soil moistures were adjusted to 60% saturation, and each incubation jar was equipped with a scintillation vial containing NaOH (0.5 N) to trap $^{14}CO_2$ evolved from atrazine mineralization. The traps were changed weekly, and $^{14}CO_2$ was quantified by using liquid scintillation techniques. Soil moistures were maintained by adding ultrapure water to the incubation jars to maintain the original weight at 60% saturation. Aerobic conditions were maintained in the incubation jars by opening the jars up during trap changing each week.

The greatest amount of atrazine mineralization occurred in soils from Bravo (both nonvegetated and rhizosphere soils) (Figure 2). After 9 weeks, 33% and 35% of the applied ^{14}C-atrazine was mineralized to CO_2 in the nonrhizosphere and rhizosphere soils, respectively. Perkovich et al. *(7)* found that atrazine was mineralized to a greater extent in rhizosphere soil than nonrhizosphere soil from this site after 36 days of incubation. In the current study, there was no statistical difference was shown between the rhizosphere and nonrhizosphere soils after 63 days, although the lag time for mineralization was less in rhizosphere soils than nonrhizosphere soils (Figure 2). One soil from the agrochemical dealership in Nebraska (Foxtrot B) mineralized 31% of the applied atrazine. Before the herbicide treatment for this study, atrazine was the only detectable contaminant of this soil (9 μg/g). The other Nebraska soil (Foxtrot A) that did not readily mineralize atrazine contained background levels of alachlor and pendimethalin (2.7 and 1.4 μg/g, respectively). Both soils from Illinois (Echo A and B) did not mineralize atrazine. As mentioned previously, Echo A had background concentrations of alachlor and metolachlor (1.4 and 3.4 μg/g, respectively) and less than 0.5 μg/g of atrazine, trifluralin, and pendimethalin. The other Echo soil (B) had extremely high levels of trifluralin (2638 μg/g) (Table 1). This study shows that the extent of atrazine mineralization in pesticide-contaminated soils from agrochemical dealerships is variable. Sites that have rapid mineralization characteristics for atrazine might be candidates for isolation of atrazine-degrading microorganisms for potential use as inoculants of soils that do not have this capacity. Several researchers have isolated atrazine-degrading microorganisms from pesticide-contaminated soils *(8-10)*. While an atrazine-mineralizing capacity has been demonstrated in several soils, most soils do not develop an atrazine-mineralizing characterization. The natural selection of atrazine-degrading populations of microorganisms may be inhibited in most soils by other pesticide contaminants or by other characteristics of the soil.

Suppression of Atrazine Mineralization

An experiment was conducted to determine if greater concentrations of herbicide mixtures would suppress atrazine mineralization in the Bravo soil. Four treating solutions were made with combinations of analytical grade atrazine, metolachlor, and trifluralin. All treating solutions contained uniformly ring-labeled ^{14}C-atrazine. The four treatments were: (1) atrazine, (2) atrazine and metolachlor, (3) atrazine and trifluralin, (4) atrazine, metolachlor, and trifluralin. Twenty-g aliquots of Bravo rhizosphere soil were treated with one of the four treating solutions (two replications), as described previously, to reach a final soil concentration of 200 μg/g per herbicide. The methods of soil moisture adjustment, $^{14}CO_2$ trapping, and incubation were identical to our study described in the previous section.

Atrazine mineralization was not suppressed by the higher concentrations and mixtures of herbicides (Figure 3). Instead, the rate of mineralization of atrazine in soils treated with herbicide mixtures at 200 μg/g each was approximately twice that in Bravo rhizosphere soils treated with 50 μg/g of the individual herbicide atrazine (Bravo B, Figure 2). After 9 weeks, there were no significant differences in the percentages of atrazine mineralized among the four treatments, with the percentages ranging from 69 to 76% of the applied ^{14}C-atrazine. Herbicide mixtures at these concentrations may be toxic to competing microbial populations, thus allowing for proliferation and increased degradation of atrazine by the atrazine-degrading population. It may also be possible that more of the ^{14}C-atrazine is available to the microorganisms at the higher concentration.

Transfer of Degradative Capabilities

Results of the screening study, which assessed the degradative capabilities of soils from several agrochemical dealer sites, revealed that not all soils could rapidly mineralize atrazine (Figure 2). An experiment was conducted to determine if a soils ability to degrade atrazine could be transferred to a soil that could not mineralize atrazine by mixing the two soil types in an attempt to "inoculate" the slow atrazine-degrading soil. The two soils used for this study were the Bravo B (rhizosphere) soil and the Echo A (nonrhizosphere) soil with the lower background herbicide concentrations (Table 1). Four combinations of soil mixtures were used for this study: (1) 100% Bravo, (2) 90% Echo: 10% Bravo, (3) 80% Echo: 20% Bravo, and (4) 100% Echo. The treating solution consisted of analytical grade and uniformly ring-labeled ^{14}C-atrazine in acetone. Soil combinations were mixed thoroughly, and 20 g of each combination (two replications) were treated in 250-ml incubation jars to obtain a soil concentration of atrazine at 50 μg/g. The methods of soil moisture adjustment, ^{14}CO$_2$ trapping, and incubation were identical to the previous studies described earlier.

Atrazine mineralization in soils of all combinations is shown in Figure 4. Thirty-six percent of the applied ^{14}C-atrazine was mineralized in the 100% Bravo soil. The 100% Echo soil did not mineralize any atrazine. The soil combination made up of 80% Echo and 20% Bravo had significantly greater mineralization than did the 100% Echo soil or the combination of 90% Echo and 10% Bravo soil. On a percentage mineralized *per gram soil* basis, the addition of Bravo soil to the Echo soil did not enhance mineralization of atrazine in the Echo soil. It is possible that characteristics of the Echo soil did not provide an optimal environment for survival of the degraders. Echo soil had a much lower pH (4.3) compared with the Bravo soil (pH = 7.5). Neutral pH is generally optimal for microbial growth, because most microorganisms cannot tolerate extreme pH *(11)*. The Echo soil was also a very saline soil with an electrical conductivity value of 8.9 mmho cm^{-1} *(12)*. Fertilizer spills at this site may have contributed to the high salt content in this soil. Generally, the success rate of enhancing biodegradation in soils by inoculation with known degraders has not been good *(13)*, although transfer has been accomplished with carbamate and certain organophosphate insecticides. Preliminary evaluations of pesticide-contaminated soils would be necessary for determining the viability of an inoculation approach.

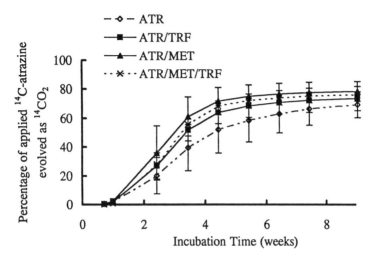

Figure 3. Herbicide concentrations in soil at 200 µg/g did not suppress atrazine degradation in Bravo rhizosphere soil. Bars represent standard deviations of the mean (n=2).

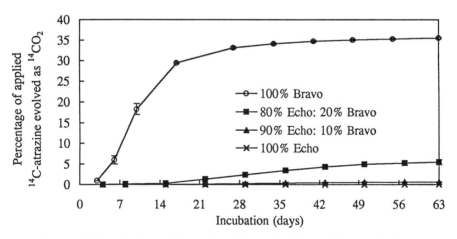

Figure 4. Mineralization of ^{14}C-atrazine (50 µg/g) in Echo and Bravo soils. Bars represent standard deviations of the mean (n=2).

Influence of Two Plant Species on the Degradation of Aged Residues

Independent composite soil samples were taken from the top 15 cm of a nonvegetated area from an Iowa agrochemical dealership. Soils were transported to the lab at 4 °C, sieved (2.4 mm), and mixed well before use in this experiment. A composite sample was sent to Midwest Laboratory (Omaha, NE) for determination of physical and chemical characteristics. Background herbicide concentrations were determined by gas chromatographic analysis, as described earlier. Concentrations of herbicides in this soil are shown in Table 1 (Alpha A). Subsamples of soil (nine replications) were treated with a solution made up of analytical grade atrazine and metolachlor, and uniformly ring-labeled ^{14}C-atrazine to achieve a final atrazine and metolachlor concentration in soil of 50 μg/g for each herbicide. One hundred-gram aliquots of treated soil were transferred to French square bottles, and soil moistures were adjusted and maintained at -33 kPa soil moisture tension. Treated soils were allowed to incubate in the dark at 24 °C for 165 days, thus allowing the herbicides to age in soil. During this aging period, the jars were opened weekly in to maintain aerobic conditions above the soil.

On day 165 posttreatment, incubation jars were opened and either left unvegetated or transplanted with *Kochia scoparia* or *Brassica napus*. The incubation jars were placed in an enclosed growth chamber in a temperature-controlled room (24 °C) with a light:dark cycle of 16:8. To ensure that ^{14}C incorporated into the plant tissue at the end of the study was the result of uptake from the soil only, a flow-through system was used to remove ^{14}CO$_2$ and ^{14}C-organic volatiles from the atmosphere of the growth chamber. Twenty-seven days after planting, the *Brassica* plants seemed very stressed, and all *Brassica*-vegetated samples along with a portion of the *Kochia*-vegetated and nonvegetated samples were extracted and analyzed at this time. *Kochia* plants were thriving, so the study was continued with the remaining set of unvegetated and *Kochia*-vegetated soils. At 240 days postherbicide treatment, 75 days after planting, the experiment was ended. At the end of each incubation period, entire plants were analyzed for ^{14}C residue by combusting pellets made up of plant tissue and hydrolyzed starch in a Packard Sample Oxidizer (Packard Instruments, Downers Grove, IL). Soils were extracted and analyzed by following the methods in Kruger et al. *(14)* which included solvent extractions of soil, liquid-liquid partitioning, thin-layer chromatography, autoradiography, and liquid scintillation techniques. Subsamples of extracted soil were combusted to determine the amount of unextractable ^{14}C-residue in soil.

After 27 days postplanting, significantly less atrazine was extractable from soils vegetated with *Kochia scoparia* compared with soils vegetated with *Brassica napus*, with 4.3% and 9.8% of the applied ^{14}C-atrazine extractable from these soils, respectively (ANOVA; p < 0.05). From the nonvegetated soil, 9.3% of the applied ^{14}C-atrazine was extractable, which was not significantly different from either of the vegetated soils. A significantly greater amount of ^{14}C was taken up by *Kochia* plants (10% of the applied ^{14}C) than by the *Brassica* plants (less than 1%) (ANOVA; p = 0.0001).

Significantly less atrazine was extractable from *Kochia*-vegetated soils than from nonvegetated soils 75 days postplanting (240 days postherbicide treatment) (ANOVA; p < 0.01). For the *Kochia*-vegetated soil 5.3% of the applied atrazine was extractable from soil, whereas 8.3% was extractable for the nonvegetated soil. There were no significant differences in quantities of degradates formed or nonextractable residues between the two treatments. Combustion of plants revealed that 6.5% of the applied ^{14}C was taken up by the *Kochia* plants.

Summary

Atrazine degradation in pesticide-contaminated soils was unaffected by the presence of metolachlor and/or trifluralin. Rapid mineralization of atrazine, applied individually, in soils from pesticide-contaminated sites was indicated in three out of eight soils tested. This degradative capability was not suppressed with the addition of even higher concentrations of herbicide mixtures. Transfer of the rapid atrazine-mineralizing capability of one soil was not achieved by mixing this soil with a soil that did not exhibit this characteristic. Plants had a positive influence on the dissipation of aged atrazine in soil, with significant uptake by *Kochia* plants, and decreased extractable atrazine from vegetated soils compared with nonvegetated soils. Overall, the results suggest that phytoremediation of soils contaminated with atrazine is a viable treatment method for soils contaminated with a mixture of some common herbicides. Coupling of phytoremediation with inoculation technology could provide a scenario in which survival of the inoculated organisms is increased. Further exploration of this along with degradation studies on the other compounds will help identify situations inshich the use of plants for cleanup of agrochemical dealer sites may be appropriate.

Acknowledgments

This research was partly funded by the Center for Health Effects of Environmental Contamination (University of Iowa, Iowa City, IA), Ciba Crop Protection (Greensboro, NC), the Great Plains-Rocky Mountain Hazardous Substance Research Center, and the U. S. Environmental Protection Agency (USEPA) (Cooperative Agreement CR-823864-01). Partial support for E.L.K. was provided in part by a USEPA Graduate Fellowship. Ciba Crop Protection also provided analytical standards and radiolabeled chemicals. This document may not reflect the views of the USEPA, and no official endorsement should be inferred. This is Journal Paper J-17162 of the Iowa Agriculture and Home Economics Experiment Station, Ames, IA, Project 3187.

Literature Cited

1. Gannon E. Environmental Cleanup of Fertilizer and Ag Chemical Dealer Sites. Iowa Natural Heritage Foundation, Des Moines, IA, **1992**, 201 pp.
2. Anderson, T.A., E.L. Kruger, and J.R. Coats. In T. A. Anderson and J. R. Coats, eds. *Bioremediation Through Rhizosphere Technology*, American Chemical Society, Washington, D. C., **1994a**, pp 199-209.
3. Alexander, M. *Biodegradation and Bioremediation*, Academic Press, Inc., San Diego, CA, **1994**, pp 226-245.
4. Roeth, F.W. Rev. Weed Sci. **1986**, 2:46-65.

5. Anderson, T.A., E.A. Guthrie, and B.T. Walton. *Environ. Sci. Technol.* **1993**, 27:2630-2636.
6. Anderson, T.A., E.L. Kruger, and J.R. Coats. Chemosphere **1994b**, 28:1551-1557.
7. Perkovich, B.S., T.A. Anderson, E.L. Kruger, and J.R. Coats. *Pestic. Sci.* **1996**, 46:391-396.
8. Mandelbaum, R.T., D.L. Allan, and L.P. Wackett. *Appl. Environ. Microbiol.* **1995**, 61:1451-1457.
9. Mandelbaum, R.T., L.P. Wackett, and D.L. Allan. *Appl. Environ. Microbiol.* **1993**, 27:1943-1946.
10. Kontchou, C.Y., and N. Gschwind. *J. Agric. Food Chem.* **1995**, 43:2291-2294.
11. Atlas, R.M. and R. Bartha. *Microbial Ecology,* Benjamin/Cummings, Menlo Park, CA, **1987**, pp 233-259.
12. Moscinski, J. Graduate student in microbiology, Iowa State University, Ames, IA.
13. Alexander, M. Biodegradation and Bioremediation. Academic Press, New York, **1994**, p 149.
14. Kruger, E.L., L. Somasundaram, R.S. Kanwar, and J.R. Coats. *Environ. Toxicol. Chem.* **1993**, 12: 1959-1967.

Chapter 5

Utilization of Plant Material for Remediation of Herbicide-Contaminated Soils

S. C. Wagner[1] and Robert M. Zablotowicz[2]

Southern Weed Science Laboratory, Agricultural Research Service, U.S. Department of Agriculture, P.O. Box 350, Stoneville, MS 38776

Biostimulation is a successful method for remediation of soils and other matrices contaminated with a wide range of xenobiotics. Use of the appropriate soil amendments can enhance the biodegrading potential of indigenous soil microbial populations. Plant materials have been studied by others as biostimulating amendments for soils contaminated with a wide range of herbicides. Our previous studies indicated that annual ryegrass (*Lolium multiforum* L.) residue was the most effective amendment for enhancing cyanazine (2-[[4-chloro-6-(ethylamino)-1,3,5-triazin-2yl]amino]-2 methyl-propanenitrile) and fluometuron (N,N-dimethyl-N'-[trifluoromethyl)-phenyl]urea) degradation in soils. Thus we pursued a comparative study of various crop residues on the degradation of fluometuron in soil. In this study we investigated the effects of amending soil with hairy vetch (*Vicia villosa* Roth), rice (*Oryza sativa* L.), or ryegrass residues on the degradation of a high concentration (500 mole kg^{-1} soil) of fluometuron (technical grade or commercial formulation). Initially, all three amendments enhanced fluometuron degradation in soil treated with technical material or commercial formulation. Hairy vetch transiently enhanced degradation, while the two grass residues stimulated degradation during the entire study (60 d). Rice straw had the greatest stimulation. In short term studies (21 d), ryegrass had the greatest effect on stimulating soil bacterial populations and several enzyme activities. Use of the appropriate plant residue is a promising approach for enhancing the remediation of herbicide-contaminated soils.

Groundwater and soils associated with agrochemical facilities (e.g. agrochemical manufacturing, formulation, and distribution facilities; farm loading/rinse sites; or abandoned waste sites) have been found to be contaminated with various pesticides and organic solvents (1-4). In order to avert damage to environmental quality, these sites must be cleaned up.

[1]Current address: Department of Biology, Stephen F. Austin State University, P.O. Box 13003, SFA Station, Nacogdoches, TX 75962–3003
[2]Corresponding author

Table I. Examples of Investigations Evaluating the Use of Amendments to
Stimulate Pesticide Degradation

Pesticide	Amendment (Reference)
alachlor	corn and soybean residue (1,6) cornmeal (5,7,8) sewage sludge (5)
aldrin, dieldrin	manure (17)
atrazine	cornmeal, sewage sludge (5) corn and soybean residue (6)
cyanazine	cornmeal, poultry litter, ryegrass residue (9)
DDT	rice straw (16)
fluometuron	cornmeal, poultry litter (9) ryegrass residue (9, 11)
metolachlor	cornmeal, sewage sludge (5) corn and soybean residue (6)
metribuzin	soybean residue (10) wheat straw and alfalfa residue (12)
pentachlorophenol	sewage sludge (13) wood chips (14)
toxaphene	manure (17)
trifluralin	cornmeal and sewage sludge (5) corn and soybean residue (6) chicken litter and soybean meal (7)

Materials and Methods

Dundee silt loam (fine-silty, mixed, thermic Aeric Ochragualf) was bulk treated with solutions of either technical grade or formulated fluometuron (Cotoran® 4L, Ciba, Greensboro, NC, 41.7% a.i., polyethylene glycol) and ^{14}C-uniformly ring-labeled fluometuron to achieve 500 mol kg^{-1} soil with 201.6 Bq ^{14}C g^{-1} soil and 25% moisture. Hairy vetch residues, Biostimulation with amendments is a promising approach that enhances the biodegrading potential of indigenous soil microbial populations.

Biostimulation optimizes environmental conditions for intrinsic degradation of xenobiotics. This can be achieved by the addition of suitable amendments, nutrients, electron acceptors or water. Soil amendments capable of biostimulating pesticide degradation, especially herbicides have been studied under laboratory and field conditions (Table I). Combinations of landfarming and biostimulation were used to demonstrate the potential of corn meal, sewage sludge, and corn and soybean residues to reduce high levels of the herbicides alachlor, atrazine, metolachlor, and trifluralin (1, 5-7). Under saturated moisture conditions, addition of soybean meal or chicken litter reduced high levels of trifluralin in soil, while under field capacity conditions, only soybean meal was effective (8). Cyanazine and fluometuron (250 mole kg^{-1}) degradation was enhanced with the addition of corn meal, poultry litter, or ryegrass residues with the greatest effect from ryegrass (9). Half-lives for cyanazine were 18.2, 21, 21, and 28.3 d for soils amended with ryegrass, cornmeal, poultry litter, and no additions, respectively, while those for fluometuron were 41.0, 57.3, 27.7, and 66.4 d for cornmeal, poultry litter, ryegrass, and no additions, respectively. In a Dundee soil, soybean residue and winter ryegrass residue stimulated the degradation of metribuzin and fluometuron, respectively (10, 11), when the herbicides were applied at agronomic concentrations. In another study, glucose and wheat straw enhanced metribuzin degradation while alfalfa residues slowed its degradation (12).

Soil amendments can also stimulate bioremediation of other pesticides, i.e., fungicides and insecticides. The addition of anaerobic sewage sludge to soil enhanced removal of 10-30 ppm pentachlorophenol (PCP) compared to untreated controls (13). Others found wood chips effective in stimulating PCP degradation (14); toxicity of PCP to the degradative microbes was reduced due to sorption of the PCP to woodchips. Under anaerobic conditions, amendment with rice straw decreased the persistence of DDT in soil (15). Degradation of the insecticides aldrin, dieldrin (16) and toxaphene (17) has been stimulated by addition of manures.

In order to adopt biostimulation as a method to remediate pesticide contamination sites, potential amendments must be screened for the ability to stimulate microbial action on pesticides. Because we observed that ryegrass residue was the most effective in stimulating cyanazine and fluometuron degradation (9, 11), we compared other crop residues available in our geographical region for biostimulation efficacy. We determined the effects of hairy vetch residue, rice stubble, and ryegrass residue on the degradation of high concentrations of fluometuron, an important herbicide in the southeastern U.S cotton production. Most sites are contaminated by formulated rather than the technical grade pesticides. We compared the effect of these amendments on the degradation of technical as well as formulated fluometuron.

rice stubble, ryegrass residues (5% w:w), or no amendment was added to 25 g soil in 250 mL Nalgene bottles and the final moisture content was adjusted to 35% (w:w). Plant residues were collected from field samples, rice stubble was from harvested rice, hairy vetch and ryegrass was from herbicide desiccated plots, and chopped to 2 to 3 cm length as described elsewhere (9, 11, 18, 19). Experimental design was a randomized complete block with four replications. The soils were incubated at 25 °C and aerated weekly by briefly uncapping each bottle, to replace head space with fresh air. Fluometuron and metabolites were extracted from each soil at 21, 40, and 60 d after initiation of the experiment (initial recovery efficiency > 98%).

Fluometuron and metabolites were extracted by ethyl acetate phase partitioning. Each soil was treated with 10 mL distilled deionized water followed by 40 mL ethyl acetate in the 250 mL Nalgene bottles with shaking for 18 h. After the soils were centrifuged for 15 min, 5860 x g, the supernatant was decanted, and the soils were extracted again with ethyl acetate/water. The two extracts were pooled and the ethyl acetate fraction containing the fluometuron and metabolites was separated from the water fraction using a separatory funnel. Extracted ^{14}C was determined in a liquid scintillation counter (Packard TriCarb 4000 series, Packard Instruments Co., Meriden, CT) using Ecolume scintillation cocktail (ICN, Inc., Costa Mesa, CA). Unextractable ^{14}C in each soil sample was determined by oxidation (Packard 306 oxidizer, Packard Instruments Co., Meriden, CT) and liquid scintillation counting.

Aliquots (20 mL) of the ethyl acetate extracts of each soil were evaporated to 3 mL under nitrogen gas at 40 °C. These extracts were analyzed for fluometuron and metabolites using thin layer chromatography. Each sample was spotted on a silica gel plate (250 m thick; 3.3 cm preabsorbent layer) and developed 10 cm with chloroform:ethanol, 95:5, v:v (20, 21). Fluometuron and metabolites were quantified using a BioScan 200 Imaging Scanner (Bioscan, Inc., Washington D.C.). The R_f values for analytical standards were: trifluoromethylphenylurea (TFMPU) = 0.13, desmethyl fluometuron (DMF) = 0.30, fluometuron = 0.58, and trifluoromethyl (TFMA) aniline = 0.76.

The effects of crop residues on microbial populations and soil enzyme activities was studied in Dundee soil treated with formulated fluometuron (500 mol kg^{-1} unlabeled). Soil (50 g in a 500 mL bottle) was amended with hairy vetch, rice and ryegrass (5 % w/w) in addition to unamended controls and the soil moisture was adjusted to 35 %. Following a 21 d incubation at 25 °C microbial populations were determined by serial dilution spiral plating on rose bengal potato dextrose agar (total fungi), 10% tryptic soy agar (total bacteria), 10% tryptic soy agar with crystal violet (Gram-negative bacteria) and S-1 media (fluorescent pseudomonads) as described elsewhere (18). Fluorescein diacetate (FDA) hydrolytic activity (20) was determined by incubating 2 g soil in 15 mL potassium phosphate buffer (0.10 M, pH 7.6) containing 0.5 mg of FDA for one h on a reciprocal shaker at 25 °C. FDA assays were terminated by acetone extraction (15 mL), clarified by centrifugation, and the optical density of supernatant determined at 490 nm. Triphenyl-tetrazolium chloride (TTC) dehydrogenase activity (23) was determined by incubating 2 g of soil in 4 ml 3 % TTC for 24 h. The assay was terminated by extraction with 12 mL methanol and clarified by centrifugation. The optical density of the supernatant determined at 485 nm. We used 2-nitroacetanilide (2-NAA) as substrate for determination of aryl acylamidase activity (24, 25) in soil. Soil (1.0 g) was incubated in 4 mL of 2 mM 2-NAA in phosphate buffer (0.05 M, pH 8.0) for 20 h on a rotary shaker. The assay was terminated by extraction with 4 mL of methanol, and clarified by centrifugation. The optical density at 410 nm determined. Enzyme activities were determined for each of the four replicates, with two substrate and one no substrate controls.

Analysis of variance of the treatments was determined with SAS (26). Means were separated at alpha = 0.05 using Fisher's Least Significant Difference test.

Results and Discussion

Fluometuron Degradation. There was a significant interaction between amendment and type of fluometuron (formulated versus technical grade). Initial discussion will compare amendments within the herbicide applied. All three plant amendments initially stimulated formulated fluometuron degradation, with rice the most effective. After 60 d, only 28%, 54%, and 61% fluometuron remained under rice, ryegrass and vetch residue treatments, respectively, compared to 78% in unamended soil (Table II). Hairy vetch provided a stimulatory effect only until 40 d, while both rice and ryegrass residues stimulated degradation throughout the study. DMF (Table III) was the major accumulating metabolite in all treatments (16 to 26% after 60 d); soils amended with plant material had significantly higher accumulation of DMF than unamended soil. TFMPU (Table IV) accumulated only in plant material-amended soils. After 60 days, 91 to 99 % of the ^{14}C applied was accounted for as either extractable or unextractable material. Unextractable ^{14}C (Table V) was significantly higher in soil amended with plant material than in unamended soil; rice-amended soil had the highest level of unextractable ^{14}C. This suggested that metabolites were perhaps incorporated into the soil organic matter via oxidative coupling reactions with TFMA and humic compounds. There was no significant increase in DMF, TFMPU or unextractable material in soils treated with hairy vetch from 40 to 60 d, indicating an inhibition of further degradation in this treatment.

Table II. Interactions of Amendment and Formulation (Technical Grade or Formulated) on ^{14}C Fluometuron Recovery from a Dundee Soil at Three Sample Times

| | Time | | | | | |
| Amendment | 21 (d) | | 40 (d) | | 60 (d) | |
	Tech.	Form.	Tech.	Form.	Tech.	Form.
	% of applied ^{14}C, recovered as fluometuron					
None	88.8	91.4	73.7	85.7	61.6	78.1
Hairy Vetch	81.7	70.1	63.5	60.0	59.1	61.3
Rice	77.4	70.0	65.8	52.0	44.5	27.7
Ryegrass	73.7	66.8	61.8	62.4	52.6	54
LSD=0.05	------- 1.9 -------		------- 6.3 ------		------- 8.6 -------	

Table III. Interactions of Amendment and Formulation (Technical Grade or Formulated Fluometuron) on ^{14}C- Desmethylfluometuron Recovery From a Dundee Soil at Three Sampling Times

Amendment	21 (d)		40 (d)		60 (d)	
	Tech.	Form.	Tech.	Form.	Tech.	Form.
	\% of Applied ^{14}C recovered as DMF					
None	8.9	8.9	18.5	12.7	19.1	15.8
Hairy Vetch	11.7	15.6	15.8	23.7	11	21.8
Rice	14.2	18.7	18.2	22.5	16.5	21.8
Ryegrass	12.4	18.3	16.1	22.9	21.8	25.6
LSD=0.05	------- 1.9 -------		------- 4.2 -------		------- 4.3 -------	

Mean of four replicates, LSD for comparison within a column and row for a given sample time.

Table IV. Interactions of Amendment and Formulation (Technical Grade or Formulated Fluometuron) on ^{14}C-Trifluoro-methylphenylurea Recovery From a Dundee Soil at Three Sample Times

Amendment	21 (d)		40 (d)		60 (d)	
	Tech.	Form.	Tech.	Form.	Tech.	Form.
	\% of applied ^{14} C, recovered as TFMPU					
None	0.0	0.0	0.0.	0.0.	4.2	0.0
Hairy Vetch	4.4	7.0	8.9	9.7	6.9	7.0
Rice	2.3	6.6	4.1	10.2	7.2	14.0
Ryegrass	6.3	6.1	9.7	7.9	5.9	6.2
LSD=0.05	------ 0.7 ------		------ 2.8 ------		- ----- 2.5 ------	

Mean of four replicates, LSD for comparison within a column and row for a given sample time.

Table V. Interactions of Amendment and Formulation (Technical Grade or Formulated Fluometuron) on Unextractable ^{14}C From a Dundee Soil at Three Sample Times

Amendment	Time					
	21 (d)		40 (d)		60 (d)	
	Tech.	Form.	Tech.	Form.	Tech.	Form.
	% of applied ^{14}C, unextractable					
None	2.4	2.4	4.1	3.7	5.5	4.1
Hairy Vetch	3.5	4.4	9.7	8.4	7.7	8.6
Rice	4.2	6.7	10.2	13.3	13.3	26.7
Ryegrass	4.3	6.8	8.6	8.5	9.9	10.3
LSD=0.05	------ 1.0 ------		------ 2.9 ------		------ 5.3 ------	

Mean of four replicates, LSD for comparison within a column and row for a given sample time.

As observed with the formulated material, degradation of technical fluometuron was stimulated by all three amendments, with the greatest effect caused by amendment with rice (Table II). After 60 d, 44% and 53% fluometuron remained under rice and ryegrass residues, respectively, compared to 62% in unamended soil. Hairy vetch transiently stimulated fluometuron degradation, but levels of this herbicide were not different from that observed in unamended soil at 60 d.

DMF was the major accumulating metabolite in all amended soils treated with technical fluometuron as observed with the formulated material (Table III). With the exception of the 21 d samples, DMF levels in unamended were similar to those observed for the amended soils. However, for both the 21 d and 40 d samples, TFMPU (Table IV) and unextractable-^{14}C levels (Table V) were lower in unamended soils than soils amended with plant material. A greater amount of ^{14}C was recovered in the unextractable fraction in rice-amended soil compared to the other treatments, although it was approximately 50% of that observed with rice amended soil treated with formulated fluometuron (Table V). Although differences in terms of fluometuron and metabolite levels between the unamended and plant material-amended soils were not as pronounced in technical fluometuron- treated soils than as in formulated fluometuron- treated soils, our results still illustrate that plant amendments stimulate technical fluometuron degradation.

A significant interaction between amendment and type of herbicide applied was observed. Differences in degradation of fluometuron when applied as the commercial formulation compared to technical material were observed in the unamended controls and rice-amended treatments. When no amendment was added, greater fluometuron degradation was observed for the technical material (61.6% remaining after 60 d) compared to the formulated material (78.1 %), while a greater degradation was observed in formulated fluometuron treated soils with rice (27.7 % recovered at 60 d) compared to technical material (44.5%). Similar recoveries of fluometuron were observed comparing formulated and technical grade fluometuron in hairy vetch and ryegrass-amended soil. These differences in degradation are also reflected in patterns of metabolite accumulation and incorporation of fluometuron into unextractable soil components.

Fluometuron degrades via sequential N-demethylation to desmethyl-fluometuron (DMF) and trifluoro-methylphenylurea (TFMPU), and/or hydrolysis of the urea moiety to trifluoromethylaniline (TFMA), [20, 21, 28]. Our studies indicated that DMF and to a lesser extent TFMPU, were the major accumulating metabolites. TFMA was infrequently observed and when found was less than 2% of the initial ^{14}C applied. TFMA is readily incorporated into insoluble bound residue and thus rarely detected (20, 24, 25). In certain microbial species N-dealkylation of the phenylurea herbicides renders them less toxic to the microbe (30). Similar degradation patterns were observed in other investigations of fluometuron at agronomic concentrations, i.e. primary degradation to DMF and accumulation of TFMPU mostly in ryegrass residues and ryegrass residue-managed soils (11). In another study, amendment with glucose enhanced the degradation of this herbicide, with TFMPU accumulating only in glucose-amended soil (17). Degradation of another phenylurea herbicide, diuron, in soil was also stimulated by the addition of glucose (31). Rice straw enhanced microbial populations and activity, which, in turn, decreased the persistence of the insecticide DDT (15) under anaerobic conditions. This study indicates rice residues may have potential for detoxifying high concentrations of other pesticides.

These results and our previous studies (9, 11) suggest that plant or other organic amendments stimulate microbial action that drives fluometuron degradation toward completion, thus substantially reducing the levels of this herbicide in soil. However it was suggested by others that organic amendments enhance organisms tolerant of high concentrations of pesticides that can co-metabolize the contaminant (7). Other factors such as sorption of the contaminant to the amendment need to be considered. Most crop residues also contain high microbial populations compared to soil (19). Use of these residues may also supplement the indigenous soil population.

A transient enhancement of fluometuron degradation in hairy vetch-amended soil compared to rice or ryegrass may be explained by several factors. Studies with the herbicide metribuzin indicated that the addition of legume residues (alfalfa) inhibited degradation (12). There may have been a higher degree of sorption of the herbicide to hairy vetch compared to the other plant amendments, rendering it unavailable for microbial degradation. Batch sorption techniques (11) indicated that fluometuron had a greater sorption to hairy vetch residues (Freundlich sorption $[K_f]$ coefficient = 28.0) compared to rye residues ($K_f = 21.8$). The sorption potential to the crop residues were several fold greater than a Dundee soil ($K_f = 2.6$). Likewise, studies with the sulfonylurea herbicide chlorimuron (19) also indicated a greater potential for sorption to hairy vetch residues ($K_f = 6.33$) compared to ryegrass and Dundee soil ($K_f = 3.95$ and 0.81) respectively. Addition of hairy vetch residue to a Dundee soil reduced chlorimuron degradation (32). Legumes such as hairy vetch, contain more nitrogen than the other residues, and the higher C:N ratio permits a more rapid decomposition of this crop residue compared to the monocot residues. Thus additional stimulating carbonaceous substrates are available for a shorter period. Fluometuron degradation also will eventually liberate inorganic nitrogen from hydrolysis of the urea group. Lowering the C:N ratio of the soil by using a monocot amendment may enhance activity of nitrogen scavenging microbes and enhance the degradation of nitrogenous contaminants. These concepts deserve further investigations when optimizing plant residues as amendments for remediation of contaminated soils.

Effects of amendments on soil microbial populations and enzyme activities. All three amendments elicited significant increases in populations of soil microorganisms studied compared to those in unamended control soil (Table VI). Hairy vetch had the greatest stimulatory effect on soil fungal propagules, while ryegrass had the greatest stimulation on the three bacterial populations. Soil fungal propagules were enhanced from about 12-fold (rice and ryegrass) to 60-fold (hairy vetch). Total bacterial populations increased about 15-fold (rice) to 35-fold (ryegrass) in residue treatments, while Gram-negative bacteria increased about 100-fold (rice and hairy vetch) to 360-fold (ryegrass). The crop residues elicited an increase in Gram-negative bacterial populations such as fluorescent pseudomonads that is similar to that observed due to a rhizosphere enrichment. The three soil enzyme activities were also significantly increased by all amendments; however ryegrass had the greatest effect (Table VII). FDA- hydrolysis represents an indication of general hydrolase activity in soil (esterases, lipases and certain proteases), while TTC-dehydrogenase activity is indicative of respiratory activity because TTC serves as an alternative electron acceptor. These two enzyme assays indicate a generic enhancement of soil microbial activity and their activities are increased up to 40-fold above that of soil without residue depending upon amendment. A relatively lower stimulation of aryl acylamidase activity was observed compared to FDA hydrolysis or TTC-dehydrogenase activity. Aryl acylamidase activity is an important factor in the degradation of phenyl urea herbicides, i.e. cleavage of the amide bond resulting in formation of the corresponding aniline (34). Although distributed in a wide variety of microorganisms, aryl acylamidase activity is not found in all genera of soil bacteria and fungi a wide range of enzyme activity is observed among isolates of a given species (35). Longer term effects of amendments of both microbial populations and enzyme assays should be addressed in future studies.

Conclusions

Our studies demonstrate that high concentrations of fluometuron were significantly reduced in soil by amendment with plant material compared to non-amended soil. All plant residues (hairy vetch, rice and ryegrass) used as amendments enhanced degradation of fluometuron with rice having the greatest effect. Differences between levels of fluometuron degradation and metabolite formation in soil treated with formulated compared to technical grade herbicide demonstrate the need to study the effect of formulation on herbicide bioremediation. All three plant residues dramatically increased soil microbial populations and enzyme activities, although at least in the short term, ryegrass had the greatest stimulation. The use of plant residues as biostimulating agents increases degradation of the contaminant via providing a generic stimulation of microbial activity. These plant residues also a have a high sorption capacity for herbicides. Reducing the level of pesticide concentration in the soil solution may reduce potential toxicity to the microorganisms, as has been demonstrated with wood chips and pentachlorophenol degradation (15). However, if the contaminant remains too tightly bound to the introduced amendment it may be rendered unavailable for further

Table VI. Effect of Crop Residue Amendments on Soil Microbial Populations in Fluometuron Treated Dundee Soil (21 d after treatment)

Treatment	Total Fungi	Total Bacteria	Gram-negative Bacteria	Fluorescent Pseudomonads
		log (10) colony forming units g^{-1} soil		
None	5.01	7.88	5.7	4.66
Rice	6.05	9.05	7.72	6.03
Hairy Vetch	6.79	9.14	7.79	6.14
Ryegrass	6.1	9.41	8.26	7.03
LSD=0.05	0.21	0.14	0.27	0.1

Mean of four samples

Table VII. Effect of Crop Residue Amendments on Enzyme Activities of Fluometuron-treated Dundee Soil (21 d after treatment)

Amendment	Triphenyl tetrazolium Chloride Dehydrogenase	Fluorescein diacetate hydrolysis	2-nitroacetanilide aryl acylamidase
		nmol g^{-1} soil h^{-1}	
None	0.3	238	19.8
Rice	10.3	1073	48.6
Hairy Vetch	9.7	2381	49.3
Ryegrass	13.8	4255	57.6
LSD=0.05	0.9	15.2	4.7

Mean of four replicates

degradation. Future studies to understand the mechanism(s) of biostimulation by plant residues need to consider sorption of the contaminant to the amendment as well as long term effects on both specific and nonspecific microbial processes. Biostimulation with plant amendments is a promising approach for the bioremediation of herbicide contamination in soil, especially if optimized for the appropriate contaminant. Biostimulation is inexpensive, environmentally friendly, and requires readily available technology for implementation

Acknowledgment

We are grateful to Ciba for providing the ^{14}C labeled fluometuron and metabolites used in this work. We appreciate the technical assistance of M.E. Smyly. Mention of a trademark or product does not constitute a guarantee or warranty of the product by the U.S. Department of Agriculture, and does not imply its approval to the exclusion of other products that may also be suitable.

Literature Cited

1. Felsot, A.S.; Dzantor, E. K. In *Enhanced Biodegradation of Pesticides in the Environment*; Racke, K. D., Coats, J. R., Eds.; ACS Books Washington, D.C., **1990**, pp 249-268.
2. Myrick, C. A. In *Pesticide Waste Management*; Bourke, J. D., Felsot, A. S., Gilding, T. J., Jensen, J. K. Seiber, J. N., Eds.; ACS Books, Washington, D.C., **1992**, pp 224-233.
3. Meharg, A. A. *Rev. Environ. Contam. Toxicol.*; **1994**, *138*, 21-48.
4. Zagula, S. J.; Risatti, J. B. In *47th Purdue Industrial Waste Conference Proc.*; Lewis Publishers, Inc., Chelsea, MI, **1992**, 93-104.
5. Dzantor, E. K.; Felsot, A. S.; Beck, M. J. *Appl. Biochem. Biotech.*; **1993**, *39/40*, 621-629.
6. Felsot, A. S.; Dzantor, E.K. *Environ. Toxicol. Chem.*; **1994**, *14*, 23-27.
7. Felsot, A., S.; Dzantor, E. K. *Proc. Air Waste Mgmt. Assoc.*; **1994**, Paper 94-TA45.05, **1994**.
8. Zablotowicz, R. M.; Dzantor, E. K. *Proc. Air Waste Mgmt. Assoc.*; **1994**, Paper 94-RA126.04;
9. Wagner, S.C.; Zablotowicz, R.M. *J. Environ. Health Sci. Sec. B.*; **1997**, *32*, 37-54.
10. Locke, M. A.; Harper, S. S. *Pestic. Sci.*; **1991**, *31*, 221-237.
11. Locke, M. A.; Zablotowicz, R. M.; Gaston, L. A. *Proc. 1995 So. Conservation Tillage Conf. for Sustainable Agric.* 1995 MAFES Special Bull. 88-7, 55-58.
12. Pettygrove, D. R.; Naylor, D. V. *Weed Sci.*; **1985**, *33*, 267-270.
13. Mikesell, M. D.; Boyd, S. A. *Environ. Sci. Technol.*; **1988**, *22*, 1411-1414.
14. Apajalahti, J.H.A., Salinoja-Salonen, M.S.; *Microb. Ecol.*; **1984**, *10*, 359-367.
15. Mitra, J.; Raghu, K. *Toxicol. Environ. Chem.*; **1986**, *11*, 171-181.
16. Hugenholtz, P.; Macrae, I.C. *Bull. Environ. Contamin. Toxicol.*; **1990**, *45*, 223-227.
17. Mirasatari, S. G.; McChesney, M. M.; Craigmill, A. C.; Winterlin, W. L.; Seiber, J. N. *J. Environ. Sci. Health*; **1987**, B22, 663-690.

18. Wagner S.C.; Zablotowicz, R.M.; Locke, M.A.; Bryson, C.T. *Proc. 1995 So. Conservation Tillage Conf. for Sustainable Agric.* **1995,** *MAFES Special Bull.88-7,* 86-89.

19. Reddy, K. N.; Locke, M. A.; Wagner, S. C.; Zablotowicz, R. M.; Gaston, L. A.; Smeda, R. J. *J. Agric. Food Chem.*; **1995,** *43,* 2752-2757.

20. Bozarth, G. A.; Funderburk, H. H., Jr. *Weed Sci.*; **1971,** *19,* 691-695.

21. Ross, J. A.; Tweedy, B. F. *Soil Biol. Biochem.*; **1994,** *5,* 739-746.

22. Schnürer, J.; Rosswall, T. *Appl. Environ. Microbiol.* **1982,** *43,* 1256-1261.

23. Casida, L.E. Jr.; Klein, D.A.; Santoro, T. *Soil Sci.* **1964,** *98,* 371-376.

24. Hoagland, R.E.; Zablotowicz, R.M. *Pestic. Biochem. Physiol.;* **1995,** *52,* 190-200.

25. Zablotowicz, R.M.; Hoagland, R.E.; Wagner, S.C. *95th Amer. Soc. Microbiol. General Meeting Abstracts;* **1995,** 331.

26. SAS Institute, Inc. *SAS Version 6.07*; SAS Institute Inc., Cary, NC, **1989.**

27. Geissbühler, H. *In Degradation of Herbicides;* P.C. Kearney, D.D. Kaufman Eds., Dekker Inc, New York, **1969,** pp 79-99.

28. Bartha, R. *J. Agr. Food Chem.*; **1971;** *19,* 385-387.

29. Chisaka, H., Kearney, P. C. *J. Agr. Food Chem.*; **1970,** *18,* 854-858.

30. Wallnöfer, P.R.; Safe, S.; Hutzinger, O. *Pestic. Biochem. Physiol.* **1973,** *3,* 253-258

31. McCormick, L. L.; Hiltbold, A. E. *Weeds;* **1965,** *13,* 77-82.

32. Wagner, S.C.; Reddy, K.N.; Zablotowicz, R.M.; Locke, M.A. *96th Amer. Soc. Microbiol. General Meeting Abstracts;* **1996,** 337.

33. Englehardt, G.; Wallnöffer, P.R., and Plapp, R. *Appl. Microbiol.* **1971,** *22,* 284-288.

34. Hoagland, R.E., Zablotowicz, R.M., Locke, M.A. In *Bioremediation Through Rhizosphere Technology,* T.A. Anderson, J.R. Coats, Eds.; ACS Books, Washington D.C. **1994,** pp 160-183.

Chapter 6

Potential of Biostimulation To Enhance Dissipation of Aged Herbicide Residues in Land-Farmed Waste

A. S. Felsot[1] and E. K. Dzantor[2]

[1]Washington State University, 100 Sprout Road, Richland, WA 99352
[2]Department of Agronomy, University of Maryland, College Park, MD 20742

One limitation to the rate of pesticide residue degradation in contaminated soil is the age of the residue. Research has suggested that "aged" residues may be less bioavailable and thus more persistent. Other research has shown that simply diluting aged, herbicide-contaminated soil with uncontaminated soil can stimulate degradation. In addition to dilution, addition of a carbon source, like dried, ground plant material, stimulated degradation. These studies suggested that landfarming as a disposal method for pesticide-contaminated soil could be made more efficient by amendments with different organic nutrients. Landfarming of aged herbicide-contaminated soils were simulated in the laboratory by diluting contaminated soil with uncontaminated soil. We present results to show that the dissipation of several herbicides was stimulated by dilution of aged, contaminated soil and amendment with different forms of corn (i.e., plant residues, ground seed, or commercial meal).

Wastewater from rinsing of spray tanks at commercial agrichemical facilities, farms, and homes can contaminate soil with high concentrations of a variety of pesticides. When in soil at high concentrations, pesticides can be unusually persistent (*1-4*), perhaps as a result of microbial toxicity causing inhibition of biodegradation (*5*). Recent research has suggested that residue aging may also contribute to prolonged persistence of pesticide residues. Sorption of herbicides can increase after prolonged residence in soil (*6,7*) and may lead to slower degradation rates (*8*). Aged residues seem to be less bioavailable than freshly applied residues, which might explain why degradation of aged residues is slower (*9*).

Contamination of soil by pesticide waste requires cleanup to protect local well water and surrounding properties from leaching and runoff of pesticide residues. Disposal of contaminated soil is expensive, and the two most used methods, landfilling or incineration, may not be practical for small-scale users like farmers and homeowners. Bioremediation of contaminants has been proposed as a lower cost solution that is applicable to ex-situ or in-situ cleanup operations (*10*).

Of the various bioremediation techniques proposed for cleanup of pesticide-contaminated soils, composting and phytoremediation are receiving increased research attention (*10,11*), while landfarming has been used in pilot studies and for cleanup of agrichemical retail sites (*12,13*). Regardless of method, however, all bioremediation techniques depend on enhancing microbial activity that may be either directly or indirectly associated with pesticide metabolism.

Some states now allow by statute landfarming of pesticide-contaminated soils (*14*). From an engineering perspective, landfarming is "a managed treatment and ultimate disposal process that involves the controlled application of a waste to a soil or soil-vegetation system (*15*)." Its objective is placement of contaminated soil within the upper A-horizon of uncontaminated soil where dilution would lower the waste concentration sufficiently to facilitate both chemical and aerobic microbial degradation. Normal agronomic management techniques like tilling, fertilization, and irrigation may be sufficient to stimulate microbial activity and truly achieve a bioremediation effect. However, dilution of contaminated soil with uncontaminated soil alone may stimulate biodegradation, perhaps in association with increased activity of soil dehydrogenase (*4*).

Research suggests landfarming may be feasible if effective rates of application are very low and the contamination is recent (*16*). Landfarming could be applied to a small pesticide spill, wherein soil would be quickly excavated and spread very thinly over uncontaminated soil. For larger areas of contamination, more land would be needed for treatment, especially if final contaminant loads had to be kept below normal field application rates. In this case, the speed at which landfarming could reduce contaminant residues to acceptable levels would be important, especially when the land was normally used for agricultural or other commercial purposes. One limitation to the speed with which pesticide residues in contaminated soil would degrade is the age of the residue; contamination may have existed for numerous years prior to excavation or in-situ treatment.

We have been investigating the idea that landfarming of aged contaminants could be made more efficient by amendments with different organic nutrients (i.e., biostimulation). In essence, in-situ biostimulation is analogous to phytoremediation because the objective is to enhance general microbial activity in the root zone, which is where landfarmed soil would be mixed. In laboratory experiments we have shown that corn meal can enhance degradation of freshly applied alachlor at concentrations two orders of magnitude above field application rates (*17*). In miniplot field tests, we have shown that sewage sludge and corn meal can also enhance dissipation of freshly sprayed alachlor and to a lesser extent aged residues (*16*). Studies reported by others have suggested that half-life of several pesticides in aerobically incubated, nutrient-amended soil was shorter when the compounds were freshly added to uncontaminated soil than when they had aged in situ at a site contaminated for many years (*18*).

It is still possible, however, that certain amendments may stimulate microbial activity towards degradation more effectively than other amendments. To further test the feasibility of the biostimulation hypothesis, we have simulated landfarming in the laboratory by diluting contaminated soils containing aged and fresh herbicide residues with uncontaminated soil followed by amendment with different organic materials (i.e., corn or soybean plant residue, ground corn seed, or commercial corn meal).

Materials and Methods

Soils. Aged, contaminated soil was obtained from an agrichemical facility in Piatt Co., IL that was undergoing remediation during 1986. The top 60 cm of soil at the facility had been contaminated since the late 1970's with a mixture of herbicides including alachlor, atrazine, metolachlor, and trifluralin (*19*). Prior to disposal by landfarming, the soils were temporarily stored in piles that are herein referred to as WASTE-PILE SOIL. Samples of this soil were stored at 2-4 °C for up to several years before use. The soil had an organic carbon content of 5.6% and a moisture content of 30.3% at 0.3 bar; mechanical analysis gave 23.3% sand, 46.9% silt, and 29.8% clay.

The soil used to dilute WASTE-PILE SOIL was collected from a cultivated field adjacent to the agrichemical facility; this site was used to test the feasibility of landfarming the contaminated soil, and several plots had been left untreated (*19,20*). Soil collected from these untreated plots was coded as CHECK SOIL. CHECK SOIL was a mixture of Ipava silt loam and Sable silty clay loam with an organic carbon content of 3.1% and a moisture content of 22.6% at 0.3 bar; mechanical analysis gave 6.1% sand, 57.1% silt, and 36.8% clay. CHECK SOIL also contained residues of the four herbicides as a result of past use. The concentrations, however, were below 0.1 ppm and at least two orders of magnitude lower than concentrations in the WASTE-PILE SOIL.

Experimental Design. Four experiments were conducted to test the hypothesis that biostimulation could enhance herbicide dissipation in aged soils, whether diluted or undiluted. Prior to starting the experiments, all soil was passed through a 3-mm mesh screen while moist and stored at 2-4 °C until used. Organic amendments included post-harvest corn and soybean plant stubble residue (ground to pass a 2-

mm mesh screen), or ground corn seed and commercial corn meal. All amendments were added to an equivalent of 2% or 10% by soil weight (i.e., 0.6 or 3.0 g per flask). All soil treatments were prepared in triplicate and 30 g oven dry weight equivalents of each was incubated in 250 mL Erlenmeyer flasks. Soil moisture was brought up to approximately 0.1 bar after all plant materials and herbicides were added. Flasks were held in unlighted, constant temperature incubators at 25 °C. For some experiments, flasks were closed with Parafilm but aerated weekly; moisture content was not adjusted. For other experiments, flasks were stoppered with polyurethane foam plugs that allowed free air flow; soil moisture (0.1 bar) was adjusted every several days.

Experiment I. WASTE-PILE and CHECK SOIL in each flask were amended with 0.6 g of air-dry ground up corn or soybean plant stubble that had been collected from a field after harvest. Unamended soils served as controls for the effects of the organic amendments. The control for the effects of aged residues on herbicide dissipation was CHECK SOIL freshly treated with a diluted solution of formulated pesticides. Lasso 4E (45.1% alachlor), Dual 8E (86.4% metolachlor), Aatrex 80W (80% atrazine), and Treflan EC (41.2% trifluralin) were combined and diluted with water. Enough herbicide solution was added to CHECK SOIL to approximate residue concentrations that had been previously determined in quadruplicate samples of WASTE-PILE SOIL. Soil flasks were covered with Parafilm and incubated for 61 days before extraction to determine the parent herbicide concentration.

Experiment II. WASTE-PILE SOIL was diluted 90% by weight with CHECK SOIL and then analyzed for residue concentration. The soil in each flask was then amended with either ground corn plant or soybean plant stubble. Unamended and undiluted WASTE-PILE SOIL was also extracted for residue concentration; this treatment served as a control for the effects of dilution and amendment. Soil flasks were covered with Parafilm and incubated for 32 days before extraction.

Experiment III. WASTE-PILE SOIL was diluted 10, 50, and 90% by weight with CHECK SOIL and then extracted for residue concentration. Each soil treatment was amended with 10% by weight ground corn seed (ground to the consistency of flour) or commercial corn meal (Quaker Oat brand). Unamended, diluted WASTE-PILE SOIL was used as a control. Soil flasks were capped with polyurethane foam plugs and incubated for 28 days before extraction.

Experiment IV. Lasso 4E was added to 3 kg of CHECK SOIL to produce an alachlor concentration of 10,000 ppm w/w. The soil was incubated with occasional stirring for 15 months at approximately 25 °C and moisture content of 30% by weight. Approximately 9000 ppm of alachlor remained in the laboratory aged CHECK SOIL after 15 months of incubation. The soil was thoroughly mixed with untreated CHECK SOIL to attain an initial concentration of 100 ppm alachlor. One batch of the diluted, laboratory-aged soil was mixed with ground corn stubble (2% by weight) and another batch was left unamended. After weighing the soils into flasks, one group was treated with propylene oxide (2.5 mL/flask) sterilant, while another group was left untreated.

The control for the laboratory-aged soil treatment was CHECK SOIL freshly treated with Lasso 4E to produce an alachlor concentration of 100 ppm. Freshly treated soil was also amended with ground corn stubble and sterilized with propylene oxide. The flasks were covered with Parafilm and aerated weekly for 56 days. In addition to determining the concentration of alachlor, soils were also assayed for microbial numbers and dehydrogenase activity as previously described (*5*).

Analytical Methods. Soils were extracted twice with glass-distilled ethyl acetate without further cleanup as described previously (*4*). Gas chromatography employed a column of 5% Apiezon + 0.13% DEGS (90 cm x 0.2 mm i.d.) at 190 °C isothermal and a nitrogen-phosphorus detector. All herbicide residue data were transformed to percentage recovery based on recoveries in each treatment on Day 0. Data were then subjected to analysis of variance and means separated by Fisher's Least Significant Difference Test ($p < 0.05$).

RESULTS & DISCUSSION

Experiment I. The effect of corn or soybean plant stubble on herbicide degradation was studied first in undiluted WASTE-PILE SOIL. Alachlor, metolachlor, and trifluralin concentrations after application to fresh soils were similar in WASTE-PILE SOIL and CHECK SOIL (Table I). Atrazine was about two-fold higher in CHECK SOIL than in the WASTE-PILE SOIL. After the start of the experiment, the container containing the Aatrex formulation was discovered to have been mistakenly labeled as having an active ingredient concentration of 40.8% rather than 80%.

Determination of initial concentration differences among treatments was important because the true kinetic function describing degradation is unknown. Unless degradation followed a first-order model, initial concentration conditions could have significantly influenced the difference in herbicide recoveries when examined 30 days after addition of the amendments. Thus, except for atrazine losses, concentration differences between aged and freshly treated soil could be eliminated as a significant factor affecting herbicide degradation.

Comparatively lower recoveries of alachlor, metolachlor, and trifluralin in corn and soybean-amended soils after 61 days suggested that dissipation had been enhanced by organic nutrients (Table II). Atrazine recoveries, however, were not significantly affected by amendment, suggesting a less important role for microbial

Table I. Average herbicide concentration (ppm ± standard deviation) over all treatments in Experiment I on day 0 [1]

Herbicide	WASTE-PILE SOIL	CHECK SOIL
alachlor	30.0 ± 3.7	27.1 ± 7.5
metolachlor	23.9 ± 2.6	22.6 ± 3.8
atrazine	2.9 ± 0.3	5.4 ± 0.8
trifluralin	1.6 ± 0.1	1.2 ± 0.7

[1] WASTE-PILE SOIL was collected during May 1986

Table II. Percentage herbicide recovery from aged WASTE-PILE SOIL and freshly treated CHECK SOIL in Experiment I after 61 days of incubation [1]

Amendment	WASTE-PILE SOIL	CHECK SOIL
	Alachlor	
None	59.0 ± 9.9 A	22.9 ± 2.8 A *
Corn	36.5 ± 8.9 B	1.2 ± 0.4 B *
Soybean	30.5 ± 3.5 B	1.0 ± 0.0 B *
	Atrazine	
None	49.7 ± 55.4 A	18.7 ± 1.0 A
Corn	34.7 ± 31.7 A	29.9 ± 2.2 B
Soybean	92.8 ± 69.7 A	13.3 ± 0.8 C
	Metolachlor	
None	63.6 ± 10.2 A	72.2 ± 4.0 A
Corn	45.7 ± 9.0 B	37.2 ± 7.2 B
Soybean	37.8 ± 4.2 B	8.6 ± 0.9 C *
	Trifluralin	
None	74.0 ± 16.5 A	67.8 ± 23.4 A
Corn	41.2 ± 10.1 B	14.9 ± 9.3 B *
Soybean	37.9 ± 5.5 B	12.3 ± 10.5 B *

[1] Percentage recovery within a soil class and herbicide type followed by different letters are significantly different, and recoveries followed by * are significantly lower than the recovery in the corresponding soil and amendment class ($p < 0.05$).

degradation in its dissipation. Such a conclusion is consistent with long-time observations suggesting that atrazine is primarily degraded through chemical mechanisms (21,22). On the other hand, a role for biodegradation is suggested by observations of >70% mineralization of atrazine incubated for 3 months in a Hanford sandy loam soil (23).

Alachlor dissipation in all CHECK SOIL treatments was significantly faster than in WASTE-PILE SOIL treatments. Trifluralin and metolachlor dissipation in

CHECK SOIL treatments was comparatively enhanced only in the presence of organic amendments. Differences in chemical and physical properties between the two soil sources suggested different microbial ecologies that may have contributed to differences in recoveries after 61 days. An alternative hypothesis that explains differences in recoveries, at least for alachlor, may be a relatively lower bioavailability of chemical in the aged soil than in freshly treated soil. We previously reported that less alachlor was desorbed from CHECK SOIL collected about one year after application of WASTE-PILE SOIL in a landfarming experiment than from CHECK SOIL freshly treated to an approximately equal concentration of alachlor (*20*).

Experiment II. In this experiment, landfarming was simulated by a 90% dilution of WASTE-PILE SOIL with CHECK SOIL before adding organic materials. With the exception of trifluralin in 90%-diluted, corn amended soil, recovery of initial herbicide residues within a dilution class did not differ significantly among amendment treatments (Table III). Initial residues were lower in this experiment than in Experiment I because WASTE-PILE SOIL was derived by mixing soil collected in May, 1986, with soil collected in June, 1987, after residues had declined during storage in the field (*20*).

After 32 days of incubation, analysis of variance indicated significant differences in recoveries of alachlor, metolachlor, and trifluralin from amended and unamended, diluted and undiluted WASTE-PILE SOIL (Figure 1). Only recoveries of the three herbicides from 90%-diluted soil, however, indicated a definite trend toward degradation (i.e., recoveries were substantially less than 100%). Recoveries greater than 100% for undiluted soil could be explained as variation caused by sampling chemical hot spots present in soil aliquots that were assigned to flasks earmarked for extraction after one month. With either no or very slow dissipation,

Table III. Initial herbicide residues (mean ± standard deviation) recovered from undiluted and 90%-diluted WASTE-PILE SOIL from Experiment II [1/]

%		*Initial Herbicide Residue, ppm*			
Dilution	Amendment	*Alachlor*	*Atrazine*	*Metolachlor*	*Trifluralin*
0	none	14.1 ± 0.1	0.9 ± 0.3	10.5 ± 1.0	1.2 ± 0.1
	corn	10.5 ± 3.1	0.6 ± 0.0	8.9 ± 2.7	1.1 ± 0.4
	bean	12.0 ± 0.0	0.8 ± 0.1	9.4 ± 0.2	1.1 ± 0.2
90	none	2.8 ± 0.5	0.1 ± 0.0	1.9 ± 0.1	0.2 ± 0.0
	corn	2.7 ± 0.1	0.1 ± 0.0	2.6 ± 0.0	0.3 ± 0.0
	bean	2.6 ± 0.0	0.1 ± 0.0	1.8 ± 0.0	0.2 ± 0.0

[1/] WASTE-PILE SOIL was a mixture of collections made during May 1986 and June 1987.

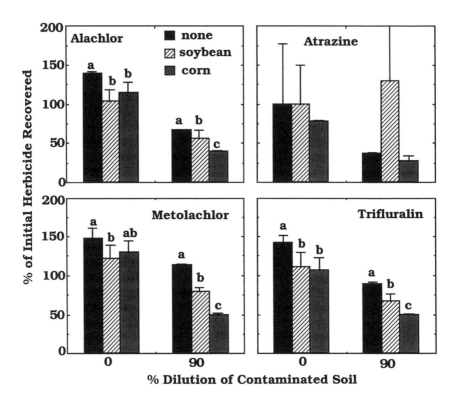

Figure 1. Effect of soil dilution and dried, ground plant residue on percentage recovery of herbicides after 32 days of incubation. Vertical lines represent standard deviation; bars within a dilution class followed by the same letter are not significantly different by the Fisher's Least Significant Difference Test ($p < 0.05$).

these chemical hot spots would result in seemingly higher residues when analyzed. The appearance of essentially no dissipation in undiluted WASTE-PILE soil relative to Experiment I may be due to the use of a shorter incubation time (32 d vs. 61 d) and a soil in a relatively more aged state (see Table III footnote). Nevertheless, in soil amended with corn and soybean plant residue, alachlor, metolachlor, and trifluralin recoveries were significantly lower than the recoveries from unamended soil. Furthermore, extensive dilution of the soil significantly stimulated dissipation of the herbicides, especially in amended soil. Thus, the results presented in Figure 1 suggested that corn and soybean plant residue actually stimulated biological degradation of the herbicides.

Atrazine was not affected by amendment; the high recovery in the soybean amendment was associated with a very large standard deviation. Thus, if atrazine is subject to predominantly chemical degradation in the soils studied, addition of organic amendments to soil might not be expected to significantly influence degradation rate. One study reported that "microbial energy sources" accelerated atrazine decomposition (*24*), which suggests that organic amendments may yet have potential for enhancing atrazine degradation in different soil types. Indeed, rice hulls and their water extracts enhanced mineralization of atrazine in a sandy loam soil (*23*).

Experiment III. This experiment was designed to determine if ground corn seed and meal were as effective in stimulating herbicide dissipation as was plant residue in the previous two experiments. Landfarming was simulated by diluting the WASTE-PILE SOIL with different percentages of CHECK SOIL. With few exceptions, the initial herbicide residue concentrations among amendments within a dilution class were not significantly different (Table IV). Alachlor, metolachlor, and trifluralin residues showed evidence of dissipation after 28 days in undiluted, amended and unamended soil (Table V). Only minor differences in percentage recoveries were observed among dilution classes. Such results were consistent with the definite trend for herbicide dissipation in Experiment I, which used undiluted WASTE-PILE SOIL

Table IV. Initial herbicide residues (mean ± standard deviation) recovered from undiluted and diluted WASTE-PILE SOIL from Experiment III [1]

% Dilution	Amendment	Initial Herbicide Residue, ppm [2]			
		Alachlor	Atrazine	Metolachlor	Trifluralin
0	none	33.4 ± 3.5	16.9 ± 13.7	21.9 ± 1.4	2.9 ± 0.3
	ground corn	39.1 ± 4.1	7.7 ± 0.5	26.0 ± 3.3	3.4 ± 0.3
	corn meal	31.6 ± 1.7	6.7 ± 0.1	21.6 ± 2.4	2.8 ± 0.2
10	none	34.4 ± 13.7	14.4 ± 8.8	21.9 ± 1.4	2.9 ± 0.2
	ground corn	29.9 ± 2.2	15.6 ± 7.9	26.0 ± 3.3	2.8 ± 0.1
	corn meal	28.5 ± 1.5	11.2 ± 3.4	21.6 ± 2.4	2.8 ± 0.1
50	none	14.2 ± 3.1	6.6 ± 4.9	8.6 ± 1.6	0.9 ± 0.1
	ground corn	13.3 ± 1.3	0.7 ± 2.0	8.2 ± 0.9	0.8 ± 0.2
	corn meal	11.2 ± 0.8	2.5 ± 0.2	6.9 ± 0.4	0.8 ± 0.0
90	none	4.3 ± 0.7	0.3 ± 0.1	2.7 ± 0.6	0.3 ± 0.0
	ground corn	3.8 ± 0.6	0.1 ± 0.2	1.4 ± 1.0	0.3 ± 0.0
	corn meal	3.4 ± 0.1	1.1 ± 1.2	1.9 ± 0.1	0.3 ± 0.0

[1] WASTE-PILE SOIL was collected during May 1986
[2] With the exception of alachlor and trifluralin residues recovered from 0% dilution in corn meal-amended soil, initial concentrations did not significantly differ among amendment type within a dilution class.

collected solely during 1986. These results also stand in contrast to the lack of herbicide dissipation in Experiment II when undiluted WASTE-PILE SOIL collected during 1986 and 1987 was mixed.

Initial recoveries of atrazine during Experiment III were highly variable (Table IV) and give more weight to the hypothesis that heterogeneous hot spots existed in the soil despite mixing and sieving. Although both pure atrazine and trifluralin are crystalline solids, atrazine is formulated as a flowable liquid rather than an emulsifiable concentrate. As a result of its high concentrations in the initially excavated WASTE-PILE soil, atrazine may have actually crystallized unevenly throughout the soil. Thus, repeated 30-g aliquots of soil weighed into individual flasks would be expected to have very high and very low concentrations of atrazine randomly distributed; for example, average recoveries of atrazine ranged from 0.7 ppm to 6.6 ppm in 50%-diluted WASTE-PILE SOIL.

Mean residue values with very high coefficients of variation precluded detection of significant trends in atrazine degradation during Experiment III (data not shown). Variability in recovery of alachlor, metolachlor, and trifluralin, which are usually formulated as emulsifiable concentrates, was low enough to establish a significant trend in degradation over a one month incubation period (Table V). Both ground corn and commercial corn meal stimulated the dissipation of the three herbicides in undiluted and diluted soil. Alachlor and metolachlor losses were

Table V. Recovery of herbicide residues (mean % of initial ± standard deviation) after 28 days of incubation in Experiment III [1]

% Dilution	Amendment	Alachlor		Metolachlor		Trifluralin	
0	none	79.2 ± 10.2	A	74.1 ± 8.6	A	39.8 ± 8.5	A
	ground corn	35.8 ± 10.6	B	49.4 ± 2.1	B	31.7 ± 4.7	A
	corn meal	36.6 ± 2.7	B	41.3 ± 1.9	B	8.6 ± 14.9	B
10	none	79.1 ± 7.5	A	77.9 ± 6.3	A	41.8 ± 7.0	A
	ground corn	30.4 ± 5.5	B	37.8 ± 4.0	B	11.5 ± 11.0	B
	corn meal	36.4 ± 5.1	B	43.8 ± 6.2	B	14.3 ± 8.1	B
50	none	88.0 ± 12.0	A	95.2 ± 16.1	A	66.6 ± 9.8	A
	ground corn	48.3 ± 15.1	B	52.2 ± 13.1	B	40.5 ± 16.7	B
	corn meal	52.3 ± 5.5	B	62.8 ± 1.3	B	48.0 ± 6.1	AB
90	none	77.4 ± 9.7	A	95.2 ± 10.2	A	78.2 ± 14.2	A
	ground corn	25.8 ± 4.0	B	55.5 ± 5.3	B	23.2 ± 3.6	B
	corn meal	25.8 ± 4.0	B	37.3 ± 4.4	C	32.1 ± 7.7	B

[1] Means followed by the same letter within a dilution class are not significantly different by the Fisher's Least Significance Difference Test at $p < 0.05$.

significantly less from unamended soil than from amended soil. In a previously reported study, addition of corn meal stimulated degradation of 10-1000 ppm concentrations of freshly applied alachlor (*17*). Corn meal has been used by others to help remediate pesticide waste by a landfarming system (*18*) and by composting (*25*).

In contrast to alachlor and metolachlor persistence in unamended soils, trifluralin losses were sometimes proportionately lower, especially in the 0 and 10% diluted soils. A distinct yellowish color that appeared on the bottom of the polyurethane foam plugs after one month suggested that dissipation of trifluralin may have also occurred by volatilization. Trifluralin, which is a bright orange crystal, has been shown to be significantly volatilized from soil (*28*).

Experiment IV. To eliminate potential confounding effects when comparing herbicide dissipation in aged and freshly treated soil arising from two different sources, CHECK SOIL was pretreated with an extremely high concentration of alachlor (10,000 ppm) and incubated in the laboratory for 15 months. We reported previously that this concentration caused microbial toxicity and was associated with essentially no herbicide dissipation (*5*). Diluting the soil with untreated CHECK SOIL simulated landfarming and presumably re-establishment of a viable microbial population. Dissipation of alachlor in the diluted aged soil approximated first-order kinetics as evidenced by straight line curves obtained when the data were plotted as semilogarithms (Figure 2); however, the curvature in the dissipation lines representing unaged (fresh) CHECK SOIL suggested non-first order kinetics.

In contrast to degradation in field-aged soil, alachlor seemed to degrade faster in laboratory-aged soil after dilution with CHECK SOIL (31% recovered after 56 d) than in freshly treated CHECK SOIL (60% recovered) (Figure 2). For both treatments, however, ground corn plant stubble significantly enhanced the rate of alachlor degradation in laboratory-aged (8% recovered) and freshly treated soil (23% recovered). Sterilization of freshly treated and aged soil, with or without corn stubble amendment, significantly inhibited degradation of alachlor throughout the 56 day incubation period (Figure 2).

Initial bacterial counts in corn stubble-amended soils were 2-3 orders of magnitude higher than in unamended soils (Day 0 data not shown), probably as a result of cold storage of the soils for several weeks before numbers were determined. By day 56, bacterial counts in amended soils freshly treated with alachlor had declined about 10-fold, but bacterial counts in aged soils remained comparatively stable (Table VI). Initial fungal counts (data not shown) in unamended soils were at least 30-fold lower than those in corn stubble-amended soils. After 56 days fungal populations in unamended soils remained 10-fold lower than in corn stubble-amended soils (Table VI).

Zero day dehydrogenase activities (data not shown) varied several-fold among treatments, but by day 56 the activities were similar in both amended and

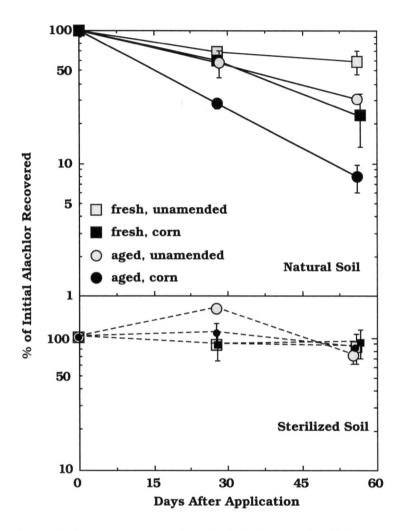

Figure 2. Percentage recovery of alachlor in freshly treated and laboratory-aged soil that was either unamended or amended with ground corn plant residue. Vertical lines through symbols represent standard deviations.

unamended soils (Table VI). Although bacteria and fungi were not detected in propylene oxide-sterilized samples during incubation, dehydrogenase activity did resurge by day 56. However, it remained at a significantly lower level in the sterilized soil than in the natural soil (Table VI).

    The absence of bacteria and fungi in sterilized soils and consequent lack of

Table VI. Estimates of bacterial and fungal populations and dehydrogenase activities in freshly treated and laboratory-aged soil after 56 days

			Microbial Numbers/g soil		
			Bacteria	Fungi	Dehydrogenase
Age	Amendment	Condition	(cells x 10^9)[1]	(cfu x 10^4)[2]	(μg TF/g)[3]
-- [4]	none	natural	0.6 (0.2-1.4)	2.6 (0.0)	125.2 (3.7)
fresh	none	natural	0.3 (0.1-0.8)	2.2 (0.1)	104.2 (13.4)
aged	none	natural	0.3 (0.1-0.8)	3.2 (0.7)	182.3 (11.4)
fresh	none	sterile	-- [5]	--	1.0 (1.0)
aged	none	sterile	--	--	10.1 (9.0)
-- [4]	corn	natural	4.0 (1.7-10.4)	30.3 (1.2)	182.5 (18.7)
fresh	corn	natural	13 (5-33)	28.0 (3.7)	138.9 (14.0)
aged	corn	natural	250 (99-605)	24.0 (0.8)	100.8 (19.4)
fresh	corn	sterile	-- [5]	--	69.8 (29.8)
aged	corn	sterile	--	--	3.5 (3.5)

[1] 95% confidence interval shown in parentheses
[2] Standard deviation shown in parentheses
[3] Dehydrogenase activity measured as μg of triphenylformazan formed from triphenyltetrazolium chloride
[4] CHECK SOIL without alachlor treatment.
[5] Bacteria or fungi not detected in propylene oxide-treated soil at the 10^{-3} dilution.

dissipation indicated that the observed dissipation of alachlor in all treatments of CHECK SOIL could be accounted for at least in part by biodegradation. Likewise, the sterilization experiments suggested that the losses of metolachlor and trifluralin also resulted from biodegradation. Past microbiological analyses of WASTE-PILE SOIL showed microbial numbers were at least 10-fold less than in CHECK SOIL, and dehydrogenase activity was nil (*20*). Such conditions may have been responsible for the prolonged persistence of herbicides in WASTE-PILE SOIL stored in the field for two years (20). Thus, biodegradation associated with increased microbial activity was likely the major cause of dissipation of alachlor, metolachlor, and trifluralin in WASTE-PILE SOIL that was diluted with CHECK SOIL or to which organic amendments were added.

   Addition of ground corn stubble augmented microbial numbers at least 10-fold compared to unamended soil (Table VI), which suggested that biostimulation may effect degradation by a general stimulation of the microbial population. Although dehydrogenase activity was not correspondingly stimulated by corn stubble amendment, shifts in specific microbial populations that can cometabolize alachlor may have taken place, rather than an overall increase in respiration (*17, 27*).

Cometabolism of alachlor rather than mineralization is recognized as its major biotransformation pathway (*4, 28*).

CONCLUSIONS

To stimulate clean up of herbicide-contaminated soil, we combined corn or soybean plant residue or corn meal amendments with soil dilution, which alone had previously been shown to stimulate herbicide degradation. Corn meal seemed to be an effective biostimulant for degradation of alachlor, metolachlor, and trifluralin even if aged soil was not diluted first. However, diluting the soil would significantly lower the concentration of the contaminants and thereby reduce the risk of adverse effects from translocation and phytotoxicity. Atrazine degradation could not be stimulated by organic amendments under laboratory conditions but has been observed to be less persistent when soil is biostimulated in field miniplots (*29*). A combination of landfarming with organic amendments might be an inexpensive option for disposal of pesticide-contaminated soils. This option would be feasible only if the waste pesticides were not classified as priority pollutants and were still legally registered for field application.

ACKNOWLEDGMENTS

This manuscript is a contribution from the Washington State University Agricultural Research Center; the experiments were conducted at the Illinois Natural History Survey, University of Illinois at Urbana-Champaign. Funding was received from the Illinois Hazardous Waste Research and Information Center, Project No. HWR 91-072, the Illinois Fertilizer and Chemical Association, and the Tennessee Valley Authority.

LITERATURE CITED

1. Wolfe, H. R.; Staiff, D. C.; Armstrong, J. F.; Comer, S. W. *Bull Environ. Contamin. Toxicol.* **1973**, *10*, 1-9.
2. Staiff, D. C.; Comer, S. W.; Armstrong, J. F.; Wolfe, H. R.. *Bull. Environ. Contam. Toxicol.* **1975**, *13*, 362-368.
3. Schoen, S. R.; Winterlin, W. L. *J. Environ. Sci. Health B* **1987**, *22*, 347-377.
4. Felsot, A. S.; E. K. Dzantor. In *Enhanced Biodegradation of Pesticides in the Environment*, K. D. Racke; Coats, J. R. Eds; Am. Chem. Soc. Symp. Ser. No. 426, Am. Chem. Soc., Washington, D. C., 1990; pp. 249-268.
5. Dzantor, E. K.; Felsot, A. S. *Environ. Toxicol. Chem.* **1991**, *10*, 649-655.
6. McCall, P. J.; Agin, G. L. *Environ. Toxicol. Chem.* **1985**, *4*, 37-44.
7. Steinberg, S. M.; Pignatello, J. J.; Sawhney, B. L. *Environ. Sci. Technol.* **1987**, *21*, 1201-1208.
8. Scribner, S. L., Benzing, T. R.; Sun, S.; Boyd, S. A. *J. Environ. Qual.* **1992**, *21*, 1150-120.

9. Alexander, M.; *Biodegradation and Bioremediation.* Academic Press, New York, NY, 1994; p. 275.
10. Bollag, J.-M; Mertz, T.; Otjen, L. In *Bioremediation through Rhizosphere Technology*; Anderson, Todd A.; Coats, J. R.; ACS Symp. Ser. 563; American Chemical Society: Washington, DC, 1994; pp 2-10.
11. Anderson, T. A.; Guthrie, E. A.; Walton, B. T. *Environ. Sci. Technol.* **1993**, *27*, 2630-2635.
12. Andrews Environmental Engineering, Inc. *Use of Landfarming to Remediate Soil Contaminated by Pesticides*; HWRIC TR-019; Hazardous Waste Research and Information Center, Champaign, IL; 42 pp.
13. Felsot, A. S. *J. Environ. Sci. Health,* , **1996**, *B31*, 365-381.
14. Bicki, T., J.; Felsot, A. S. In *Mechanisms of Pesticide Movement into Ground Water*; Honeycutt, R. C.; Schabacker, D. J., Eds.; Lewis Publishers, Boca Raton, FL, 1994; pp 81-99.
15. Loehr, R., M. and M. R. Overcash. *J. Environ. Engin.* 1985, 111, 141-159.
16. Felsot, A. S.; Mitchell, J. K.; Dzantor, E. K. In *Bioremediation: Science and Applications.* Skipper, H. D.; Turco, R. F., Eds. SSSA Special Publication Number 43; Soil Science Society of America: Madison, WI, 1995; pp 237-257.
17. Felsot, A.; Dzantor, E. K. *Environ. Toxicol. Chem.* 14, **1995**, 23-28.
18. Winterlin, W. L.; Seiber, J. N.; Craigmill, A.; Baier, T.; Woodrow, J.; Walker, G. *Arch. Environ. Contam. Toxicol.* **1989**, *18*, 734-747.
19. Felsot, A. S.; Liebl, R.; Bicki, T. *Feasibility of land application of soils contaminated with pesticide waste as a remediation practice*; HWRIC RR 021, Hazardous Waste Research and Information Center, Champaign, IL, 1988, 55 pp.
20. Felsot, A.; Dzantor, E. K.; Case, L.; Liebl, R. *Assessment of problems associated with landfilling or land application of pesticide waste and feasibility of cleanup by microbiological degradation*; HWRIC RR-053, Hazardous Waste Research and Information Center, Champaign, IL, 1990, 68 pp.
21. Armstrong, D. E.; Chesters, G.; Harris, R. F. *Soil Sci. Soc. Am. Proc.* **1967**, *31*, 61-66.
22. Avidov, E.; Aharonson, N.; Katan, J.; Rubin, B.; Yarden, O. *Weed Sci.* **1985**, *33*, 457-461.
23. Alvey, S.; Crowley, D. E. *J. Environ. Qual.* **1995**, *24*, 1156-1162.
24. McCormick, L L.; Hiltbold, A. E. *Weeds* 1966, *14*, 77-82.
25. Berry, D. F.; Tomkinson, R. A.; Hetzel, G. H.; Mullins D. E.; Young, R.W. *J. Environ. Qual.* **1993**, *22*, 366-374.
26. Grover, R.; Smith, A. E.; Shewchuk, S. R.; Cessna, A. J.; Hunter, J. H. *J. Environ. Qual.* **1988**, *17*, 543-550.
27. Stevenson, I. L. *Can. J. Microbiol.* **1962**, *8*, 501-509.
28. Novick, N. J.; Mukherjee, R.; Alexander, M. *J. Agric. Food Chem.* **1986**, *34*, 721-725.
29. Felsot, A. S.; Mitchell, J. K.; Dzantor, E. K. *Use of Landfarming to Remediate Soil Contaminated by Pesticide Waste*; HWRIC RR-070, Hazardous Waste Research and Information Center, Champaign, Illinois, 1994; 53 pp.

Chapter 7

An Integrated Phytoremediation Strategy for Chloroacetamide Herbicides in Soil

Robert E. Hoagland, Robert M. Zablotowicz, and Martin A. Locke

Southern Weed Science Laboratory, Agricultural Research Service, U.S. Department of Agriculture, P.O. Box 350, Stoneville, MS 38776

We have tested an integrated system for phytoremediation using corn (*Zea mays* L.), a safener specific for chloroacetamides (benoxacor), and an inoculum of a rhizosphere-competent *Pseudomonas fluorescens* strain UA5-40rif capable of catabolizing these herbicides. Initial growth chamber studies with a Bosket sandy loam soil (organic matter content < 1%), benoxacor (0.75 kg ha^{-1}) and inocula of this bacterium provided protection to corn seedlings at herbicide application rates of up to 45 and 54 kg ha^{-1} of alachlor and metolachlor, respectively. Satisfactory root colonization, i.e. log (10) 6.2 to 7.4 cfu g^{-1} root by UA5-40rif, was observed at concentrations up to 12x of these herbicide rates. Following 12 days of plant growth, alachlor concentrations in soil from safened and inoculated corn seedlings were about 25% and 35 % of those observed in unplanted soil at the 12x alachlor and metolachlor rate, respectively. Additional experiments studied applications of up to 36x of formulated metolachlor. At this higher metolachlor rate, normal physiological development was observed in benoxacor-safened seedlings, although there were slight reductions in root and shoot biomass. Metolachlor residues in soil treated with the 36x rate were 89%, 80%, 75%, respectively, for the unplanted soil, corn, and corn + benoxacor treatments, while only 54% remained in corn + benoxacor + UA5-40 treatment. Results indicate that use of a combination of chemical and biological safeners (competent herbicide-detoxifying rhizobacteria) is a novel and useful approach for increasing herbicide tolerance in an agronomic crop plant for enhanced phytoremediation.

Contamination of some soils with herbicides has become a serious environmental problem. The chloroacetamide herbicides, especially alachlor [2-chloro-*N*-(2,6-diethylphenyl)-*N*-(methoxymethyl) acetamide] and metolachlor [2-chloro-*N*-(2-ethyl-6-methylphenyl)-*N*-(2-methoxy-1-methylethyl) acetamide] are common contaminants in agricultural chemical dealerships, (1) and on farm loading and rinse sites (2). In many cases, it is desirable to remediate these contaminated areas. Phytoremediation is an attractive option to reduce levels of certain pesticide contaminants (3, 4). Some researchers have utilized plants that naturally colonize such contaminated sites for remediation (3, 4), since they are intrinsically tolerant to the contaminants. These plants may also harbor a microbial community that is tolerant or that has adapted to the level and/or nature of the contaminants. Domesticated crop plants may be limited in utility for remediation of pesticide- contaminated soils due to limited contaminant tolerance. Under landfarming conditions with a mixture of herbicides {alachlor, atrazine [6-chloro-N-ethyl-N'-(1-methylethyl)-1,3,5-triazine-2,4-diamine], metolachlor, and trifluralin [2,6-dinitro-N,N-dipropyl-4-(trifluoromethyl)benzenamine]}, corn (*Zea mays* L.) has been found to be more tolerant than soybeans (*Glycine max* Merr.) (5). Corn was capable of adequate establishment in plots treated with soil contaminated with these herbicides at rates up to 5x above the recommended application rates. Corn is tolerant to the triazine herbicides, and many sites contaminated with the chloroacetamides are also contaminated with atrazine and or cyanazine.

Safeners are compounds used to protect crop plants from herbicide injury. Their chemistries and modes of action are varied (6). Numerous safeners have been developed to protect several crop plants (typically monocots) against chloroacetamide phytotoxicity (7, 8). Benoxacor [CGA-154281; 4-(dichloroacetyl)-3,4-dihydro-3-methyl-2H-1,4-benzoxazine] was introduced to protect corn against commercial application of metolachlor (9). Corn is sometimes injured by chloroacetamide herbicides, especially at high rates early in the growing season and under cool and wet conditions (10). The chloroacetamides are traditionally detoxified via GSH-conjugation catalyzed by the enzyme glutathione S-transferase (GST) (11, 12). Benoxacor enhances GST in corn and is responsible for the major detoxification step of metolachlor to a non-phytotoxic metabolite (13, 14). Safeners such as benoxacor may have utility in phytoremediation strategies to enhance tolerance of the remediation crop to high herbicide levels.

The chloroacetamides are metabolized in soil by various microbial processes. The major metabolic route in soil is dechlorination, resulting in formation of several acidic polar metabolites (15, 16). We have identified glutathione conjugation as a detoxification mechanism of alachlor by various rhizosphere bacteria (17, 18), especially fluorescent pseudomonads. Laboratory studies of soil treated with high levels (100 ppm) of alachlor indicated that inoculation with *Pseudomonas fluorescens* strain UA5-40 plus cornmeal as an amendment could substantially accelerate initial degradation of alachlor (19). The use of microorganisms that transform contaminants by co-metabolism as bioaugmentation agents alone may have limited utility. However, these microorganisms may have greater utility when combined with other bioremediation strategies. We are developing approaches to bioremediation of herbicide-contaminated areas using crop safeners (for herbicides), microorganisms, and plants.

Our objectives in this study were to examine an integrated system for phytoremediation of soils containing high levels of chloroacetamide herbicides, particularly alachlor and metolachlor (Figure 1). Our studies used the moderately tolerant plant species corn, the safener benoxacor (Figure 1) and one of our superior rhizosphere-competent chloroacetamide-detoxifying fluorescent pseudomonads, strain UA5-40.

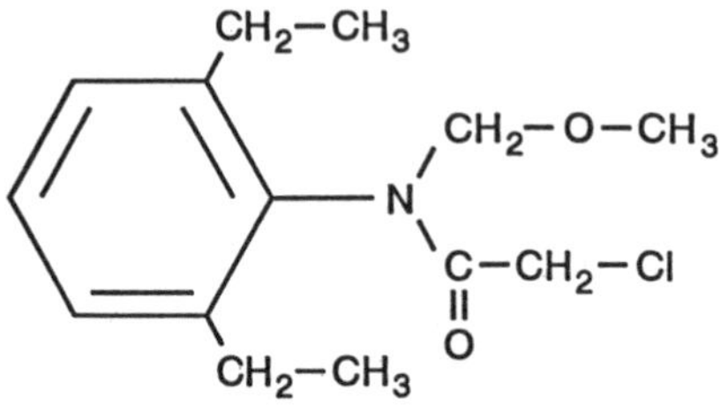

Alachlor

2-Chloro-N-(2,6-diethylphenyl)-N-(methyoxymethyl)acetamide

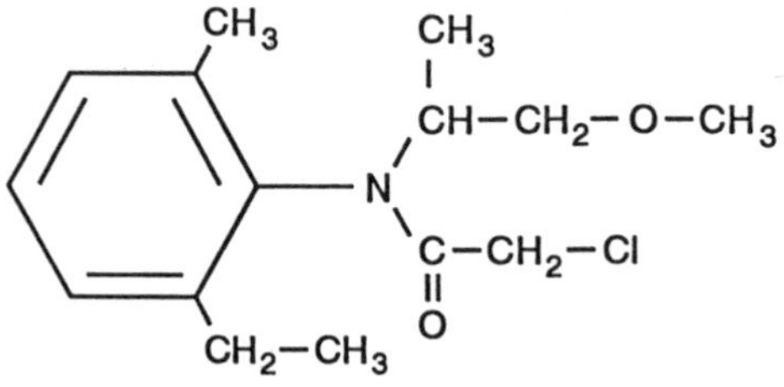

Metolachlor

2-Chloro-N-(2-ethyl-6-methylphenyl)-
N-(2-methoxy-1-methylethyl)acetamide

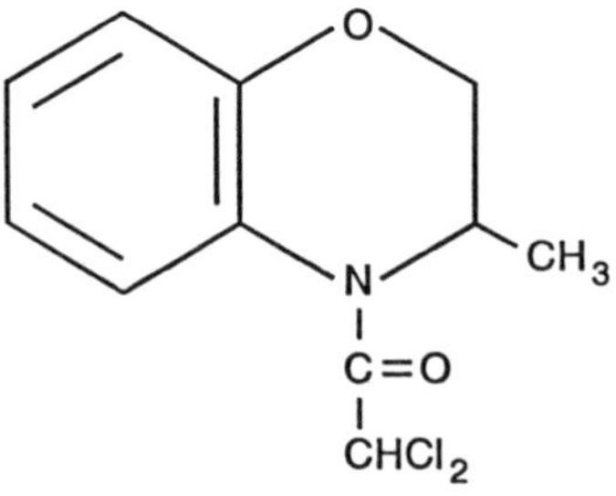

Benoxacor
(CGA-154281)

4-(Dichloroacetyl)-3,4-dihydro-3-methyl-2H-1,4-benzoxazine

Figure 1. Structures of two chloroacetamide herbicides (alachlor and metolachlor) and a safener (benoxacor) used in these studies.

Materials and Methods

Catabolism of Chloroacetamide Herbicides by *P. fluorescens* UA5-40. Strain UA5-40 was grown on nutrient glucose broth, washed 3 times, and resuspended to a final cell density of 6.0 (optical density at A_{660}). Cell suspensions or buffer (1.9 ml) were treated with 100 μl of 4 mM ethanolic solutions of alachlor, dimethenamid {2-chloro-N-[(1-methyl-2-methoxy)ethyl]-N-(2,4-dimethyl-thienyl-3-yl)acetamide}, DIMM [des-isopropyl-methoxy-metolachlor = 2-chloro-N-(2-ethyl-6-methylphenyl)acetamide], DMMA [des-methylmethoxy-alachlor = 2-chloro-N-(2,6-diethylphenyl)-acetamide], metolachlor, or propachlor [2-chloro-N-(1-methylethyl)-N-phenylacetamide]. After 20 h incubation, the assay was terminated with 4 ml acetonitrile, extracted by sonication, and centrifuged (15,000 xg, 10 min). Supernatants were analyzed by HPLC: μBondapak C18 reverse phase column, 1% aqueous acetic acid/acetonitrile gradient (initial 60% acetic acid, final 30% acetic acid) monitored as UV absorbance at 250 and 262 nm, as described elsewhere (20). Retention times were: alachlor 13.0 min, DIMM 6.8 min, DMMA 9.0 min, dimethenamid 11.4 min, metolachlor 13.0 min, and propachlor 8.7 min. Activity was calculated based upon recovery of parent substrate from cell suspensions compared to that recovered from controls.

Growth Chamber Studies Assessing Interactions of Herbicide Level, Safener, and Bacterial Inoculation on Corn Growth and Herbicide Persistence. Initial studies evaluated several rates of alachlor and metolachlor on corn growth, herbicide persistence and root colonization by strain UA5-40. A Conetainer® system similar to that described elsewhere (21) was used for these studies. Cones were filled with 175 g of a Bosket sandy loam (no herbicide history, pH 6.5, 0.9 % organic matter). The experimental design consisted of 5 treatments: no plant, corn alone, corn inoculated with UA5-40rif (rifampicin resistant), corn treated with benoxacor, and corn treated with benoxacor and with UA5-40. A randomized complete block design that also included four rates of herbicide and one no-herbicide treatment, with five replicates, was used. Technical grade alachlor and metolachlor (Chem Service, PA), were dissolved in water at appropriate dilutions so that application of 30 ml resulted in rates of 1x, 3x, 6x, and 12x the maximum label rates of application (3.0 kg ha^{-1} alachlor and 4.5 kg ha^{-1} metolachlor). Soils were initially moistened with the first 20 ml of appropriate solution. When benoxacor was applied at 5x the recommended rate (0.75 kg ha^{-1}), it was incorporated into the last 10 ml of herbicide solution. Strain UA5-40rif was grown in tryptic soy broth for 24 h. Cells were recovered by centrifugation, washed twice in phosphate buffer (0.05 M, pH 7.0), and diluted to a final cell density of log (10) 9.0 cells ml^{-1}. Untreated (i.e., no fungicide) Pioneer 3394 corn seeds were planted (2 per cone); and if inoculated, 1.0 ml of UA5-40 was placed on each seed. The seeds were covered with an additional 15 g of soil and then the remaining 10 ml of herbicide, or herbicide + safener solution were added. Plants were grown in an environmental chamber [light:dark cycle of 14:10 h at 30:24° C]. Light (425 μE m^{-2} s^{-1}, PAR) was provided by sodium halide and incandescent lamps. Seedlings were watered only sparingly (via pipette) so that no leaching occurred. Plants were fertilized with dilute N:P:K commercial fertilizer.

Twelve days after treatment (DAT), plants were evaluated for herbicide damage, and shoot and root fresh weights were determined. Root colonization was only examined in UA5-40-inoculated treatments (all with safener and certain alachlor rates without safener). For root colonization measurements, closely adhering soil was removed, and two 5 cm root segments (2 cm distal to the seed) were placed in 10 ml potassium phosphate buffer and serially diluted. Rhizosphere suspensions were then spiral plated on 10% tryptic soy agar containing 100 µg ml^{-1} cycloheximide and 500 µg ml^{-1} rifampicin. Following 48 h incubation (30° C in the dark), colonies were counted and populations were expressed as log (10) colony forming units (cfu) per g of root fresh weight.

In this initial study the persistence of alachlor and metolachlor was evaluated in soils treated at the 12 x application rate. Soils were extracted using ethyl acetate phase partitioning. Twenty-five g of soil was placed in high-density polypropylene bottles, 25 ml of acidified water (pH 3.0) and 50 ml of ethyl acetate were added. The bottles were capped and placed on a platform shaker for 24 h. Soil was sedimented by centrifugation (7,500 rmp, 10 min) and the supernatant decanted into separatory funnels. The aqueous phase was discarded and the ethyl acetate fraction was concentrated under nitrogen. Alachlor and metolachlor were determined by HPLC using the method previously described (20). Initial recoveries exceeded 98% (data not shown).

A second growth chamber study examined corn growth and persistence of metolachlor at higher application rates. Four treatments were used: no plant, untreated corn, corn treated with benoxacor, and corn treated with benoxacor and UA5-40. Commercially formulated herbicide (Dual 8E; 960 g a.i. l^{-1}) was used to attain rates of 12x, 24x and 36x. Methodology similar to that described for the first experiment was used, except for the following modification. In this study, the seeds were placed on dry soil, and either 5 ml of potassium phosphate buffer (0.05 M; pH 7.0) or 5 ml of UA5-40 (log 9.0 cfu ml^{-1}) were pipetted on top of the seed, followed by application of 25 ml of the herbicide solution. This alternate inoculation method was used so that greater movement of UA5-40 through the soil could be achieved. Plants were harvested 17 DAT, at which time shoot and root fresh weights were measured and the soil was examined for herbicide persistence. Data from both studies were subjected to ANOVA and means separated using Fisher's Protected LSD.

Results and Discussion

Chloroacetamide Catabolism by *P. fluorescens* Strain UA5-40. The catabolism of six chloroacetanilides by the most active strain, UA5-40, was investigated. Metabolism was based upon dissipation of the parent compound 20 hours after treatment (HAT) as determined via HPLC analysis (Table I). Propachlor and alachlor were the most rapidly transformed. Metolachlor was transformed more rapidly than dimethenamide. There was no difference between the dissipation rate of metolachlor and its N-dealkylated metabolite, DIMM. The rate of dissipation of the N-dealkylated metabolite of alachlor, DMMA, was lower than that of alachlor. Accompanying the dissipation of the chloroacetamides were metabolites with a retention time of approximately 5.5 min (20). This was not observed in the herbicide controls or in cells incubated without substrate. These are most likely the glycylcysteine- and cysteine-conjugates of these substrates. All six chloroacetamides were catabolized by strain UA5-40. Other researchers have shown that the non-enzymatic conjugation of chloroacetamides varied in the following order: propachlor > alachlor > metolachlor, and that plant enzymes could cause significant conjugation that differed with enzyme source (11).

Table I. Catabolism of Chloroacetamides by *P. fluorescens* UA5-40 Suspensions

Substrate	nmol Transformed 20 HAT[1]
Alachlor	54.4
Dimethenamid	27.2
DIMM	33.6
DMMA	44.2
Metolachlor	37.2
Propachlor	56.8
Fisher's Protected LSD (0.05)	9.9

[1] Initial concentration = 200 μM; mean of four replicates.

Corn Growth as Affected by Herbicide Concentration, Benoxacor and Inoculation.
Corn grown in soil treated with alachlor or metolachlor at 3x the recommended rate or greater, without safener, exhibited visual symptoms of phytotoxicity, with greater phytotoxicity occurring at the highest rates (data not shown). At the 12 x rate of both herbicides, leaves of unsafened plants did not expand or unfold in a normal manner (i.e., abnormal whorl formation and development). Inoculation with UA5-40 caused no reduction of visual phytotoxicity, but plants treated with benoxacor exhibited normal leaf development. The phytotoxicity was reflected in shoot fresh weight, and to a lesser extent, root fresh weight (Table II). Shoot fresh weight was reduced about 35 % at 3x alachlor and 50% at the 12x rate, while root weights were reduced 20% at the 3x rate and only 35% at the 12x rate, compared to the no-herbicide control plants. Similar patterns of fresh weight reduction were observed for metolachlor (Table III). Even at the highest rate (12x), benoxacor-safened seedlings attained 80% or greater of the shoot or root fresh weight of the no-herbicide treatment for both herbicides.

In the second study, normal plant development was observed at metolachlor concentrations as high as 324 μmoles kg^{-1} soil (36x rate) in both safened seedlings and safened seedlings inoculated with UA5-40. A significant effect (P=.001) of herbicide concentration and treatment (P=.001) on corn root and shoot fresh weight was observed, but no treatment x rate interaction was found. Overall, shoot and root fresh weight decreased with increasing herbicide concentration (Figure 2.).

There were significant differences in shoot fresh weight accumulation among all three treatments (corn alone, corn + benoxacor, and corn + benoxacor + UA5-40) averaged over three metolachlor concentrations (Figure 3). The greatest shoot fresh weight accumulation was in the corn + benoxacor + UA5-40 treatment. Since the UA5-

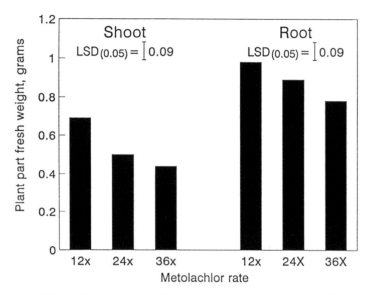

Figure 2. Effects of three metolachlor rates on corn shoot and root fresh weight accumulation, 17 DAT. Bars = pooled means of 3 treatments (corn, corn + benoxacor, and corn + benoxacor + UA5-40).

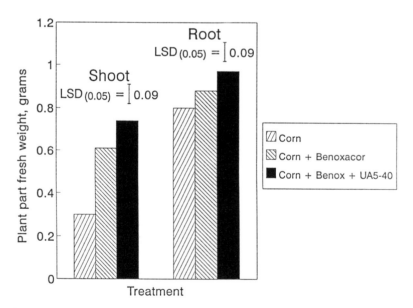

Figure 3. Effects of several treatments on corn shoot and root fresh weight accumulation, 17 DAT. Bars = pooled means from 3 metolachlor rates (12x, 24x, and 36x).

40 treatment alone was not able to counteract the growth inhibition caused by metolachlor (Table III), the increased fresh weight accumulation in the corn + benoxacor + UA5-40 treatment above that of corn + benoxacor suggests a synergistic effect between the safener and the microorganism. This interaction (synergy?) was not observed in root fresh weight accumulation, where the only significant difference was between the corn and the corn + benoxacor + UA5-40 treatments (Figure 3).

Table II. Interaction of Benoxacor and UA5-40 on the Growth of Corn Seedlings in Soil Treated with Several Alachlor Rates, 12 DAT

Treatment	\	\	Alachlor Rate	\	\
	0	1x	3x	6x	12x
	\	\	Shoot fresh weight (g^{-1} plant)	\	\
Control	1.13[a]	1.09	0.73	0.70	0.55
UA5-40	1.16	1.02	0.86	0.74	0.62
Benoxacor	1.13	1.02	0.98	1.05	1.08
Benox. + UA5-40	1.18	1.18	1.02	1.03	0.98
LSD (0.05)[b]	\	\	0.17	\	\
	\	\	Root fresh weight (g^{-1} plant)	\	\
Control	0.81	0.81	0.64	0.54	0.52
UA5-40	0.75	0.64	0.63	0.58	0.51
Benoxacor	0.83	0.80	0.72	0.83	0.74
Benox. + UA5-40	0.89	0.76	0.75	0.70	0.70
LSD (0.05)	\	\	0.12	\	\

[a] Means of 5 reps.
[b] LSD valid for comparison of effects of herbicide rates within a given treatment (rows) or of treatments within a given herbicide rate (column) for each plant part.

Root Colonization by UA5-40. A satisfactory degree of root colonization by *P. fluorescens* UA5-40 was observed in all roots sampled (>log 6 cfu g^{-1} soil) (Table IV). Efficient root colonizing bacteria have been defined as introduced strains that can displace native microflora and persist at propagule densities of log 3 to 6 cfu/g root fresh weight (22).   At the 12x rate of both herbicides, a significantly greater degree of root colonization was observed compared to the lower herbicide rates and the untreated controls. Perhaps slower root development at the highest levels (12x) of alachlor and metolachlor resulted in higher colonization by UA5-40. No significant effect of safener on root colonization by UA5-40 was noted (data not shown).

Herbicide Persistence/Degradation. In the first study, about 50% of the initial levels of alachlor and metolachlor were recovered in soil without corn (Figure 4). Significant (P=0.01) effects of treatment on alachlor or metolachlor recovery were observed. About 50% of the initial herbicide was recovered from soil without corn (herbicide control), while in soil with corn, 13 to 22 % of initially-applied alachlor and metolachlor was recovered. Herbicide recovery in the corn + benoxacor treatment did not differ significantly from the corn alone or corn + UA5-40 treatments for either herbicide. However, the corn + benoxacor + UA5-40 treatment had significantly (P=0.05) lower residual herbicide levels than all other treatments except corn + benoxacor for both herbicides. This suggested a possible interaction on herbicide dissipation, so studies with increased metolachlor concentrations were conducted.

Table III. Interaction of Benoxacor and UA5-40 on the Growth of Corn Seedlings in Soil Treated with Several Metolachlor Rates, 12 DAT

	Metolachlor Rate				
Treatment	0	1x	3x	6x	12x
	——————— Shoot fresh weight (g^{-1} plant) ———————				
Control	1.14[a]	1.03	0.79	0.73	0.57
UA4-50	1.16	1.12	0.89	0.67	0.55
Benoxacor	1.13	1.10	1.06	0.85	1.20
Benox. + UA5-40	1.18	1.15	1.10	0.95	1.08
LSD (0.05)[b]	——————————— 0.17 ———————————				
	——————— Root fresh weight (g^{-1} plant) ———————				
Control	0.81	0.76	0.67	0.61	0.60
UA5-40	0.75	0.79	0.62	0.63	0.49
Benoxacor	0.83	0.73	0.84	0.65	0.77
Benox. + UA5-40	0.89	0.75	0.78	0.72	0.70
LSD (0.05)	——————————— 0.11 ———————————				

[a] Means of 5 reps.
[b] LSD valid for comparison of effects of herbicide rates within a given treatment (rows) or of treatments within a given herbicide rate (column) for each plant part.

Table IV. Colonization of Corn Roots by UA5-40 in Soil Treated with Several Rates of Alachlor or Metolachlor, 12 DAT

Herbicide Rate	Alachlor	Metolachlor
	log (10) cfu g^{-1} root	
0	6.27[a]	6.26
1x	6.58	6.51
3x	6.65	6.50
6x	6.36	6.64
12x	7.48	7.23
LSD (0.05)	0.54	0.40

[a] Means of 5 reps.

In the second study, in which metolachlor was applied at 12, 24 and 36x rates, a significant effect of herbicide (P=.001) and treatment (P=.001) was observed, but there was no rate x treatment interaction (P=0.189). In this study, metolachlor recovery was directly proportional to the application rate (Figure 5). Considering all herbicide rates, lower recovery of metolachlor (P=0.01) was observed in soil planted with corn (Figure 6). Recovery of metolachlor in soil planted with corn was not significantly different than corn + benoxacor. However recovery of metolachlor from soil treated with UA5-40 and benoxacor were significantly lower (P=0.05) compared to corn alone or corn treated with benoxacor. Metolachlor residues in soil treated with the 36x rate were 89%, 80%, 75%, respectively, for the unplanted soil, corn, and corn + benoxacor treatments, while only 54% remained in corn + benoxacor + UA5-40 treatment (data not shown). No metabolites were observed in any soil extracts because we used an ethyl acetate-phase partitioning and the polar metabolites of these herbicides would not have been extracted. Since we used unlabelled herbicide, we were not able to separate these metabolites from soil components. Likewise, no residual benoxacor (retention time = 10.5 min) was observed.

Summary and Conclusions

We have shown that corn can be used to accelerate dissipation of very high levels of alachlor (36 kg ha^{-1}) and metolachlor (54 to 162 kg ha^{-1}) in soil. Furthermore, although presently based only on a model system (growth chamber conditions, short duration study), this domesticated plant (corn) can be modified chemically and microbiologically to enhance its tolerance to chloroacetamide herbicides. Chemical safeners are well accepted tools in agricultural production to minimize herbicide injury to crops, and the search for more efficient compounds continues. Recently, compounds of different chemical classes have been found superior to some of the standard commercial safeners

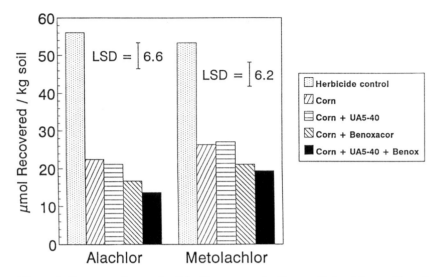

Figure 4. Recovery from soil of alachlor or metolachlor applied at 12x as affected by corn, benoxacor, and UA5-40, 12 DAT. LSDs at the 0.05 confidence level are presented.

used to protect against chloroacetamide herbicide injury (23). These same chemicals may have utility in developing phytoremediation as a practical technology for remediation of herbicide-contaminated soils that will allow the use of traditional crops for phytoremediation. Furthermore, bacteria that degrade the herbicide dicamba (3,6-dichloro-2-methoxybenzoic acid) have been used in soils to protect susceptible crop species from herbicide toxicity (24). Detoxification has been proposed as a potential mode of action of plant-growth-promoting rhizobacteria (25). It has also been suggested that various strains of *Pseudomonas* might be used as seed inoculants to protect crops against herbicides (17, 22). Recently, microbial strains that rapidly degrade the sulfonylurea herbicide, chlorsulfuron {2-chloro-N-[[(4-methoxy-6-methyl-1,3,5-triazin-2-yl)-amino]carbonyl]benzenesulfonamide}, have been used to inoculate flax (*Linum* sp.) and beet (*Beta vulgaris* L.) seeds for protection against herbicide injury (26). Phytoremediation is a young technology and a greater understanding of plant-soil-microbial relationships affecting the fate of particular xenobiotics is needed. We have shown that a bacterial strain capable of only co-metabolic transformation of alachlor and metolachlor can have additive benefits on reducing unwanted herbicide residues in soil when integrated with a suitable plant and a chemical safener.

Future studies will require scale-up and design of an experimental system that will enable determination of a mass balance of the fate of the contaminant before implementation of our phytoremediation strategy into a real world situation. We used a sandy loam soil with very low organic matter, thus a high proportion of the added herbicide would be readily available for both phytotoxicity via plant uptake and microbial degradation in the soil. Herbicides that are more tightly adsorbed to different soil types would perhaps be more easily remediated due to reduced availability for phytotoxicity. However, such increased concentration in the soil solution may limit microbial

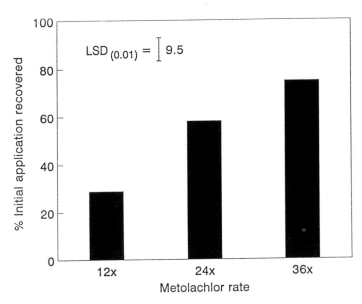

Figure 5. Effect of metolachlor rate on its recovery from soil, 17 DAT. Bars = pooled means of 4 treatments (metolachlor alone, corn, corn + benoxacor, and corn + benoxacor + UA5-40).

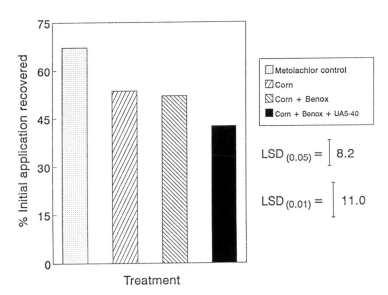

Fig. 6. Effects of several treatments on the recovery of metolachlor from soil, 17 DAT. Bars = pooled means from 3 metolachlor rates (12x, 24x, and 36x).

degradation, depending on kinetics of sorption and of metabolism. Assessment of herbicide uptake by the plant versus degradation in the soil/rhizosphere needs to be addressed. Relationships between, plant genotype and species, plant and microbial nutrition , soil type, and environmental conditions are also important considerations. Chemical safeners are also available for other herbicides in addition to the chloroacetamide herbicides. The role of these compounds for providing bioremediation potential of other herbicides by other crops is also being investigated.

Acknowledgements

We thank Ciba Crop Protection Corp., Greensboro, NC, for generously supplying the high-purity benoxacor, DIMM, DMMA, and metolachlor. Sandoz Agro, Inc., Des Plaines, IL, generously provided the technical grade dimethenamid. Our thanks to Pioneer Hi-Bred International, Des Moines, IA, for providing the Pioneer 3394 seeds used in this study. The valuable technical assistance of Liz Smyly, Velma Robertshaw, and Earl Gordon during this project is appreciated. R.W. Hoagland's help with graphics preparation is also acknowledged.

Literature Cited

1.   Habecker, M. A. *Environmental contamination at Wisconsin pesticide mixing/loading facilities: case study, investigation and remedial action evaluation;*   Wisconsin Dept. Agric., Trade and Consumer Protection, Agricultural Resource Div., Madison, WI, **1989**; 80 pp.

2.   Long, T. *Proc. Illinois Agric. Pesticides Conf. '89.* Coop. Ext. Serv. Univ. IL, Urbana-Champaign, IL, **1989**, pp. 139-149.

3.   Anderson, T. A.; Krueger, E. L.; Coats, J. R. *Chemosphere,* **1994**, *28,* 1551-1557.

4.   Anderson, T. A.; Krueger, E. L.; Coats, J. R. In *Bioremediation Through Rhizosphere Technology;* Anderson, T.A.; Coats, J. R., Eds.; ACS Books, Washington, DC, **1994**, pp 199-209.

5.   Felsot, A.S., Dzantor, E.K. In *Pesticides in the Next Decade: The Challenges Ahead;* Weigmann, E. L., Ed.; Proc. Third National Res. Conf. on Pesticides: Va. Polytechnical Inst., Blacksburg, VA, **1991**, pp 532-551.

6.   Hatzios, K. K.; Hoagland, R. E., Eds.; *Crop Safeners for Herbicides;* Academic Press: San Diego, CA, **1989**; 400 pp.

7.   Erza, G.; Stephenson, G. R. In *Crop Safeners for Herbicides;* Hatzios, K. K., Hoagland; R. E., Eds.; Academic Press: San Diego, CA, **1989**; pp 147-161.

8.   Ebert, E.; Gerber, H. R. In *Crop Safeners for Herbicides;* Hatzios, K. K., Hoagland; R.E., Eds.; Academic Press: San Diego, CA, **1989**; pp 177-193.

9.   Peek, J. W.; Collins, H. A.; Propiglia, P. J.; Ellis, J. F.; Maurer, W. *Proc. Weed Sci. Soc. Amer.* **1988**, *28,* pp. 13-14.

10.   Viger, P. R.; Eberlein, C. V.; Fuerst, E. P. *Weed Sci.* **1991**, *39,* 227-231.

11.   Scarponi, L.; Perucci, P.; Martinetti, L. *J. Agric. Food Chem.,* **1991**, *39,* 2010-2013.

12.   Breaux, E. J. *Weed Sci.,* **1987**, *35,* 463-468.

13. Viger, P. R.; Eberlein, C. V.; Fuerst, E. P.; Gronwald, J. W. *Weed Sci.* **1991**, *39*, 324-328.
14. Dean, J. V.; Gronwald, J. W.; Anderson, M. P. *Z. Naturforsch.* **1991**, *46c*, 850-855.
15. Lamoureux, G. L.; Rusness, D. G. *Pestic. Biochem. Physiol.*, **1989**, *34*, 187-204.
16. Feng, P. C. C. *Pestic. Biochem. Physiol.*, **1991**, *40*, 136-142.
17. Zablotowicz, R. M.; Hoagland, R. E.; Locke, M. A. In *Bioremediation Through Rhizosphere Technology;* Anderson, T. A.; Coats, J. R., Eds.; ACS Symp. Ser. No. 563; Amer. Chem. Soc.: Washington, DC, **1994**; pp 184-198.
18. Zablotowicz, R. M.; Hoagland, R. E.; Locke, M. A.; Hickey, W. J. *Appl. Environ. Microbiol.* **1995**, *61*, 1054-1060.
19. Zablotowicz, R. M.; Dzantor, E. K. *Proc. Air Waste Mgmt. Assoc. Annu. Meet.*, **1994**, Paper No. 94-RA126.04.
20. Locke, M. A.; Gaston, L. A.; Zablotowicz, R. M. *J. Agric. Food Chem.*, **1996**, *44*, 1128-1134.
21. Hoagland, R. E.; Zablotowicz, R. M.; Reddy, K. N. In *Saponins Used in Food and Agriculture;* Waller, G. R.; Yamasaki, K. Eds.; Plenum Publishing Corp.: New York, **1996**; pp 57-73.
22. Kloepper, J. W.; Zablotowicz, R. M.; Tipping, E. M.; Lifshitz, R. In *The Rhizosphere and Plant Growth;* Keister, D. L.; Cregan, P. B., Eds.; Kluwer Academic Publ, The Netherlands, **1991**; pp. 315-326.
23. Repasi, J.; Hulesch, A.; Suvegh, G.; Dutka, F. *J. Environ. Sci. Health, Part B*, **1996**, *B31*, 567-571.
24. Krueger, J. P.; Butz, R. G. *J. Agric. Food Chem.* **1991**; *39*, 1000-1003.
25. Kloepper, J. W.; Lifshitz, R. L.; Zablotowicz, R. M. *Trends Biotechnol.* **1989**, *7*, 39-44.
26. Gavrilkina, N. V.; Filipshanova, L. I.; Zimenko, T. G. *Vestsi Akad. Navuk. Belarusi, Ser. Biyal. Navuk.*, **1995**, *3*, 55-58.

Chapter 8

Ascorbate: A Biomarker of Herbicide Stress in Wetland Plants

T. F. Lytle and J. S. Lytle

Institute of Marine Sciences, Gulf Coast Research Laboratory, University of Southern Mississippi, P.O. Box 7000, Ocean Springs, MS 39564

In laboratory exposures of wetland plants to low herbicide levels (<0.1 μg/mL), some plants showed increased total ascorbic acid suggesting a stimulatory effect on ascorbic acid synthesis occurred; at higher herbicide concentrations ($\geq$0.1 μg/mL) a notable decline in total ascorbic acid and increase in the oxidized form, dehydroascorbic acid occurred. *Vigna luteola* and *Sesbania vesicaria* were exposed for 7 and 21 days respectively to atrazine (0.05 to 1 μg/mL); *Spartina alterniflora* 28 days at 0.1 μg/mL trifluralin; *Hibiscus moscheutos* 14 days at 0.1 and 1 μg/mL metolachlor in fresh and brackish water. The greatest increase following low dosage occurred with *S. alterniflora*, increasing from <600 μg/g wet wt. total ascorbic acid to >1000 μg/g. Ascorbic acid may be a promising biomarker of estuarine plants exposed to herbicide runoff; stimulation of ascorbic acid synthesis may enable some wetland plants used in phytoremediation to cope with low levels of these compounds.

Tissue or plasma levels of ascorbic acid (AA) and dehydroascorbic acid (DAA), the oxidized form of AA, have been considered as possible biomarkers of oxidative stress in animals (*1*). We have examined several estuarine plants in laboratory bioassays to see whether AA and DAA can serve in the same capacity in wetland plants. In plants it is thought that AA serves directly to scavenge superoxide and hydrogen peroxide (*2,3*) both harmful to cell structures and functions, and indirectly to oxidant species through reduction of oxidized forms of vitamin E. AA can protect plants from effects of atmospheric ozone (*4,5*) and nitric oxide (*6*) furthermore may postpone senescence in plants through antioxidant effects. Plants, like many animals, can regenerate AA from the oxidized from, DAA, through the action of glutathione and glutathione reductase (*7*), but unlike most animals, plants may synthesize additional AA in response to oxidative stress (*4,8*). If herbicides undergo conversion to radicals they may produce oxidative stress to plants (*9*), this action may be offset by the antioxidant properties of AA.

A number of estuarine wetland plants have been cultured in our laboratory and exposed to several herbicides under a variety of conditions and examined for indicators of stress. Because of runoff from agricultural fields into coastal estuaries following pesticide application, a series of studies were designed to determine effects of these pesticides on the indigenous plant communities bordering these estuaries. Seedlings of four estuarine wetland plants, *Vigna luteola*, *Sesbania vesicaria*, *Spartina alterniflora* and *Hibiscus moscheutos* were exposed to either atrazine, trifluralin or metolachlor, herbicides commonly used in the Southeastern U.S., and then examined for tissue levels of total ascorbic acid (TAA) and DAA. At low herbicide exposures enhanced levels of TAA were found. At higher herbicide exposure concentrations significant loss of TAA and increased DAA were found, suggesting that TAA and DAA may be useful biomarkers of herbicide exposure in estuarine plants. Furthermore, the stimulation of synthesis of ascorbic acid at low herbicide concentrations may enable these wetland plants to withstand some herbicide components of agricultural runoff in areas where they were used for phytoremediation.

Methods

Test Exposure Conditions. Seeds of the four test plants were collected in either Weeks Bay, Alabama, a National Estuarine Research Reserve, or in Ocean Springs, Mississippi and germinated and raised on quartz sand as growth media with nutrients (in this and all tests) from 10% Hoaglands solution (*10*) for 3-6 weeks or until plants were about 10cm tall. Seedlings of *V. luteola* were grown for exposure to the triazine herbicide, atrazine in distilled water. Three replicate bioassay chambers contained 30 plants each with quartz sand as a growth media. The atrazine was added every second day as an aqueous solution of a commercial formulation at nominal concentrations of 0.05, 0.1, 0.5 and 1 μg/mL. At the end of 7 days exposure, plants were harvested and analyzed.

S. vesicaria seedlings were also exposed to atrazine in distilled water. Four replicate bioassay chambers contained 30 plants each. The atrazine was added every second day as an aqueous solution of a commercial formulation at concentrations of 0.05 and 0.1 μg/mL. At the end of 21 days plants were harvested and analyzed.

Seedlings of *S. alterniflora*, a dominant marsh plant of the Gulf of Mexico and Atlantic estuaries, were exposed to the pre-emergent herbicide, trifluralin, in seawater diluted to 4‰. Three replicate bioassay chambers contained 60 plants each in quartz sand. Trifluralin was added every second day as an aqueous solution of a commercial formulation at concentration of 0.1 μg/mL. Because of the difficulty in obtaining sufficient numbers of seedlings of the same age for this test, three groups of seedlings were tested as follows: control plants were 1 week older than treatment group B and 1 week younger than treatment group A at initiation of exposure. Plants were harvested one week after a 28-day exposure to trifluralin.

H. moscheutos seedlings were exposed to the chloroacetamide, pre-emergent herbicide, metolachlor, both in fresh distilled water and in brackish water (seawater adjusted to 15 g/kg salinity with distilled water). Duplicate bioassay chambers for

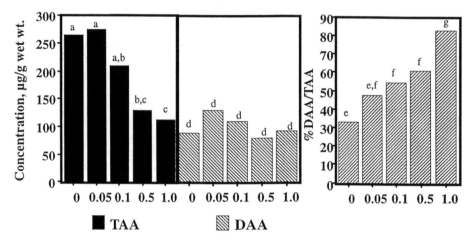

Figure 1. Ascorbate in *V. luteola* exposed to atrazine. Nominal concentrations of atrazine, μg/mL, are shown along X-axis with 0 being control. Means for replicate exposure groups are shown for each variable, level of total ascorbic acid (TAA), dehydroascorbic acid (DAA) and %dehydroascorbic acid/total ascorbic (%DAA/TAA); bars in each variable group not sharing a letter above bar are significantly different (ANOVA, $p < 0.05$); bars sharing letter are statistically indistinguishable.

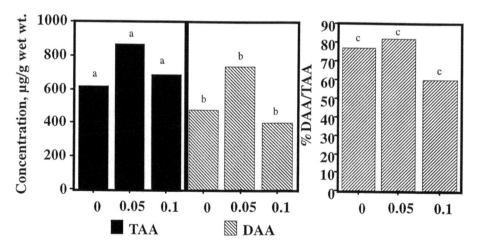

Figure 2. Ascorbate in *S. vesicaria* exposed to atrazine. Nominal concentrations of atrazine, μg/mL, are shown along X-axis with 0 being control. See Figure 1 caption for explanation of variables plotted and bar letters.

each exposure condition and controls contained 30 plants each in quartz sand. Metolachlor was added every second day as an aqueous solution of a commercial formulation at nominal concentrations of 0.1 and 1 μg/mL. At the end of 14 days plants were harvested and analyzed for TAA and DAA.

Ascorbic Acid/Dehydroascorbic Acid Analysis. Analysis followed the AOAC technique (*11*) with following summary. At the end of each exposure leaf tissue was collected from 10 plants in each of the bioassay chambers and analyzed immediately. These leaves were transferred to a glove box, flushed with nitrogen, and cut into small pieces. One g samples were weighed and extracted with HPO_3-$HOAc$-H_2SO_4. Extracts were filtered and oxidized with Norit (Fisher), converting original AA to DAA after which the reaction of DAA with o-phenylenediamine (Sigma) created a highly fluorescent derivative. This reaction was the basis of a fluorometric determination of TAA using a Perkin Elmer MPF44 fluorescence spectrofluorometer with excitation λ at 353 nm and emission λ at 426 nm with both slits set at 10 nm. In a separate analysis, DAA was analyzed directly in an acid extract not treated to the oxidation step. Standards were ascorbic acid (Sigma) prepared at 100 μg/L and 10 μg/L. With each set of samples actual plant extracts were spiked with ascorbic acid and the recovery of these spikes measured between 85 to 95%. All sample handling was done in a darkened laboratory and conducted with latex gloves for those steps outside the glove box.

Data Treatment. The three variables computed for each treatment were: TAA in μg/g wet tissue, DAA in μg/g wet tissue and %DAA/TAA (the exception being the *H. moscheutos* test in which only TAA was measured). Mean values for each of these variables were compared in each exposure experiment by one-way analysis of variance (ANOVA) followed by least significant difference (LSD) multiple range test at 95% confidence limit (*12*). For each variable, statistically significant differences were noted and presented.

Results

V. luteola **and** *S. vesicaria* **Exposed to Atrazine.** Figures 1 and 2 display results of exposure of *V. luteola* and *S. vesicaria* to atrazine. Tissue levels of TAA and DAA in *S. vesicaria* were more than double that of *V. luteola* in control plants. At the three exposure levels of atrazine that were identical for both plants, DAA in *S. vesicaria* represented a higher % of TAA than it did in *V. luteola*. Slight enrichment of TAA and DAA occurred following exposure at a nominal concentration of 0.05 μg/mL atrazine in both plants though neither was statistically significant. At this exposure level the relative contribution of the oxidized form of DAA was greater in both plants than in controls. After exposure at 0.1 μg/mL atrazine, levels of TAA dropped in both plants compared to the 0.05 levels with a prominent and significant increase in %DAA/TAA occurring in the *V. luteola*. At the higher dose levels of 0.5 and 1 μg/mL, levels of TAA continued to decrease in *V. luteoma* in a dose response accompanied by increasing fraction of TAA existing in the oxidized DAA form.

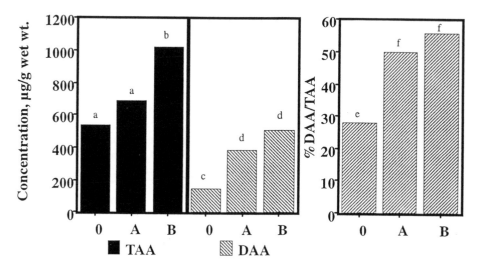

Figure 3. Ascorbate in *S. alterniflora* exposed to trifluralin. Plant group A (one week older than controls, shown as 0 on graph) and group B (one week younger than controls) were exposed to 0.1 μg/mL nominal level of trifluralin. See Figure 1 caption for explanation of variables plotted and bar letters.

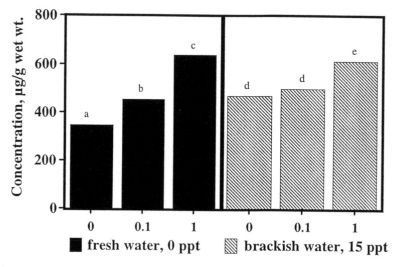

Figure 4. Total ascorbic acid in *H. moscheutos* exposed to metolachlor. Nominal concentrations of metolachlor, μg/mL, shown along X-axis with 0 being control. Plants were exposed in fresh water, salinity = 0 g/kg, and in brackish water, salinity = 15 g/kg. See Figure 1 caption for explanation of bar letters (note: only total ascorbic acid determined).

S. alterniflora **Exposed to Trifluralin.** Figure 3 indicates that more TAA occurred in *S. alterniflora* plants exposed to 0.1 μg/mL trifluralin than in controls with significantly more occurring in the Group B plants that were younger than controls. Both groups of trifluralin-exposed plants experienced a significant enhancement of DAA with the DAA/TAA ratio almost doubling that found in the control plants, rising from <30% to >50%.

H. moscheutos **Exposed to Metolachlor.** TAA levels in *H. moscheutos* following exposure to metolachlor shown in Figure 4 demonstrate a positive dose response in plants growing in either fresh water or saline water. There were statistically significant differences in TAA levels at both exposure levels compared to controls in fresh water with statistically significant differences exhibited only at the 1 μg/mL level for saline water exposure.

Discussion

Low Level Exposures. In each of the four plant experiments there seemed to be a stimulatory response on TAA observed at the very lowest dose level regardless of the pesticide or plant studied, though it was only with *H. moscheutos* that this trend proved to be statistically significant. Further testing will be required to see whether what seemed to be a trend in stimulatory response at very low doses of herbicides will withstand statistical scrutiny for these wetland plants. If these herbicides induce an oxidative stress, an oxidative stimulatory effect would not be surprising since Ranieri et al (*4*) and Machler et al. (*5*) showed that levels of both AA and DAA increase as a response to the oxidative stress of atmospheric ozone in agricultural plants.

Results of *S. alterniflora* exposed to trifluralin emphasizes the need for consistency in toxicity bioassays with plants. The Group B plants (see Figure 3) had substantially higher amounts of TAA than did either the controls or Group A plants. As noted earlier, the Group B plants were a week younger than the controls which in turn were a week younger than Group A. Dinakin and Ayanlaja (*2*) observed a three-fold increase in TAA in 8-week jute (*Corchorus olitorius*) seedlings compared to 5-week seedlings grown under identical conditions. It is quite likely then that differences in TAA enhancement seen in Group A and B *S. alterniflora* seedlings were due in part to the age difference of plants used in this study. Nevertheless, both the plants older and younger than controls showed the same response but to a different degree. Furthermore, %DAA/TAA was almost double in both sets of exposed plants compared to controls. These results are further testament to the resilience of *S. alterniflora* to herbicide damage shown by Lytle and Lytle (*13*).

Higher Exposure Levels. Above 0.1 μg/mL atrazine, both *V. luteoma* and *S. vesicaria* experienced loss of TAA that behaved in a negative dose response in the *V. luteoma* exposure from 0.1 to 1 μg/mL. Under normal circumstances plants can regenerate AA from reduction of DAA by glutathione and glutathione reductase (*7*). The decline in TAA and increasing relative levels of the oxidized form, DAA (>80% of TAA following 1 μg/mL exposure), in this plant indicates that insufficient AA regeneration via DAA reduction and/or irreversible loss of DAA through further oxidation occurred at these higher dose levels. Apparently the AA/DAA defense system has been compromised by atrazine at these higher concentrations.

Multiple Stress Exposures. It is noteworthy that in spite of salt stress to one group of *H. moscheutos*, these plants responded to 0.1 μg/mL and 1 μg/mL of metolachlor in a manner almost identical to plants grown in fresh water. Because a serious salt stress did not affect the ascorbic acid response to herbicide stress, this plant may be a good candidate for use as an indicator plant in estuaries exposed to herbicide run-off under conditions of varying salinities and may also serve in certain phytoremediation applications.

Conclusions. Though these experimental procedures differed in their design and execution, it appears clear that atrazine, trifluralin and metolachlor representing three classes of herbicides can elicit a measurable response in levels of TAA and DAA, from what seemed to be a stimulatory response at very low levels to pronounced loss of TAA and increased %DAA/TAA at higher levels. Since seeds of all four plants are readily available and can, with methodologies developed in our laboratory and elsewhere, be germinated in sufficient quantities for study, it is recommended that more future work with plant bioassays be directed from agricultural species and use indigenous plant species such as those used in this study. *S. alterniflora* and *H. moscheutos* showed no evidence of losing ability to produce ascorbic acid with exposure to low levels of herbicides and there are indications that these and other estuarine wetland plants may have the resilience to herbicide damage to be used in buffer zones around agricultural fields in coastal areas.

Acknowledgments

The authors thank Faye Mallette, Nghe Nguyen, Lea Sharpe and Hong Cui for technical assistance and the U.S. Environmental Protection Agency for partial financial support under Grant No. CR820666-10.

Literature Cited

1.   Lykkesfeldt, J.; Loft S.; Poulsen, H. E. *Anal. Biochem.* **1995**, 229, pp. 329-335.
2.   Dinakin, M. J.; Ayanlaja, S.A. *HortScience* **1996**, *31*, p. 164.
3.   Maellaro, E.; Bellow, B. D.; Sugherini, L.; Pompella, A.; Casini, A. F.; Comporti, M. *Xenobiotica* **1994**, *24*, pp. 281-289.
4.   Ranieri, A.; Lencioni, L.; Schenóne, G.; Soldatini, G. F. *J. Plant Physiol.* **1993**, *142*, pp. 286-290.
5.   Machler, F.; Wasescha, M. R.; Krieg, F.; Oertli, J. J. *J. Plant Physiol.* **1995**, *147*, pp. 469-473.
6.   Kashiba-Iwatsuki, M.; Yamaguchi, M.; Inoue, M. *FEBS Letters* **1996**, *389*, pp. 149-152.

7.   Mehlhorn, H.; Cottam, D. A.; Lucas, P. W.; Wellburn, A. R. *Free Rad. Res. Comm.* **1987**, *3*, pp. 193-197.

8.   Lee, E. H.; Jersey, J. A.; Gifford, C.; Bennett, J. *Envir. exp. Bot.* **1984**, *24*, pp. 331-341.

9.   Kappus, H. *Arch. Toxicol.* **1987**, *60*, pp. 144-149/

10.   Hoagland, D. R.; Arnon, D. I.; Circular 347, California Experiment Station, Berkeley, 1950.

11.   *Official Methods of Analysis*; method 43.069, Williams, S. Ed.; Association of Official Analytical Chemists, Arlington, VA, 1984.

12.   *Statgraphics Plus*; Manugistics, Inc., Rockville, MD, 1995.

13.   Lytle, J. S.; Lytle, T. F. In *Environmental Toxicology and Risk Assessment: Biomarkers and Risk Assessment*; D. A. Bengtson; D. S. Henshel, Eds; ASTM STP 1306, Philadelphia, PA, 1996, Vol. 5; pp 270-284.

Chapter 9

Degradation of Persistent Herbicides in Riparian Wetlands

D. M. Stoeckel, E. C. Mudd, and James A. Entry

Department of Agronomy and Soils, College of Agriculture, 202 Funchess Hall, Auburn University, Auburn, AL 36849-5412

Modern agricultural practices make extensive use of herbicides to increase crop yields. Persistent herbicides (recalcitrant to degradation) are often preferentially used for season-long protection. The persistence of these herbicides makes them environmentally hazardous if they leach or are carried by surface runoff and erosion to pollute surface- or ground-waters. Three heavily used persistent herbicides are presented for illustration: atrazine (a triazine), fluometuron (a substituted urea), and trifluralin (a dinitroaniline).

Vegetated border strips between agricultural fields and adjoining streams are sometimes cleared and protected from flooding to increase the amount of cultivable land. These areas, left in their natural state as seasonally-flooded riparian wetlands, contain micro-environments conducive to immobilization and degradation of persistent herbicides. While natural riparian wetlands should not be used to treat point-source herbicide pollutants, the literature indicates that maintenance of riparian wetlands can help to slow migration of and to enhance degradation of herbicides from non-point sources.

Herbicides are economically essential to agriculture in the United States. Production of corn (*Zea mays*), cotton (*Gossypium hirsutum*), soybean (*Glycine max*), and other crops, which was once labor intensive, has become more efficient with chemical weed control. Agricultural herbicides are formulated to be somewhat persistent to confer season-long protection to a crop. While agriculture has become more efficient with the use of herbicides, some risk has been conferred to the environment. Herbicides can disrupt non-target plant physiological processes and may adversely affect human health and natural ecosystems (*1*).

The United States encompasses about 42 million hectares of wetlands in the conterminous states (*2*). They fall into a number of categories including tidal salt marshes, tidal freshwater marshes, mangrove swamps, freshwater marshes, southern deepwater swamps, and riparian wetlands. Approximately 27 million hectares (64%) of these wetlands are Southern deepwater swamps and riparian wetlands (*3*). In agricultural watersheds these areas are often forested zones of varying width between agricultural land and surface waters. They are characterized by dense vegetation with flood-tolerant species, high primary productivity, and a seasonal flooding regime. Their soils are usually hydric, characterized by low pH with a high organic matter content and a shallow oxidized surface layer underlain by deeper reduced layers (*4*).

A frequently ignored benefit of riparian wetlands is that they have characteristics which can act to retain and assist in the degradation of recalcitrant compounds. The basis for this characteristic is in the chemical conditions caused by frequent flooding and plant adaptations to these conditions. This combination results in an accumulation of organic matter in the soil and partially oxidized zones surrounding plant roots. The result is a buffer area which can slow the migration of herbicides and provide conditions conducive to biological and chemical degradation of recalcitrant compounds.

Riparian wetlands are sometimes drained for logging or conversion to field. Conversion of wetlands to agricultural use can result in high-yield croplands for several seasons, but is followed by a steady decline in innate productivity as the soils revert to a nutrient-poor state (*2*). When left in their natural state these lands have other values including buffering natural waters against runoff of agricultural chemicals (*5*), runoff of sediment (*6*), provision of hydrologic flood damping (*2*), and maintenance of wildlife habitat (*7, 8*). It is hoped that this paper will demonstrate the utility of keeping riparian wetlands intact for their long-term benefits in the face of pressure to clear them for short-term gain.

Fate of Herbicides in the Environment

There are 26.3 million hectares of corn-producing land, four million hectares of cotton-producing land, and 21 million hectares of soybean-producing land in the respective major-producing regions of the United States (*9*). In 1993 the aggregate use of chemical herbicides on these lands was estimated at 117 million kilograms (*9*). The fate of these herbicides is a question of some concern to the environment. The literature indicates that while most of these applied herbicides do not leave the field, some portion washes off with surface flow (dissolved in water or attached to sediment particles) or leaches to the subsurface (*1,10-13*). Another portion is degraded by sunlight in the process of photolysis (*14*), and yet another portion is volatilized (*15*). The protective value of riparian wetlands is associated with the fraction which leaves the site of application with surface flow.

Water insoluble herbicides leave the point of application associated with erosion sediment. The sediment will either run into the drainage basin or it will settle out short of running water. The initial benefit of riparian wetlands is based on hydrology. A riparian wetland on the border of a field can act as a sediment filter by slowing storm water before it hits the stream. This effect can be caused by topography or by physical barriers such as vegetation which can slow water flow rates.

Once herbicide-carrying sediment is deposited in the wetland, the herbicide may leave by the same mechanisms by which it could leave an agricultural field. The wetland environment is, however, very different from that of the field. The remainder of this paper describes mechanisms of herbicide removal from soil and how riparian wetlands can immobilize and enhance the degradation of agricultural herbicides. Soil characteristics which enhance immobilization and degradation will be introduced and illustrated with three examples of commonly used herbicides.

Illustrative Herbicides

Three herbicides will be used to illustrate the ability of riparian wetlands to immobilize and degrade agricultural chemicals. The herbicides were chosen because they are commonly used in the United States (9) and they encompass a wide spectrum of mobilities and degradation pathways (16). Table I is a compilation of some critical characteristics of each herbicide.

Atrazine. Atrazine (2-chloro-4-ethylamino-6-isopropylamino-1,3,5-triazine) is a commonly used triazine-class herbicide. Its structure and degradation pathway are shown in Figure 1. Atrazine is used against broadleaf weeds and grassy weeds in monocot (primarily corn) production (16). It has a wide range of half lives reported in the literature, ranging from two weeks (17) to four months (18) after field application. The average half-life value is around two months (18). Atrazine is not very soluble in water, which limits its mobility in soil. Despite this moderate mobility atrazine is frequently found contaminating groundwater resources, especially in the midwest (19). It is not volatile, so little is lost to the atmosphere. The major mechanisms of atrazine disappearance from a point of deposition are chemical and biological degradation. About 20 million kilograms of atrazine were applied in the major corn-producing area of the United States in 1993 (9).

Fluometuron. Fluometuron (1,1-dimethyl-3-(α,α,α-trifluoro-m-tolyl) urea) is a commonly used substituted urea herbicide. Its structure and degradation pathway are shown in Figure 2. Fluometuron is used to control broadleaf weeds in the production of cotton and other crops (16). The major mechanisms of disappearance of fluometuron are biological degradation (20) and leaching (10, 21). Fluometuron is a moderately persistent herbicide with reported half-life values from 26 days (22) to twelve weeks after summer field application, and up to 52 weeks upon fall field application (23). One million kilograms of fluometuron were used in the cotton-producing region of the United States in 1993 (9).

Table I: Compilation of important characteristics in the dissipation of the three study herbicides. References are listed in the text.

Characteristic	Atrazine	Fluometuron	Trifluralin
Molecular Wt.	215.7 g/mole	232.2 g/mole	335 g/mole
Common name	AAtrex	Cotoran	Treflan
Solubility	33 ppm in water Moderate mobility	90 ppm in water Moderate mobility	0.3 ppm in water Slight mobility
Photolysis susceptibility	Minimal	Slight	Moderate
Volatility	$P_v = 3.0 \times 10^{-7}$ mm Hg at 20 °C 8 g ha^{-1} d^{-1} at 0.5 day, 25 °C lost when 2.5 kg ha^{-1} applied 20% lost over 5 mos from field	$P_v = 5 \times 10^{-7}$ mm Hg at 20 °C no field loss rate information available in the literature	$P_v = 1.1 \times 10^{-4}$ mm Hg at 25 °C 150 g ha^{-1} d^{-1} at 0.5 day, 25 °C when 2.5 kg ha^{-1} applied 36% lost over 5 mos from field
Half-life in soil	14 to 120 days	49 to 364 days	154 to 420 days
Effect of Redox condition	Degradation rate decreases under anaerobic conditions	No information available under anaerobic conditions.	Degradation rate increases under anaerobic conditions.
Degradation	Initial steps are dominated by chemical reactions though microorganisms are responsible for mineralization	Moderate rate of biological degradation under aerobic conditions, with mineralization the end result	Probably dominated by chemical processes though microorganisms accelerate the degradation rate

Trifluralin. Trifluralin (α,α,α-trifluoro-2,6-dinitro-N,N-dipropyl-p-toluidine) is a dinitroaniline-class herbicide. It is applied preemergence for protection against a variety of shallow-rooted weeds (*16*). The structure and degradation pathway are shown in Figure 3. Its half-life ranges from 22 weeks to over a year after field application (*24*). This molecule has a very complex network of potential chemical and biological degradation pathways (*24-26*) and photolytic degradation (*14*). The trifluralin molecule is moderately volatile and it is nearly insoluble, which causes it to have low mobility through soils. Trifluralin was heavily applied in cotton and soybean production in 1993 in the United States with a combined use of 6.4 million kilograms (*9*).

Volatilization of Herbicides

In general, volatilization losses of field-applied herbicides increase with partial pressure. This phenomenon has been demonstrated in field studies in which 2.5 kg/ha of atrazine or trifluralin were applied. Atrazine, which has a relatively low partial pressure (Table I), was lost at a rate of only 8 g/ha-day while trifluralin was lost at a rate of 150 g/ha-day (*15*).

Volatilization may play a minor role in atrazine removal from soil. In a microcosm study, zero to eight percent of ^{14}C-radiolabel from atrazine was lost (not accounted for as ^{14}C-carbon dioxide nor as residues in the solid or liquid phases) due to volatilization as either parent compound or volatile degradation products (*17*). In the afore-mentioned field study Nash and Gish (*15*) showed a 20% loss of atrazine due to volatility over five months. Fluometuron has a partial pressure similar to that of atrazine (*16*) and is expected to show similar characteristics. Volatilization may be a major pathway in the disappearance of trifluralin from a soil over time (*24*). Nash and Gish (*15*) showed that both trifluralin and a combination of its degradation products were volatile. After 154 days, 20% to 50% of field-applied trifluralin was lost to volatilization at temperatures ranging from 5 °C to 35 °C.

Migration through Soil

Movement of organic molecules through soil is explained by hydrophobic bonding and ion exchange chemistry. Hydrophobic bonding is simply the tendency of a hydrophobic molecule to remove itself from a solution onto a neutral surface. Ion exchange chemistry is the same process which controls movement through an ion exchange column. Sorption of low solubility herbicides can be partially explained by each process (*27*).

The soil matrix is composed of sand, silt, clay, and organic matter. Most of these components hold a net neutral or negative charge at near-neutral pH. The amount of solute bound to the matrix from solution varies with the charge densities of the matrix and of the solute. The charge density of the soil matrix is analogous to its cation exchange capacity (CEC), the total of exchangeable bases a soil can hold. In soil the highest CEC is held by the organic material, though a substantial portion is held by clay particles. Compared with organic matter and clay, sand and silt have minimal CEC.

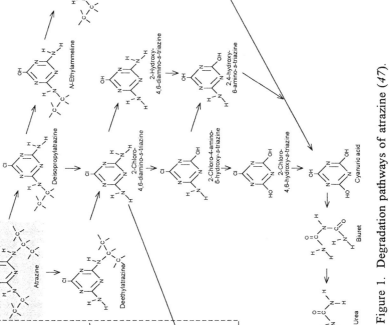

Figure 1. Degradation pathways of atrazine *(47)*.

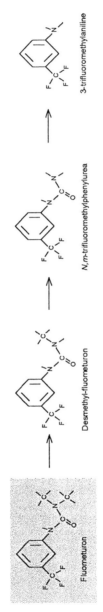

Figure 2. Degradation pathways of fluometuron (20).

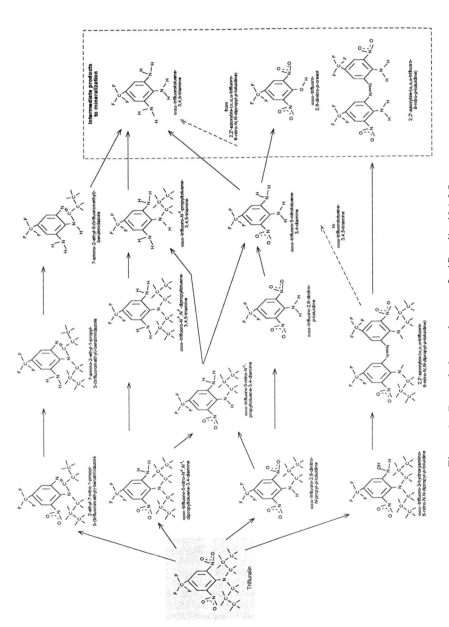

Figure 3. Degradation pathways of trifluralin (24-26).

Numerous studies have shown that movement of organic herbicides through soil is dominated by the organic content of the soil (28-30). As with an ion exchange column, pH is the driving criterion for the rate of elution through a soil of given CEC. The three herbicides discussed all have amino or nitro constituents and therefore become more positively charged as acidity increases. For example, atrazine has easily protenated amino groups which enhance its adsorption to negatively charged soil constituents under acidic pH. As the pH decreases the proportion of protenated amino constituents increases resulting in more binding to the matrix (31). However, as the pH decreases the negative charges of the soil matrix neutralize leaving fewer active sites for binding, causing the solute to elute faster. The overall effect of these competing trends is dependent upon the charge characteristics of the matrix and the solute. In the neutral to moderately acidic range, decreasing pH causes a net trend to slower elution rates for the weakly basic herbicides under discussion.

Freundlich Equation. The relationship between solute and matrix has often been described by the Freundlich equation (equation 1)

$$\frac{mass_{adsorbed}}{mass_{soil}} = K_f \cdot C_{eq}^{\alpha} \qquad (1)$$

where $mass_{adsorbed}$ over $mass_{soil}$ is the concentration of soil-bound herbicide at the soil solution equilibrium concentration C_{eq}, K_f is an empirical constant describing the adsorption characteristics of the herbicide, and alpha is an empirical constant describing the adsorption characteristics of the soil matrix (32). The resulting isotherm plots herbicide mobility in a soil as a function of the CEC of the soil and the charge on the herbicide at a given pH (28, 33).

Atrazine. When applied to field soils most atrazine stays in the top 10 cm of soil, and very little is lost to leaching below the top 20 cm of soil (18). The major path of transport to surface waters is through runoff-carried sediment. In soils with high organic matter content the mobility of atrazine decreases (34).

Fluometuron. Fluometuron easily leaches from soils. As was the case with atrazine, adsorption coefficients for fluometuron are strongly correlated with soil organic content (21, 29). One field study found that adsorption of fluometuron in the field was a function of depth, tillage practice, and cover crop (10). A sterilization-concentration application of fluometuron (40 kg/ha) traveled to the groundwater through a soil with an unknown organic content within two months (12). Migration of fluometuron may be through lateral subsurface migration, but will slow as it approaches the riparian wetland due to the increase in organic content.

Trifluralin. An increase in soil organic matter can contribute to a drastic increase in trifluralin retention (28). Minimal if any radiolabeled trifluralin or intermediate product leached during one year after application to a soil containing 4.4% organic matter (*24*).

Darcy's Equation. Accumulations of organic matter can have physical effects on herbicide movement through soil as well as the chemical effects noted above. The second-dominant characteristic of an ion exchange column (first being adsorption-desorption based on pH and charge density) is the flow rate. Flow rates through soil and resulting transport of contaminants have been described by Darcy's law (equation 2).

$$v = -k \cdot s \tag{2}$$

where v is the velocity of flow, k is the hydraulic conductivity, and s is the hydraulic gradient (*35*). Organic material can coat soil particles, increasing their diameters and decreasing interstitial pore volumes, which will decrease the hydraulic conductivity. The resulting flow will be slower, making herbicides less mobile in undisturbed soils high in organic matter (*27*). All three illustrative herbicides would be expected to follow this general trend.

Photolysis

Herbicides are subject to photolytic degradation when exposed to sunlight. This factor may be important upon field application, but is less important when the herbicides are transported to the riparian wetland. Sediment-transported herbicides will be protected from direct sunlight by vegetative cover and soil particles. Of the three illustrative herbicides, only trifluralin is susceptible to significant photolysis (*14, 16*).

Degradation

Degradation rates of herbicides have been reported in the literature by two methods. In the first method, degradation is indicated by disappearance of the parent compound or the appearance of known degradation products. This approach is complicated by the continued presence of the degradation products, some of which may also be phytotoxic. The second method is to measure the mineralization (evolution of ^{14}C-carbon dioxide) from a ^{14}C labeled substrate. This approach measures the complete destruction of the labeled component, which may be either a ring-structure or a side-chain on the parent molecule. The first method measures the initiation of degradation while the second measures the ultimate mineralization of the labeled constituent (*36*).

Degradation can occur either by chemical or biological methods. Chemical degradation relies upon entropy, such that the products are at a lower energy state than the reactant(s). These reactions are spontaneous and may be mediated by metal ions or salts (*37*). Biological degradation is enzyme mediated and may be spontaneous or may require the input of energy (*38*).

Soil microorganisms use organic compounds anabolically as carbon sources and sometimes catabolically as energy sources. Easily degraded constituents are utilized preferentially for both carbon and energy. Factors which decrease the degradative potential of a compound include the presence of aromatic ring structures, nitrogen- or sulfur- substitution, and halogen substitution (39, 40). Herbicides tend to incorporate one or more of these characteristics, making them recalcitrant to degradation. Microorganisms have not developed metabolic pathways for many pesticides if their structures are not similar to compounds found in the natural environment (38).

Biological degradation of recalcitrant compounds may coincide with other degradation pathways. Microbial communities degrading a carbon- or energy-yielding compound secrete enzymes which may incidentally degrade non-energy yielding compounds of similar structure. This phenomenon is known as cometabolism (38, 41, 42). Communities can also 'learn' degradation pathways. Exposure to xenobiotic chemicals can enhance degradation of compounds of similar structure upon re-exposure (43).

While it is impossible to predict what will happen in the field, degradation of recalcitrant compounds has been enhanced under nutrient-limiting conditions in the laboratory. In one study, incubation of an atrazine-degrading microbial community in media with atrazine as the sole nitrogen source stimulated degradation rates (44). In another, nitrogen fertilization of a grassland soil resulted in inhibition of atrazine degradation rates (45).

The redox state of the soil solution plays an important part in degradation pathways. Microorganisms degrade most organic compounds faster under aerobic conditions than anaerobic, though many compounds are only partially degraded at a given redox state. Cycling of aerobic and anaerobic conditions is a common procedure in process-based wastewater treatment for rapid and complete degradation (35, 46).

Atrazine. The degradation pathways for atrazine are shown in Figure 1 (47). While degradation of a major portion of the parent compound is fairly rapid, traces of atrazine parent compound were found in soils nine years after application and up to 50% of applied compound was recovered as soil-bound degradation product residues after nine years (11). Winkelmann and Klaine (17) found that about only 12% of ring-labeled ^{14}C-atrazine degraded to ^{14}C-carbon dioxide over six months.

The initial products of degradation are the hydrolytically dechlorinated product hydroxyatrazine and the dealkylated products deisopropylatrazine and deethylatrazine. Deethylatrazine is either formed rapidly or it is not as susceptible to degradation as deisopropylatrazine, since the deethyl product is commonly found in soil extracts at three to six times the concentration of the deisopropyl product (17, 43). Hydroxyatrazine similarly either has a higher flux or is less readily degraded since it is at least an order of magnitude higher concentration than the dealkylated products (17).

Degradation of atrazine has both chemical or biological aspects. Mineralization seems to be possible only with biological activity, however. Less than 0.1% of total radioactivity from ring-labeled atrazine evolved as ^{14}C-carbon dioxide using a sterile soil sample (17). In the same study, however, nearly all of the applied atrazine degraded biologically within six months and only a small portion remained in the form of the primary degradation products (*17*). A review by Erickson and Lee (*47*) described the conversion to hydroxyatrazine as a chemical process with rate minima at pH 3 and pH 11, while dealkylation was predominantly an energy-yielding microbial process.

Many atrazine-degrading microorganisms, both bacteria and fungi, have been isolated from field soils with histories of atrazine exposure. Most of the bacterial degraders tended to be gram-negative, with six genera represented in one study (*43*). Atrazine application is known to stimulate aerobic respiration in microbial communities (*48*), though whether this is a result of atrazine acting as a substrate or atrazine killing some organisms which are then available as a food source is unknown (*17*).

Levanon (*49*) labeled alkyl-group carbon and ring-carbon in tandem with selective inhibition of fungal and/or bacterial constituents. The results showed that ^{14}C-carbon dioxide evolved from alkyl-labeled atrazine enrichments with active fungi but not active bacteria. Labeled carbon dioxide was measured from ring-labeled atrazine enrichments only when neither bacteria nor fungi were inhibited. This was taken to indicate that in the environment fungi catabolically N-dealkylate atrazine while combined activity of fungi and bacteria is required for mineralization of the triazine ring (*49*).

Degradation of atrazine is also affected by the redox condition of soils. Under aerobic conditions no metabolites accumulated in broth culture, indicating full degradation (*44*). Under anaerobic conditions the atrazine degradation rate decreased (*47*). In one field study, no degradation of low concentrations (35 μM) of atrazine could be detected under anaerobic conditions over two years (*50*). Wilber and Parkin (*51*) studied metabolism of atrazine with acetate as the primary substrate under various redox conditions and found that degradation virtually ceased at redox levels equivalent to nitrate- and sulfate-reducing conditions, but increased dramatically at very low redox potentials corresponding to methanogenic conditions.

In anaerobic broth culture the only known degradation product which accumulated was hydroxyatrazine, indicating that hydroxyatrazine conversion was the major rate-limiting step in mineralization (*52*). Other studies have measured accumulation of the dealkylated initial products at low concentrations but there is no evidence that any of these is rate limiting.

Fluometuron. The path of microbial degradation for fluometuron is shown in Figure 2 (*20, 53*). Chemical degradation is insignificant, with over 92% of radiolabeled fluometuron remaining as parent compound over 72 days in sterile soil (*20*). The pathway is composed simply of two sequential demethylations followed by removal of the carboxyamine. The resulting intermediate compound, 3-trifluoromethylaniline, does not accumulate under aerobic conditions and is presumably easily degraded (*20*). No references were found in the literature to the activity of fluometuron in soil under anaerobic conditions, though one of the laboratory degradation studies was almost certainly carried out under anoxic conditions (*23*). No influence of organic matter content was found on fluometuron degradation rates (*10, 22, 53*) indicating that microorganisms can use fluometuron as a primary substrate rather than relying upon secondary metabolism or cometabolism. Degradation rates decreased with depth in one field study, correlating with microbial biomass decrease (29).

Trifluralin. Degradation pathways for trifluralin are shown in Figure 3 (24-26). There are four mechanisms for the disappearance of the parent compound. The first involves formation of a heterocycle caused by partial oxidation of the alpha carbon of one propyl group, partial reduction of a nitro group, then condensation. This sequence results in the C-N double bond-containing product 2-ethyl-7-nitro-1-propyl-5-(trifluoromethyl)-benzimidazole (24). The reactions occur under aerobic soil conditions and the resulting product builds up as a major intermediate in the mineralization pathway (24). The second and third conversion reactions on the parent compound are reduction of a nitro group and removal of a propyl group. The former reaction, which results in α,α,α-trifluoro-5-nitro-N^4,N^4-dipropyltoluene-3,4-diamine, is a major pathway under anaerobic and anoxic conditions (24, 25) and a minor pathway under aerobic conditions (25). The latter reaction results in α,α,α-trifluoro-2,6-dinitro-N-propyl-p-toluidine and is a major aerobic pathway (24-26). The fourth mechanism for trifluralin disappearance is the formation of the dimer 2,2'-azoxybis-(α,α,α-trifluoro-6-nitro-N,N-dipropyl-p-toluidine) upon partial reduction and condensation of two nitro groups (24).

Rate-limiting steps for the mineralization of trifluralin are indicated by the accumulation of intermediate compounds. In an anaerobic environment the rate limiting steps were the reduction of the second nitro group of α,α,α-trifluoro-5-nitro-N^4,N^4-dipropyltoluene-3,4-diamine and the first dealkylation of α,α,α-trifluoro-N^4,N^4-dipropyltoluene-3,4,5-triamine (24, 25) while in an anoxic environment only the first step was rate limiting. In an aerobic soil environment, rate limiting steps were the second dealkylation of α,α,α-trifluoro-2,6-dinitro-N-propyl-p-toluidine, dealkylation and splitting of the toluidine dimer, further degradation of the dealkylated toluidine dimer, and reduction of the nitro group of 2-ethyl-7-nitro-1-propyl-5-(trifluoromethyl)-benzimidazole (24, 25).

The initial reactions and those which follow to mineralization in Figure 3 are carried out by a combination of biotic and abiotic influences. Zayed et al. (26) showed that three fungal strains, independently or in consortium, catabolically degraded ^{3}H-radiolabeled trifluralin substrate under aerobic conditions with a half-life in broth culture of less than 10 days. Comparisons between sterile and non-sterile soils showed definitively that microbial degradation enhances the rate of disappearance of trifluralin (25). Other studies showed that the rate of trifluralin disappearance was enhanced by incubation in the presence of plants. This result indicates that either trifluralin may serve as a secondary substrate in the presence of plant exudates or that rhizosphere bacterial communities are better able to metabolize trifluralin (25, 54).

Either the halogenated constituent of trifluralin is recalcitrant to degradation or there is no rate-limiting degradation step after the dehalogenation step. Golab et al. (24) showed that of thirty-three detectable degradation products, only two were defluorinated. It appears probable that complete mineralization of trifluralin does occur in the natural environment, but none of the studies cited followed the fate of the parent compound through ring cleavage.

Fate of Herbicides in Riparian Wetland Soils

The essential differences between the soils of riparian wetlands and upland soils are the chemical reduction state and the organic content. Both oxidation state and organic content in riparian wetland soils are strongly influenced by the vegetation adapted to growth in wetland soils. The reduced environment affects both chemical and biological reactions with herbicides. Soil microbial populations in reduced environments show different community structure and metabolic activity from those in more oxidized environments (*55*). Riparian wetland soils are not uniformly reduced, however, since wetland-adapted plants frequently are able to transport oxygen to the roots. Transport of environmental oxygen below-ground to roots results in a partially oxidized rhizosphere associated with root systems (*56, 57*).

There are three reasons for the high organic content of wetland soils. First is the periodic import of organic material associated with flood wrack. This material accumulates in riparian wetlands and can increase the organic load. Second is the sclerophyllous nature of some flood-adapted vegetation (often seen as rigid, waxy leaves). The high lignin content of sclerophyllous litter is more recalcitrant than upland litter and results in accumulation of partially decomposed matter enriched in aromatic constituents (*58, 59*). Third is the anaerobic condition of the soil. Partially decomposed organic molecules tend to aggregate into the long-chain recalcitrant material known generally as humus under anaerobic conditions (*38*). Taken together, these characteristics lead to a soil containing a high content of humic material enriched in phenolic and other aromatic components. This high organic content imparts a high cation exchange capacity to the soil and slows the migration of positively charged herbicides through the soil matrix (*34, 60*).

Volatilization. Herbicide volatility in riparian wetland soils should be lower than that at comparable concentrations in upland soils. The high organic content of the soil will result in more adsorption of herbicide and lower volatility (*61*). Of the three sample herbicides, only trifluralin is volatile (*16*). By the time an herbicide residue has migrated from the point of application to a riparian wetland area volatilization is not expected to be a major mechanism of disappearance.

Migration. Increased organic matter in riparian wetland soils increases the CEC and slows the migration of herbicides. In a study comparing migration rates through riparian wetland soil cores with varying organic contents, Mudd (*34*) showed that soils with high organic content had slower movement of atrazine. Fluometuron infiltration was held to 5 cm on a soil with 4.4% soil organic carbon as opposed to 10 cm in a soil with 1.5% soil organic carbon (*21*). Trifluralin is only minimally mobile in soil, and no study showed trifluralin migration past the upper several centimeters of a soil (*28*).

Photolysis. Photodecomposition in riparian wetlands should be minimal since sediment-associated herbicides are shaded and not exposed to direct sunlight. Of the three sample herbicides, only trifluralin is subject to photolysis (*14, 16*).

Degradation. Chemical and biological degradation of herbicides are affected by the presence of oxygen. Biological degradation is further affected by the presence of alternate sources of organic material and the characteristics of those alternate sources. Riparian wetlands contain micro-environments which exhibit a wide range of redox potentials and organic carbon concentrations and types.

Reduction State. The redox state of the bulk riparian wetland soil (underlying a thin oxidized crust) is very low. This highly reduced environment leads soil microorganisms to metabolize by anaerobic or anoxic pathways (*38*). The products of these metabolic pathways are all acidic, leading to a reduction in soil pH (*62*). Low pH decreases the mobility of aminated and nitrosylated herbicides through the soil matrix, giving more time for degradation to occur.

There are pockets of less reduced soil around wetland plant roots. One adaptation of flood-tolerant plants is the ability to transport oxygen to their roots. Vegetated constructed wetlands for wastewater treatment take advantage of this phenomenon to provide oxygen to the subsurface (*46*). Many studies have been performed to measure the oxygen transport rate, assess the environmental impact, and elucidate the mechanism of this phenomenon (*63-65*). Many species, including *Phragmites australis* (*63*), *Spartina alterniflora* (*65*), *Pinus* spp. (*66*), *Typha latifolia*, *Carex lacustris*, *Scirpus acutus* (*56*), *Pontederia cordata* (*57*) and *Alnus glutinosa* (*67*) possess some ability to aerate their root zones. Oxygen transport rates as high as 1.54 grams of oxygen per kg dry root weight per hour have been measured (*57*). The resulting effect is one of aerobic micro-sites in a generally anaerobic matrix, providing conditions for both aerobic and anaerobic degradation processes within a small volume of soil. Because a gradient exists within a small scale in riparian wetlands, as rate-limiting reaction substrates accumulate they will diffuse to areas of other redox conditions and be degraded (*68*), or the redox condition of that area will change with time and the product will be degraded.

Co-substrates. The organic matter in riparian wetland soil affects the extent and type of microbial activity (*69, 70*). In soil the major carbon and energy source available to microorganisms is soil organic matter. Microbial population densities are thought to be fundamentally limited by the amount of available carbon (*71*), though in frequently flooded wetlands denitrification causes nitrogen to be the limiting nutrient (*2*). Nitrogen stress enhances degradation of atrazine under aerobic conditions, indicating that microorganisms are capable of utilizing both the amino groups and the ring nitrogens of atrazine anabolically (*44, 47*).

Humic material is composed of a wide variety of chemical types, including aromatic rings and heterocycles (*38*). This fact is important when considering microbial degradation of herbicides since all three herbicides considered in this paper contain aromatic rings or heterocycles. Organic material can influence the degradation of dilute solutions of recalcitrant molecules by providing a primary substrate for secondary metabolism or by inducing the production of degradative enzymes which cometabolize the recalcitrant molecule.

Atrazine degradation rates have been shown to be enhanced in vegetated soils. Mineralization of atrazine as a secondary substrate may be stimulated by plant root exudates (*72*). Rhizosphere soil from *Kochia* sp. was shown to elicit higher degradation rates of atrazine than the surrounding bulk soils (*54*). Addition of sucrose to broth culture containing atrazine also increased mineralization rates (*47, 51*).

Entry and Emmingham (*70*) showed that the type of vegetation affects atrazine mineralization rates. Microbial communities of coniferous forest soils were more capable of mineralizing atrazine than those from deciduous or grassland soils. Mudd (*34*) similarly showed that biologically active forested soils exhibited higher mineralization rates than grassland or cornfield soils. Forest litter, especially wetland forest litter, is more lignified and thus more capable of providing a template for cometabolism (*70*).

The major inhibitors of atrazine degradation appear to be anoxic or moderately anaerobic conditions, or a lack of organic substrate. Riparian wetland soils may accumulate atrazine while saturated, then upon drying the microbial communities may degrade it. The high humic content of these soils is conducive to atrazine cometabolism both because humic acids can serve as a primary carbon and energy source and because they are thought to be degraded by the same ring-cleaving enzymes as atrazine (*73-75*). As the organic content increases above 2%, atrazine degradation rates begin to increase (*43*).

Conclusions

The soils of riparian wetlands are not homogeneous. They contain micro-habitats which vary in time and space with edaphic features and a constantly shifting mosaic of weather and water content. The driving characteristics of these micro-environments are limited to redox potential, organic matter, and pH. These characteristics can act to slow herbicide migration or enhance degradation by chemical and biological processes.

The three herbicides evaluated in this review encompass a range of mobilities and chemical and biological degradation potentials. All three are known to be moderately persistent in soils because of their low degradation potential. We have shown that soils of vegetated riparian wetlands contain the chemical characteristics and biological constituents to enhance the degradation of these compounds.

Accumulations of herbicides with known sources should be dealt with by traditional waste-treatment methods. The literature and our research with atrazine indicate that riparian areas at field edges have a unique ability to slow migration and enhance degradation of herbicides from non-point sources. Riparian wetland areas are valuable as "nature's kidneys", especially in the southeastern U.S. where wetland areas are abundant and buffer waterways from agricultural runoff. They should be preserved for this and other ecological functions.

Literature Cited

1. Ribaudo, M.O.; Bouzaher, A. Research and Technology Division, Economic Research Service, USDA. *Agricultural Economic Report Number 699*, 1994.
2. Mitsch, W. J.; Gosselink, J. G. *Wetlands*, 2nd ed.; Van Nostrand Reinhold: New York, 1993.
3. Dahl, T. E.; Johnson, C. E. Wetland Status and Trends in the Conterminous United States Mid 1970s; US Department of the Interior, Fish and Wildlife Service, Washington DC 28 pg.
4. Kimble, J. M. Characterization, classification, and utilization of wet soils; USDA, Soil Conservation Service, National Soil Survey Center, Lincoln NB.
5. Neary, D. G.; Bush, P. B.; Michael, J. L. *Environmental Toxicology and Chemistry* **1993**, *12*, 411-428.
6. Haertel, L.; Duffy, W. G.; Kokesh, D. E. *Journal of the Minnesota Academy of Science* **1995**, *60*, 1-10.
7. Johnson, S. R. *Environmental Entomology* **1995**, *24*, 832-834.
8. Thomas, D. H. L. *Environmental Conservation* **1995**, *22*, 117-126.
9. USDA. *Agricultural Chemical Usage: 1993 Field Crops Summary*; USDA National Agricultural Statistics Service, Washington DC, 1993.
10. Brown, B. A.; Hayes, R. M.; Tyler, Donald D.; Mueller, T. C. *Weed Science* **1994**, *42*, 629-634.
11. Capriel, P.; Haisch, A.; Khan, S. U. *J. Agric. Food Chem.* **1985**, *33*, 567-569.
12. LaFleur, K. S.; Wojec, G. A.; McCaskill, W. R. *J. Environ. Qual.* **1973**, *2*, 515-518.
13. Miller, J. H.; Keeley, P. E.; Thullen, R. J.; Carter, C. H. *Weed Science* **1978**, *26*, 20-27.
14. Leitis, E.; Crosby, D. G. *J. Agric. Food Chem.* **1974**, *22*, 842-848.
15. Nash, R. G.; Gish, T. J. *Chemosphere* **1989**, *18*, 2353-2362.
16. Humberg, N. E.; Colby, S. R.; Lym, R. G.; Hill, E. R.; McAvoy, W. J.; Kitchen, L. M.; Prasad, R., Eds. *Herbicide Handbook of the Weed Science Society of America*, 6th ed.; Weed Science Society of America: Champaign, IL, 1989.
17. Winkelmann, D. A.; Klaine, S. J. *Soil Biol. Biochem.* **1991**, *10*, 335-345.
18. Weed, D. A. J.; Kanwar, R. S.; Stoltenberg, D. E.; Pfieffer, R. L. *J. Environ. Qual.* **1995**, *24*, 68-79.
19. Hallberg, G.R. *Agricultural Ecosystems and the Environment* **1989**, *26*, 299-367.
20. Bozarth, G. A.; Funderburk, H. H. *J. Weed Science* **1971**, *19*, 691-695.

21. Hance, R. J.; Embling, S. J.; Hill, D.; Graham-Bryce, I. J.; Nicholls, P. *Weed Research* **1981**, *21*, 289-297.
22. Rogers, C. B.; Talbert, R. E.; Mattice, J. D.; Lavy, T. L.; Frans, R. E. *Weed Science* **1985**, *34*, 122-130.
23. Bouchard, D. C.; Lavy, T. L.; Marx, D. B. *Weed Science* **1982**, *30*, 629-632.
24. Golab, T.; Althaus, W. A.; Wooten, H. *J. Agric. Food Chem.* **1979**, *27*, 163-179.
25. Probst, G. W.; Golab, T.; Herberg, R. J.; Holtzer, F. J.; Parka, S. J.; van der Schans, C.; Tepe, J. B. *J. Agric. Food Chem.* **1967**, *15*, 592-599.
26. Zayed, S. M. A. D.; Mostafa, I. Y.; Farghaly, M. M.; Attaby, H. S. H.; Adam, Y. M.; Mahdy, F. M. *Journal of Environmental Science and Health part B* **1983**, *18*, 253-267.
27. Domenico, P. A.; Schwartz, F. W. *Physical and Chemical Hydrogeology*; John Wiley & Sons: New York, 1990.
28. Kanazawa, J. *Environmental Toxicology and Chemistry* **1989**, *8*, 477-484.
29. Mueller, T. C.; Moorman, T. B.; Snipes, C. E. *J. Agric. Food Chem.* **1992**, *40*, 2517-2522.
30. Savage, K. E.; Wauchope, R. D. *Weed Science* **1974**, *22*, 106-110.
31. McGlamery, M.D.; Slife, S.W. *Weeds* **1966**, *14*, 237-239.
32. Wauchope, R. D.; Koskinen, W. C. *Weed Science* **1983**, *31*, 504-512.
33. Carringer, R. D.; Weber, J. B.; Monaco, T. J. *J. Agric. Food Chem.* **1975**, *23*, 568-572.
34. Mudd, E. C. M.S. Thesis, Auburn University, 1996.
35. Tchobanoglous, G.; Burton, F. L., Eds. *Wastewater engineering: treatment, disposal, and reuse*, 3rd ed.; McGraw Hill, Inc: New York, 1991.
36. Fomsgaard, I. S. *International Journal of Environmental Analytical Chemistry* **1995**, *58*, 231-243.
37. Fessenden, R. J.; Fessenden, J. S. *Organic Chemistry*, 3rd ed.; Brooks/Cole Publishing Company: Monterey, CA, 1982.
38. Atlas, R. M.; Bartha, R. *Microbial Ecology: Fundamentals and Applications*, 3rd ed.; The Benjamin/Cummings Publishing Company, Inc.: New York, 1993.
39. Berry, D. F.; Francis, A. J.; Bollag, J. M. *Microbiol. Rev.* **1987**, *51*, 43-59.
40. Chaudhry, G. R.; Chapalamadugu, S. *Microbiol. Rev.* **1991**, *55*, 59-79.
41. Alexander, M. *Agriculture and the Quality of Our Environment*; pages 331-342; Brady, N.C. ed.; AAAS:Washington DC. 1967.
42. Horvath, R. S. *Bacteriological Reviews* **1972**, *36*, 146-155.
43. Mirgain, I.; Green, G. A.; Monteil, H. *Environmental Toxicology and Chemistry* **1993**, *12*, 1627-1634.
44. Ro, K. S.; Chung, K. H. *Journal of Environmental Science and Health part A* **1995**, *30*, 121-131.
45. Entry, J.A.; Mattson, K.G.; Emmingham, W.H. *Biology and Fertility of Soils* **1993**, *16*, 179-182.
46. Reed, S. C.; Crites, R. W.; Griffes, D. A.; Knight, R. L.; Kreissl, J. F.; Kruzic, A. P.; Middlebrooks, E. J.; Otis, R.; Smith, R. G.; Tchobanoglous, G.; Vickers, K. D.; Wallace, A. T.; Zirschky, J., Eds. *Natural Systems for*

132   PHYTOREMEDIATION OF SOIL AND WATER CONTAMINANTS

3222

132   PHYTOREMEDIATION OF SOIL AND WATER CONTAMINANTS

Wastewater Treatment Manual of Practice FD-16; Water Pollution Control Federation: Alexandria, VA, 1990.

47. Erickson, L. E.; Lee, K. H. *Critical Reviews in Environmental Control* **1989**, *19*, 1-14.

48. Tu, C. M. *Journal of Environmental Science and Health part B* **1992**, *27*, 695-709.

49. Levanon, D. *Soil Biol. Biochem.* **1993**, *25*, 1097-1105.

50. Adrian, N. R.; Suflita, J. M. *Environmental Toxicology and Chemistry* **1994**, *13*, 1551-1557.

51. Wilber, G. G.; Parkin, G. F. *Environmental Toxicology and Chemistry* **1995**, *14*, 237-244.

52. Chung, K. H.; Ro, K. S.; Roy, D. *Journal of Environmental Science and Health part A* **1995**, *30*, 109-120.

53. Ross, J. A.; Tweedy, B. G. *Soil Biol. Biochem.* **1973**, *5*, 739-746.

54. Anderson, T. A.; Kruger, E. L.; Coats, J. R. *Chemosphere* **1994**, *28*, 1551-1557.

55. Bossio, D. A.; Scow, K. M. *Appl. Environ. Microb.* **1995**, *61*(11), 4043-4050.

56. Bedford, B. L.; Bouldin, D. R.; Beliveau, B. D. *Journal of Ecology* **1991**, *79*, 943-959.

57. Moorhead, K. K.; Reddy, K. R. *J. Environ. Qual.* **1988**, *17*, 138-142.

58. Day, F. P. *J. Ecology* **1982**, *63*, 670-678.

59. Day, F. P. J.; Megonigal, J. P. *Wetlands* **1993**, *13*(2), 115-121.

60. Weitersen, R. C.; Daniel, T. C.; Fermanich, K. J.; Girard, B. D.; McSweeney, K.; Lowery, B. *J. Environ. Qual.* **1993**, *22*, 811-818.

61. Diaz Diaz, R.; Gaggi, C.; Sánchez-Hernández, J. C.; Bacci, E. *Chemosphere* **1995**, *30*, 2375-2386.

62. Ponnamperuma, F. N. *Advances in Agronomy* **1972**, *24*, 29-96.

63. Armstrong, J.; Armstrong, W. *New Phytologist* **1990**, *114*, 121-128.

64. Dacey, J. W. H.; Klug, M. J. *Physiologia Plantanum* **1982**, *56*, 361-366.

65. Howes, B. L.; Teal, J. M. *Oecologia* **1994**, *97*, 431-438.

66. Armstrong, W.; Read, D. J. *New Phytologist* **1972**, *71*, 55-62.

67. Buchel, H. B.; Grosse, W. *Tree Physiology* **1990**, *6*, 247-256.

68. Montgomery, J.M. *Water Treatment Principles and Design*; John Wiley & Sons, 1985.

69. Entry, J. A.; Donelly, P. K.; Emmingham, W. H. *Biology and Fertility of Soils* **1994**, *18*, 89-94.

70. Entry, J. A.; Emmingham, W. H. *Can. J. Soil Sci.* **1996**, *76*, 101-106.

71. Powlson, D. S.; Brookes, P. C.; Christensen, B. T. *Soil Biol. Biochem.* **1987**, *19*, 159-164.

72. Anderson, T. A.; Coats, J. R. *Journal of Environmental Science and Health part B* **1995**, *30*, 473-484.

73. Entry, J. A.; Donelly, P. K.; Emmingham, W. H. *Applied Soil Ecology* **1995**, *2*, 77-84.

74. Hammel, K.E. *Enzyme Microbiology and Technology* **1989**, *11*, 776-777.

75. Lamar, R.T.; Dietrich, D.M. *Applied and Environmental Microbiology* **1990**, *56*, 3093-3100.

Chapter 10

Phytoremediation of Herbicide-Contaminated Surface Water with Aquatic Plants

Pamela J. Rice[1], Todd A. Anderson[2], and Joel R. Coats[1]

[1]Pesticide Toxicology Laboratory, Department of Entomology, Iowa State University, 112 Insectary Building, Ames, IA 50011
[2]The Institute of Wildlife and Environmental Toxicology, Department of Environmental Toxicology, Clemson University, Pendleton, SC 29670

There is current interest in the use of artificial wetlands and macrophyte-cultured ponds for the treatment of agricultural drainage water, sewage, and industrial effluents. Aquatic plant-based water treatment systems have proved effective and economical in improving the quality of wastewaters containing excess nutrients, organic pollutants, and heavy metals. This investigation was conducted to test the hypothesis that herbicide-tolerant aquatic plants can remediate herbicide-contaminated waters. The addition of *Ceratophyllum demersum* (coontail, hornwort), *Elodea canadensis* (American elodea, Canadian pondweed), or *Lemna minor* (common duckweed) significantly ($p \leq 0.01$) reduced the concentration of [14C]metolachlor (MET) remaining in the treated water. After a 16-day incubation period, only 1.44%, 4.06%, and 22.7% of the applied [14C]MET remained in the water of the surface water systems containing *C. demersum*, *E. canadensis*, or *L. minor* whereas 61% of the applied [14C]MET persisted in the surface water systems without plants. *C. demersum* and *E. canadensis* significantly ($p \leq 0.01$) reduced the concentration of [14C]atrazine (ATR) in the surface water. Only 41.3% and 63.2% of the applied [14C]ATR remained in the water of the vegetated systems containing *C. demersum* and *E. canadensis*, respectively. Eighty-five percent of the applied [14C]ATR was detected in the water of the *L. minor* and nonvegetated systems. Our results support the hypothesis and provide evidence that the presence of herbicide-tolerant aquatic vegetation can accelerate the removal and biotransformation of metolachlor and atrazine from herbicide-contaminated waters.

Herbicides in Surface and Subsurface Waters. Runoff/erosion of pesticides from agricultural fields is believed to be the largest contributor to water quality degradation in the midwestern United States. Atrazine, alachlor, cyanazine, and metolachlor are the major herbicides used in Iowa and the Midwest (*1, 2*). The intense use of these relatively water soluble and mobile compounds threatens the integrity of surface and subsurface waters (*3, 4*). Approximately 1 to 6% of the applied herbicides may be lost to the aquatic environment by runoff and drainage depending on the slope of the field, tillage practices, presence or absence of subsurface drains, and the quantity and timing of rainfall after application (5-7). Monitoring studies have detected herbicides in surface waters (3, 8, 9), tile-drain water and groundwater (5,10,11). Goolsby et al. (3) and Thurman et al. (8) reported frequent detection of metolachlor, alachlor, cyanazine, atrazine, and the atrazine degradation products deethylatrazine and deisopropylatrazine in rivers and streams of the midwestern United States. Atrazine and metolachlor were the two most frequently detected herbicides. Measurable amounts of atrazine were reported in 91%, 98%, and 76% of the preplanting, postplanting, and harvest surface waters sampled. Metolachlor was detected in 34%, 83%, and 44% of the preplanting, postplanting and harvest-season waters sampled, respectively.

Problems Associated with Pesticide-Contaminated Water. The presence of pesticides in surface water is a concern for human health and the health of aquatic ecosystems (12). Contamination of surface waters with pesticides exposes nontarget microorganisms, plants, and animals to compounds that may have an adverse effect on individual organisms or biotic communities. Aquatic insects and other aquatic arthropods are particularly susceptible to insecticides, whereas herbicides may suppress the growth of aquatic vegetation (13-16). The primary concern involving human exposure to pesticide-contaminated waters involves long-term exposure to low concentrations through drinking water (13). Conventional water treatment processes (filtration, clarification, chlorination, softening, and recarbonation) do little to reduce the levels of pesticides in drinking water (13, 17, 18). Pesticide concentrations are significantly reduced only when advanced processes such as ozonation, reverse osmosis, or granular activated carbon are used. In areas where water treatment facilities lack advanced treatment processes, the concentration of pesticides in the finished drinking water will be similar to the concentrations found in the surface water or groundwater source (17).

Phytoremediation of Contaminated Water. There is current interest in the use of artificial wetlands and macrophyte-cultured ponds for treating wastewater (agricultural drainage water, sewage, and industrial effluents) (19-23). Aquatic plant-based water treatment systems have proved to be effective and economical in improving the quality of wastewater effluents (24-27). Floating and emergent aquatic plants including water hyacinth (Eichhornia crassipes Mart.), elodea (Egeria densa P.), duckweed (Lemna and Spirodela spp.), pennywort (Hydrocotyle umbellata L.), common arrowhead (Sagittaria latifolia L.), common reed (Phragmites australis), and pickerelweed (Pontederia cordata L.) reduce the levels of total suspended solids and nutrients (N and P) in wastewater by solid filtration, nutrient assimilation, and microbial transformation (19-28). In addition, aquatic plants and their associated microbiota have contributed to the removal and biotransformation of xenobiotic compounds from contaminated waters and sediments. Microbiota of cattail roots (Typha latifolia L.) and duckweed plants (L. minor) accelerate the biodegradation of surfactants (29). Curly leaf pondweed (Potamageton crispus L.), common duckweed (L. minor), and their epiphytic microbes contributed to the removal and degradation of pentachlorophenol from a stream, and various duckweed plants (Lemna and Spirodela spp.) have been shown to accumulate metals (aluminum, cadmium, copper, lead, and mercury) from aqueous solutions (30-32).

Previous research provides evidence that aquatic plants can remediate wastewaters containing excess nutrients, organic pollutants, and heavy metals. This investigation was conducted to test the hypothesis that herbicide-tolerant aquatic plants can remediate herbicide-contaminated waters. Experiments were setup to evaluate the ability of two submerged aquatic plants (*Ceratophyllum demersum* L. and *Elodea canadensis* Rich.) and one floating aquatic plant (*Lemna minor* L.) to remediate metolachlor or atrazine contaminated waters. Metolachlor [2-chloro-N-(2-ethyl-6-methylphenyl)-N-(2-methoxy-1-methylethyl)acetamide] controls annual grass weeds and broadleaf weeds in corn, soybeans, peanuts, and potatoes. Atrazine [6-chloro-N-ethyl-N'-(1-methylethyl)-1,3,5-triazine-2,4-diamine] inhibits photosynthesis of susceptible grassy and broadleaf weeds in corn, sorghum and turf grass (*33*). Our results support the hypothesis and demonstrate the presence of herbicide-tolerant aquatic vegetation can accelerate the removal and biotransformation of metolachlor and atrazine from herbicide-contaminated waters.

Materials and Methods

Chemicals. Metolachlor [2-chloro-N-(2-ethyl-6-methylphenyl)-N-(2-methoxy-1-methylethyl)acetamide] (CGA 24705, 97.3 % pure); [U-ring-^{14}C]metolachlor ([^{14}C]MET) (98.9% pure); the metolachlor degradates N-(2-ethyl-6-methylphenyl)-2-hydroxy-N-(2-methylethyl)-acetamide (CGA 40172, 98.4% pure) and 4-(2-ethyl-6-methylphenyl)-5-methyl-3-morpholinone (CGA 40919, 99.8% pure); [U-ring-^{14}C]atrazine ([^{14}C]ATR) (98.2% pure); [U-ring-^{14}C]deethylatrazine (94.8% pure); [U-ring-^{14}C]deisopropylatrazine (92.9% pure); [U-ring-^{14}C]didealkylatrazine (98.8% pure); and [U-ring-^{14}C]hydroxyatrazine (97.5% pure) were gifts from the Ciba-Geigy Corporation, Greensboro, NC.

Surface Water and Aquatic Plant Sample Collection. Surface water and aquatic plants *Lemna minor* L. (common duckweed), *Elodea canadensis* Rich. (American elodea, Canadian pondweed), and *Ceratophyllum demersum* L. (coontail, hornwort) were collected from the Iowa State University Horticulture Station Pond, Ames, Iowa. The aquatic plants were selected as a result of their abundance and availability. Pond water samples were collected in sterile 4-L bottles and stored at $4 \pm 2°C$. Aquatic plants were collected and maintained, at $25 \pm 2°C$, in aquaria containing distilled water and Hoagland's nutrient solution with a 14:10 (L:D) photoperiod.

Experimental Design. Experiments were conducted to evaluate the degradation of metolachlor or atrazine in vegetated- and nonvegetated-surface-water incubation systems. Each experimental variation [herbicide (metolachlor, atrazine) x aquatic plant (*L. minor*, *E. canadensis*, *C. demersum*) x the duration of the incubation period (0-16 days)] was replicated a minimum of three times. Analysis of variance and least square means determined significance between treatments.

Surface Water/Plant Incubation Systems. French square bottles were filled with 150 ml of a water solution containing pond water/Hoagland's nutrient solution/ultra-pure water (1:1:4 v/v/v). A pond water/Hoagland's nutrient solution/ultra-pure water mixture was used rather than 150 ml of pond water in order to make the study more reproducible for other researchers. [^{14}C]MET or [^{14}C]ATR was added to the water at a rate of 200 μg/L. This rate was chosen to represent a runoff concentration and to ensure there was enough radioactivity for the detection of metabolites. Aquatic plants (3 g) were added to 150 ml of the treated water solutions and placed in a temperature-controlled room set at $25 \pm 2°C$ with a 14:10 (L:D) photoperiod. Three replicate vegetated- and nonvegetated-incubation systems were dismantled on each of the designated incubation days. The herbicides and their degradates were extracted from the water solutions and the plant tissues, and a mass balance was determined.

Water Extraction and Analysis. At the completion of each test, the aquatic plants were removed from the water solutions by using vacuum filtration and were rinsed with ultra pure water. The plant rinsate was added to the filtrate. A portion of the treated water was counted with a liquid scintillation spectrometer to determine the quantity of radioactivity remaining in the water. The herbicides and herbicide degradates were removed from the remaining water with a solid phase extraction (SPE) process. Supelclean Envi-18 6-cc solid phase extraction cartridges (Supelco, Inc., Bellefonte, PA) were positioned on a 12-port Visiprep Solid Phase Extraction Vacuum Manifold (Supelco, Inc., Bellefonte, PA) and activated with 18 ml (3 column volumes) of certified ethyl acetate followed by 18 ml of certified methanol and finally 18 ml of ultra-pure water. The water samples were drawn through the activated cartridges by using an applied vacuum (50 kPa). Once the entire sample had been drawn through the extraction cartridge, the packing was dried by drawing air through the cartridge for approximately 15 minutes. The cartridges were eluted with 10 ml certified methanol followed by 5 ml of certified ethyl acetate. The radioactivity of the effluent (post-SPE water sample) and the methanol and ethyl acetate eluates was determined with liquid scintillation techniques. The quantity of metolachlor, atrazine, and their degradates in the methanol eluates were characterized by thin layer chromatography (TLC).

Plant Extraction and Analysis. Plant tissues were extracted three times with certified methanol. The volume of the extract was reduced with a rotary evaporator and the plant extracts were characterized by TLC. Dry-extracted plant tissues were mixed with hydrolyzed starch and combusted in a Packard sample oxidizer (Packard Instrument Co.) to determine the activity of the nonextractable residues. The $^{14}CO_2$ produced from the combusted plant material was trapped in Carbo-Sorb E and Permafluor V (Packard Instruments Co.). Liquid scintillation spectroscopy was used to quantify the radioactivity in the plant extracts and the combusted plant tissues.

Thin-Layer Chromatography. A portion of the methanol eluates from the water samples or plant extracts, representing 70,000 dpm (0.03 μCi), was concentrated under nitrogen in a warm-water bath. [^{14}C]MET, N-(2-ethyl-6-methylphenyl)-2-hydroxy-N-(2-methylethyl)-acetamide, 4-(2-ethyl-6-methylphenyl)-5-methyl-3-morpholinone and the water and plant extracts from the metolachlor-treated systems were spotted on 20-cm by 20-cm glass plates containing a 250-μm layer of normal-phase silica gel 60 F-254. The TLC plates were developed in a hexane/methylene chloride/ethyl acetate (6:1:3 v/v/v) solvent system (*34*). [^{14}C]ATR, [U-ring-^{14}C]deethylatrazine, [U-ring-^{14}C]deisopropylatrazine, [U-ring-^{14}C]didealkylatrazine, [U-ring-^{14}C]hydroxyatrazine, and the reduced water and plant extracts from the atrazine-treated systems were spotted on normal phase silica gel plates and developed in a chloroform/methanol/formic acid/water (100:20:4:2 v/v/v) solvent system (Ciba-Geigy). An ultraviolet lamp (254 nm) was used to locate the nonradiolabeled standards and the location of the radiolabeled standards and extracted compounds was determined by autoradiography using Kodak X-Omat diagnostic film (Eastman Kodak Co., Rochester, NY). The silica gel of each spot was scrapped into vials containing 5 ml of Ultima Gold scintillation cocktail (Packard Instrument Co., Downers Grove, IL) and the radioactivity in each sample was quantified on a liquid scintillation spectrometer. The ^{14}C mass balance was determined for each system. Percentage of applied ^{14}C in the degradation products was summed and reported as the percentage of applied ^{14}C associated with total degradation products in the water or plant extracts. A report of the individual degradation products and the percentage of applied ^{14}C associated with the individual degradation products will not be discussed in this chapter. Information regarding the degradation products will be written in a paper to be submitted to the journal of *Environmental Toxicology and Chemistry*.

Results

Reduction of Metolachlor and Atrazine in the Water of the Vegetated Incubation Systems.
The concentrations of [^{14}C]MET and [^{14}C]ATR were significantly reduced ($p \leq$ 0.01) in the water of the vegetated surface water incubations systems. After 16 days, 22.7%, 4.06%, and 1.44% of the applied [^{14}C]MET remained in the water of the vegetated incubation systems that contained *L. minor* (common duckweed), *E. canadensis* (American elodea), and *C. demersum* (coontail), respectively (Figure 1). Sixty-one percent of the applied [^{14}C]MET was detected in the water of the nonvegetated incubation systems. The quantity of the [^{14}C]ATR that remained in the water of the atrazine-treated *E. canadensis* (63.2%) and *C. demersum* (41.4%) vegetated incubation systems were significantly ($p \leq 0.01$) reduced compared with the nonvegetated incubation systems (85.0%) (Figure 2). The water of the *L. minor* incubation systems (84.9%) contained levels of [^{14}C]ATR comparable to the concentrations found in the water of the nonvegetated incubation systems (85.0%). Half-lives of [^{14}C]MET and [^{14}C]ATR in the water of the vegetated and nonvegetated incubation systems were calculated assuming first-order reaction kinetics (Table I). The significant reduction in the concentration of [^{14}C]MET and [^{14}C]ATR in the water of the vegetated incubation systems may be the result of 1) the herbicide attaching to the surface of the plant, 2) the accumulation, sequestering, and degradation of the herbicide in the plant, or 3) the degradation of the herbicides in the water.

Plant Uptake of ^{14}C.
Replicates of the metolachlor- or atrazine-treated vegetated incubation system containing either *L. minor* or *C. demersum* were extracted and analyzed immediately following the herbicide treatment (day 0) and 4, 8, 12, and 16 days after the addition of the herbicide. Vegetated incubation systems containing *E. canadensis* were extracted and analyzed on day 0, 4, and 16. After 16 days, less than 25% of the applied ^{14}C was detected in the *L. minor*, *E. canadensis*, or *C. demersum* plants of the metolachlor- or atrazine-treated vegetated incubation systems (Tables II & III). Significantly greater quantities of ^{14}C were associated with the plant tissues of the metolachlor-treated systems compared with the atrazine-treated systems ($p \leq 0.01$), which may be the result of the greater water solubility of metolachlor (metolachlor = 530 mg/L at 20°C, atrazine = 33 mg/L at 27°C). Metolachlor may be more bioavailable and more readily absorbed and translocated in plants than atrazine as a result of its increased water solubility. Plants of the metolachlor-treated *L. minor*, *E. canadensis*, and *C. demersum* systems contained 7.57 ± 0.09%, 20.3 ± 3.07%, and 23.2 ± 0.02% of the applied ^{14}C after 16 days. Aquatic plants from the atrazine-treated systems contained 1.21 ± 0.05%, 11.7 ± 1.06%, and 9.23 ± 1.17% of the applied ^{14}C in the *L. minor*, *E. canadensis*,

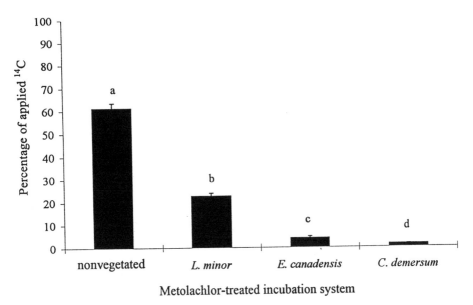

Figure 1. Percentage of applied [^{14}C]metolachlor remaining in the water of the nonvegetated- and vegetated-surface water incubation systems after 16 days.

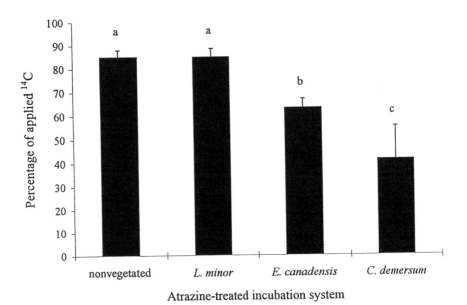

Figure 2. Percentage of applied [^{14}C]atrazine remaining in the water of the nonvegetated- and vegetated-surface water incubation systems after 16 days.

Table I. First-order half-lives of metolachlor and atrazine in the water of the nonvegetated- and vegetated-surface-water incubations systems

Incubation system	Half-life (days)[a]	
	Metolachlor	Atrazine
Nonvegetated	32 ($r^2 = 0.83$)	144 ($r^2 = 0.76$)
Vegetated - *Lemna minor*	8 ($r^2 = 0.95$)	78 ($r^2 = 0.45$)
Vegetated - *Elodea canadensis*	3 ($r^2 = 0.99$)	25 ($r^2 = 0.93$)
Vegetated - *Ceratophyllum demersum*	3 ($r^2 = 0.61$)	12 ($r^2 = 0.73$)

[a]The half-lives were calculated assuming first-order reaction kinetics. The natural log of the percentage of applied [14C]metolachlor or [14C]atrazine remaining in the water was plotted with time (days). Half-lives were calculated based on the percentage of applied [14C]metolachlor or [14C]atrazine remaining in the water. The rates of metolachlor or atrazine metabolism, plant uptake, and the release of the parent compound or its metabolites from the plant are not known. The calculated half-lives do not reflect this rate data.

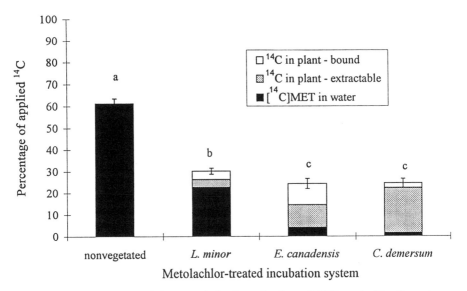

Figure 3. Significance of plant uptake in the reduction of [14C]metolachlor from the water of the vegetated incubation systems. A comparison of the percentage [14C]metolachlor remaining in the water of the nonvegetated incubation system with the summation of the percentage [14C]metolachlor in the water of the vegetated incubation system and the percentage 14C in the plant.

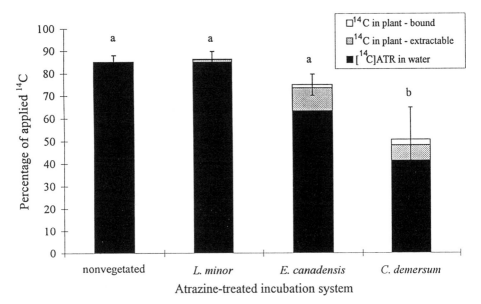

Figure 4. Significance of plant uptake in the reduction of [14C]atrazine from the water of the vegetated incubation systems. A comparison of the percentage [14C]atrazine remaining in the water of the nonvegetated incubation system with the summation of the percentage [14C]atrazine in the water of the vegetated incubation system and the percentage 14C in the plant.

and *C. demersum* systems, respectively. Based on the results of our investigation and the assumption that there was no rapid and significant plant uptake, metabolism, and release of the herbicide degradation products from the plant into the water between the extraction intervals (days 0, 4, 8, 12, and 16), plant uptake of ^{14}C by the aquatic vegetation did not, by itself, account for the significant reduction in the concentrations of [^{14}C]MET detected in the water of the vegetated incubation systems. Examination of the data presented in Figure 3 shows the summation of the percentage of applied [^{14}C]MET remaining in the water of the vegetated incubation systems plus the percentage of the applied ^{14}C associated with the plant tissues (extractable and nonextractable) represents a significantly smaller ($p \leq 0.01$) portion of the applied herbicide than the percentage of applied [^{14}C]MET remaining in the water of the nonvegetated-incubation systems. Similar results were seen in the atrazine-treated *C. demersum* system (Figure 4). These results suggest the significant ($p \leq 0.01$) reductions of [^{14}C]MET in the water of the *L. minor, E. canadensis,* and *C. demersum* systems and [^{14}C]ATR in the water of the *C. demersum* system did not occur predominantly as the result of plant uptake and the sequestering of the herbicide in the plant. Additional factors such as the degradation of the herbicide in the water or the degradation of the herbicide in the plant and the subsequent release of the herbicide and degradates into the water seem to be more important. Addition of the ^{14}C percentage in the *E. canadensis* to the percentage of [^{14}C]ATR in the water of the *E. canadensis* system was not significantly different from the percentage of [^{14}C]ATR remaining in the water of the nonvegetated system. This suggests that the accumulation of ^{14}C in the *E. canadensis* and the degradation of [^{14}C]ATR in the water were equally important to the significant reduction of [^{14}C]ATR.

Degradation of Metolachlor and Atrazine in the Water and Plant Tissues. Metolachlor, atrazine, and a number of the degradation products of metolachlor and atrazine were detected in the water extracts and plant extracts of the metolachlor- or atrazine-treated vegetated incubation systems. In the metolachlor-treated *L. minor, E. canadensis,* and *C. demersum* systems, the plant extracts and the water extracts contained significantly ($p \leq 0.01$) greater quantities of total ^{14}C (metolachlor and degradates) (*lines*) than [^{14}C]MET (*bars*) (Figure 5). The significantly reduced quantities of [^{14}C]MET relative to the total ^{14}C measured in the water and plant extracts and the detection of metolachlor degradates in these extracts indicate that the significant reduction ($p \leq 0.01$) of the [^{14}C]MET in the water of the vegetated systems occurs, in large part, as a result of degradation. The presence of herbicide degradates in the water and plant extracts may result from 1) the degradation of the herbicide in the water, 2) the degradation of the herbicide in the plant, 3) the degradation of the herbicide in the water and the accumulation of the herbicide degradates in the plant, or 4) the degradation of the herbicide in the plant and the release of the herbicide degradates into the water. Results from these vegetated incubation studies cannot definitively determine the location of the herbicide degradation. Our data (Table II) show significantly greater quantities of metolachlor degradates were found in the water fraction of the vegetated-incubations systems compared with the quantity of total ^{14}C detected in the plants ($p \leq 0.01$). The percentage of applied ^{14}C associated with the metolachlor degradates in the water of the vegetated incubation systems were at least 2.5 times greater than the percentage of applied ^{14}C detected in the plants (extractable and nonextractable) throughout the duration of the incubation. Less than twelve percent of the ^{14}C associated with the plant extracts was identified as [^{14}C]MET. This represents less than one percent of the total applied [^{14}C]MET. These results suggest that either 1) the majority of the

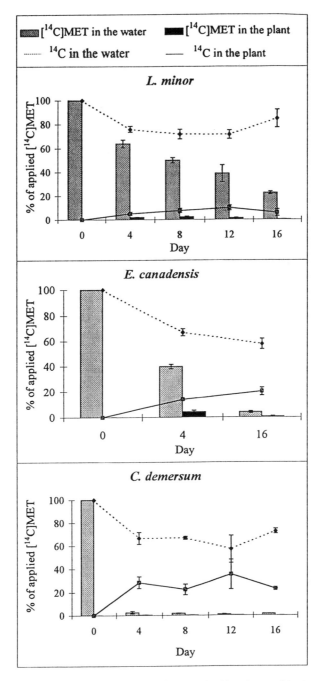

Figure 5. Percentage of applied ^{14}C and [^{14}C]metolachlor detected in the water and plant extracts of the vegetated and nonvegetated metolachlor-treated surface water incubations systems. ([^{14}C]MET = [^{14}C]metolachlor; ^{14}C = total radioactivity (parent + degradates).

Table II. Mass balance of the metolachlor-treated nonvegetated- and vegetated-surface-water incubation systems after sixteen days

Incubation system	Water			Plant				Mass balance
				Extractable		Unextractable		
	[14C]MET[a] (%)[b]	[14C]Degradates (%)[b]	Total [14C] (%)[b]	[14C]MET[a] (%)[b]	[14C]Degradates (%)[b]	[14C]Bound (%)[b]	Total [14C] (%)[b]	Total [14C] (%)[b]
Nonvegetated	61.2 ± 2.29	19.9 ± 2.20	81.1 ± 0.10	----	----	----	----	81.1 ± 0.10
Vegetated - *L. minor*	22.7 ± 1.30	56.2 ± 0.37	78.8 ± 1.64	0.43 ± 0.47	3.19 ± 2.18	3.95 ± 1.80	7.57 ± 0.09	86.4 ± 1.73
Vegetated - *E. canadensis*	4.06 ± 0.79	53.7 ± 3.51	57.7 ± 4.15	0.60 ± 0.52	9.83 ± 1.41	9.91 ± 1.71	20.3 ± 3.07	78.1 ± 7.22
Vegetated - *C. demersum*	1.44 ± 0.07	71.4 ± 2.09	72.9 ± 2.16	0.02 ± 0.02	20.9 ± 2.34	2.20 ± 0.25	23.2 ± 2.18	96.0 ± 4.34

[a][14C]MET = [14C]metolachlor.
[b](%) = percentage of applied 14C.

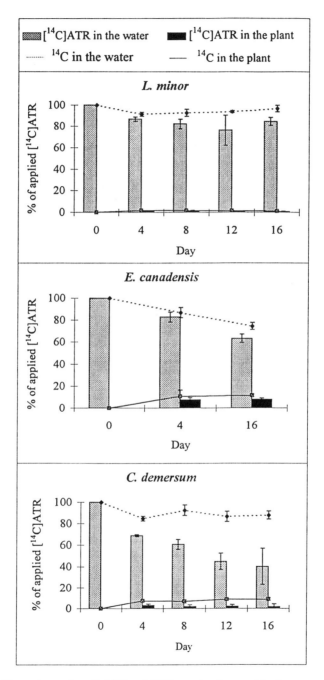

Figure 6. Percentage of applied ^{14}C and [^{14}C]atrazine detected in the water and plant extracts of the vegetated and nonvegetated atrazine-treated surface water incubation systems. ([^{14}C]ATR = [^{14}C]atrazine; ^{14}C = total radioactivity (parent + degradates).

[^{14}C]MET degradation occurred in the water of the metolachlor-treated vegetated incubation system or 2) the herbicides were rapidly taken up into the plants, metabolized, and released into the water solution within the 4-day intervals between the extraction and analysis of the incubation systems. Additional experiments need to be conducted in order to determine if the herbicides are degraded by microorganisms in the water or transformed in the plant and released into the water. In the vegetated and nonvegetated incubation systems we did not account for the mineralization of metolachlor or atrazine to CO_2. Between 78% and 98% of the applied radioactivity was recovered in the metolachlor- and atrazine-treated systems (Tables II & III).

The degradation of [^{14}C]ATR in the vegetated incubation systems primarily occurred in the water phase. With one exception (the day-four water extract in the atrazine-treated *E. canadensis* systems), the percentage of applied [^{14}C]ATR (*bars*) remaining in the water of the vegetated incubation systems was significantly less than the percentage of the total ^{14}C (atrazine and degradates combined) (*lines*) remaining in the water ($p \geq 0.02$) (Figure 6). Less than 12% of the applied ^{14}C was found in the *L. minor, E. canadensis*, and *C. demersum* plants throughout the duration of the incubations. The levels of [^{14}C]ATR detected in the plant extracts were not significantly different from the total ^{14}C (extractable and nonextractable) measured in the plants. This indicates that the degradation of [^{14}C]ATR in the plants was minimal, assuming the plant uptake, metabolism, and release of atrazine transformation products was minimal during the 4-d time intervals between the extraction and analysis of the 0, 4, 8, 12, and 16-d incubation systems. With the exception of the *E. canadensis* system, the water of the atrazine-treated vegetated incubation systems contained a significantly (p < 0.01) greater quantity of atrazine degradates than the total quantity of ^{14}C that was detected in the plants (extractable and nonextractable) (Table III). The quantity of atrazine degradates in the water of the *L. minor* and *C. demersum* systems was ten times and five times greater, respectively, than the quantity of ^{14}C detected in the *L. minor* and *C. demersum* plants. These data suggest [^{14}C]ATR was predominately degraded in the water rather than in the aquatic plants. The absence of a large accumulation of ^{14}C into the plants preceding a significant decrease in the quantity of radioactivity detected in the plant (extractable and nonextractable) suggests that the degradation of atrazine and metolachlor occurred mostly in the water phase of the incubation system rather than in the plant.

Atrazine Versus Metolachlor. When we compare the atrazine-treated vegetated and nonvegetated systems with the metolachlor-treated vegetated and nonvegetated systems, a greater percentage of the applied herbicide ([^{14}C]ATR or [^{14}C]MET) persisted in the atrazine systems compared with the metolachlor systems (Figures 1 & 2, Tables II & III). A greater percentage of the applied herbicide was characterized as degradates in the water and the plant extracts of all three metolachlor-treated vegetated systems relative to the corresponding atrazine-treated systems (Tables II & III). In addition, metolachlor and/or metolachlor degradates were more readily taken up into the plant or attached to the surface of the plant (total ^{14}C in the plant) than atrazine and its degradates. Based on this investigation, metolachlor was more readily degraded than atrazine. These results agree with the monitoring studies of Goolsby et al. (*3*) and Thurman et al. (*8*); they reported that atrazine was more persistent than metolachlor, alachlor, or cyanazine in the surface waters of the midwestern United States.

Table III. Mass balance of the atrazine-treated nonvegetated- and vegetated-surface-water incubation systems after sixteen days

| Incubation system | Water | | | Plant | | | | Mass balance |
| | | | | Extractable | | Unextractable | | |
	[14C]ATR[a] (%)[b]	[14C]Degradates (%)[b]	Total [14C] (%)[b]	[14C]ATR[a] (%)[b]	[14C]Degradates (%)[b]	[14C]Bound (%)[b]	Total [14C] (%)[b]	Total [14C] (%)[b]
Nonvegetated	85.0 ± 2.98	8.83 ± 3.71	93.9 ± 5.19	----	----	----	----	93.9 ± 5.19
Vegetated - *L. minor*	84.9 ± 3.73	12.0 ± 1.87	97.0 ± 3.00	0.64 ± 0.48	0.35 ± 0.42	0.22 ± 0.11	1.21 ± 0.05	98.2 ± 3.05
Vegetated - *E. canadensis*	63.2 ± 3.84	11.4 ± 2.78	74.6 ± 3.31	7.90 ± 1.24	2.33 ± 1.03	1.47 ± 0.42	11.7 ± 1.06	86.3 ± 4.37
Vegetated - *C. demersum*	41.3 ± 14.0	46.1 ± 11.3	87.5 ± 3.19	1.82 ± 2.75	4.97 ± 2.38	2.44 ± 0.86	9.23 ± 1.17	96.7 ± 4.36

[a] [14C]ATR = [14C]atrazine.
[b] (%) = percentage of applied 14C.

C. demersum Versus *E. canadensis* Versus *L. minor.* The presence of plants and the type of plant can make a significant difference in the quantity of metolachlor or atrazine that remains in the water. Our investigations demonstrated, with the exception of the atrazine-treated *L. minor* system, that the presence of aquatic plants significantly ($p \leq 0.01$) reduced the concentration of [^{14}C]MET and [^{14}C]ATR in the herbicide-contaminated waters (Figures 1 & 2). Lack of a significant difference in the concentration of [^{14}C]ATR in the *L. minor* incubation systems compared with the nonvegetated system may be attributed to the phytotoxicity of atrazine to the *L. minor* (*35, 36*). *C. demersum* was superior in the remediation of the metolachlor- and atrazine-contaminated waters. The herbicide-reduction efficiencies of the aquatic plants were, from most efficient to least efficient, *C. demersum* > *E. canadensis* > *L. minor* for both the metolachlor- and atrazine-treated systems. Degradation seems to be the predominant factor involved in the high herbicide-reduction efficiency of the *C. demersum* system. The quantities of atrazine and metolachlor degradates detected in the water of the vegetated incubation systems were, in descending order, *C. demersum* > *L. minor* = *E. canadensis.* The accumulation of the herbicides in *C. demersum* seemed to play a secondary role to degradation. Herbicide accumulation in the plants followed the order of *C. demersum* = *E. canadensis* > *L. minor* for the metolachlor- and atrazine-treated systems. This may be related to the surface area of the plant exposed to the herbicide-contaminated water. Both the *C. demersum* and *E. canadensis* are submerged aquatic plants whereas *L. minor* is a free-floating aquatic plant. The submerged aquatic plants would have a greater surface area exposed to the herbicide in relation to the floating *L. minor.*

Discussion

The purpose of our investigation was to evaluate the ability of aquatic plants to remediate herbicide-contaminated waters. Our results demonstrated the presence of herbicide-tolerant aquatic plants contributed to the accelerated dissipation of metolachlor and atrazine in the surface water incubation systems.

Aquatic plants can contribute directly or indirectly to the removal of pollutants from water and sediment. Direct interaction of the plant and contaminant would include the uptake and accumulation or metabolism of the xenobiotic compound within the plant. Research has shown that plants contain enzymes that transform and conjugate organic contaminants (*37-39*). Herbicides that are absorbed by herbicide-resistant plants can be transformed and conjugated by these enzymes to degradation products that may be stored in the vacuoles or cell walls of the plant cells (*37, 40*) or released from the plant back into the water. The tolerance of plants to metolachlor is often dependent on the plants' ability to rapidly conjugate metolachlor. In most cases, atrazine-resistant plants contain a different amino acid in the photosynthetic protein that will interfere with atrazine's ability to disrupt electron flow (*33*).

The dissipation of contaminants from water or sediment can be indirectly affected by plants as a result of the accelerated biodegradation of the compound in the phyllosphere or rhizosphere. Plants provide a favorable surface for the attachment of microorganisms (*41-43*), and they supply organic nutrients to epiphytic microorganisms, in the form of photosynthates and exudates, which stimulate microbial growth in the phyllosphere and rhizosphere (*43, 44*). In addition, certain plants can transport oxygen to anaerobic sediments and anoxic waters, which create oxidized microenvironments that stimulate the microbial degradation of organic substances (*45, 46*).

The presented data provide evidence that enhanced degradation is the predominant factor involved in the significant reduction of metolachlor and atrazine from the waters of the vegetated incubation systems. The sequestering of the atrazine or metolachlor or their degradation products in the plant was minimal. Additional experiments need to be conducted to determine if the accelerated degradation occurs as the result of degradation in the plant or as a result of enhanced biodegradation associated with epiphytic microorganisms in the phyllosphere or rhizosphere. Results of this investigation are similar to other phytoremediation studies that report the major mechanism of pollutant removal to be enhanced degradation (29, 47).

Metolachlor was more readily degraded than atrazine in the nonvegetated and vegetated systems. Atrazine may be more recalcitrant to degradation as a result of its chemical structure or bioavailability to microorganisms or plants. Metolachlor has been shown to be primarily degraded by microorganisms in sediments (10) and a number of metolachlor degradation products were detected in microbial cultures (48, 49). Laboratory studies have shown that atrazine, in surface water samples or aquatic solutions, was recalcitrant to microbial degradation (50). This may be the result of the resistance of the s-triazine ring to microbial attack (51 as cited by 12). Metolachlor is more water soluble than atrazine and therefore more bioavailable to plants and microorganisms. The greater solubility of metolachlor may account for the increased percentage of applied ^{14}C detected in the plants of the [^{14}C]metolachlor treated systems compared with the [^{14}C]atrazine treated systems. Greater plant uptake and bioavailability of metolachlor to the plants and epiphytic microorganisms contributes to the more rapid degradation of metolachlor compared with atrazine.

Conclusions

Our research has demonstrated that aquatic vegetation may be used to remediate herbicide-contaminated waters. With the exception of the atrazine treated *L. minor* system, concentrations of [^{14}C]MET or [^{14}C]ATR were significantly ($p \leq 0.01$) reduced in the water of the vegetated incubation systems after 16 days. In both the metolachlor- and atrazine-treated systems, the herbicide-reduction efficiencies of the aquatic plants were, from most efficient to least efficient, *C. demersum* > *E. canadensis* > *L. minor.* The results of our investigation suggest the significant ($p \leq 0.01$) reductions of [^{14}C]ATR in the water of the *C. demersum* system and [^{14}C]MET in the water of the *L. minor, E. canadensis,* and *C. demersum* systems did not occur predominantly as the result of the absorption and sequestering of the herbicides and their transformation products in the plants. Accelerated biodegradation seems to be more important than plant accumulation and storage to the enhanced dissipation of metolachlor and atrazine from the water of the vegetated systems. Additional experiments need to be conducted with surface-sterilized and non-sterilized plants to confirm whether the accelerated degradation of the herbicides was the result of xenobiotic metabolism in the plant or of enhanced biodegradation of the herbicides in the water do to increased microbial populations in the phyllosphere or rhizosphere of the aquatic plants.

Practical application of this research would be the construction of wetlands and macrophyte-cultured ponds for the phytoremediation of agricultural-drainage effluents from field runoff and tile drains. These aquatic macrophyte systems would provide a relatively maintenance-free and cost-effective means of remediating contaminated effluents before their release into streams, rivers, and lakes. Phytoremediation of wastewater effluents can reduce the levels of contaminants that enter natural waters, which would lessen the adverse impact of pollutants on aquatic ecosystems, remove unwanted nitrates and pesticides from surface drinking water sources, and help meet public demands for higher water quality.

Acknowledgments

The authors thank Jennifer Anhalt, Karin Tollefson, Carla McCullough, Piset Khuon, John Ramsey, Kara Wedemeyer, and Brett Nelson for their technical support and Ellen Kruger and Patricia Rice for their assistance in the collection of samples. Radiolabeled chemicals and analytical standards were provided by Ciba Crop Protection. This research was funded by US Department of Agriculture Management Systems Evaluation Area grant, Great Plains-Rocky Mountain Hazardous Substances Research Center at Kansas State University and the US Environmental Protection Agency, Cooperative Agreement No. CR-823864. This journal publication is Journal Paper No. J-17161 of the Iowa Agriculture and Home Economics Experiment Station, Ames, Iowa, Project No. 3187, and supported by Hatch Act and State of Iowa funds.

Literature Cited

1. *Agricultural Chemical Usage 1992 Field Crops Summary;* United States Department of Agriculture National Agricultural Statistics Service: Washington, D.C., **1993**.
2. Squillace, P. J.; Thurman, E. M. *Environ. Sci. Technol.* **1992**, *26*, 538-545.
3. Goolsby, D. A.; Thurman, E. M.; Kolpin, D. W. *Geographic and temporal distribution of herbicides in surface waters of the upper Midwestern United States, 1989-90;* 91-4034; U.S. Geological Survey Water-Resources Investigation Report: Denver, CO, 1990; pp 183-188.
4. Hall, J. K.; Mumma, R. O.; Watts, D. W. *Agric. Ecosyst. and Environ.* **1991**, *37*, 303-314.
5. Bengtson, R. L.; Southwick, L. M.; Willis, G. H.; and Carter, C. E. *Transactions of the ASAE* **1990**, *33*, 415-418.
6. Triplett, G. B.; Conner, B. J. Jr.; Edwards, W. M. *J. Environ. Qual.* **1978**, *7*, 77-84.
7. Leonard, R.A.; Langdale, G. W.; Fleming, W. G. *J. Environ. Qual.* **1979**, *8*, 223-229.
8. Thurman, E. M.; Goolsby, D. A.; Meyer, M. T.; Mills, M. S.; Pones, M. L.; Kolpin, D. W. *Environ. Sci. Technol.* **1992**, *26*, 2440-2447.
9. Wauchope, R.D. *J. Environ. Qual.* **1978**, *7*, 459-472.
10. Chesters, G.; Simsiman, G. V.; Levy, J.; Alhajjar, B. J.; Fathulla, R. N.; Harkin, J. M. *Reviews of Environ.Contamin.and Toxicol.* **1989**, *110*, 2-74.
11. Mote, C. R.; Tompkins, F. D.; Allison, J. S. *Transactions of the ASAE* **1990**, *33*, 1083-1088.
12. Solomon, K. R.; Baker, D. B.; Richards, R. P.; Dixon, K. R.; Klaine, S. J.; La Point,T. W.; Kendall, R. J.; Weisskopf, C. P.; Giddings, J. M.; Giesy, J. P. *Environ. Toxicol. and Chem.* **1996**, *15*, 31-76.

13.   Fawell, J.K. *Pesticides in Soils and Water: Current Perspectives;* British Crop Protection Council Monograph No. 47; British Crop Protection Council, Farnham, UK, 1991; pp 205-208.

14.   Leonard, R.A. In *Pesticides in the Soil Environment: Processes, Impacts and Modeling;* Cheng, H. H., Ed.; Soil Science Society of America, Inc.; Madison, WI, 1990, pp 303-349.

15.   Jones, T. W.; Winchell, L. *J. Environ. Qual.* **1984,** *13,* 243-247.

16.   Forney, D. R.; Davis, D. E. *Weed Sci.* **1981,** *29,* 677-685.

17.   Baker, D.B.; Richards, R. P. In *Long Range Transport of Pesticides;* Kurtz, D. A., Ed.; Lewis Publishers: Chelsa, MI, 1990, pp 241-270.

18.   Miltner, R.J., Baker, D. B.; Speth T. F.; Fronk, C. A. *J. AWWA* **1989,** *81,* 43-52.

19.   Reddy, K.R. *J. Environ. Qual.* **1983,** *12,* 137-141.

20.   Reddy, K.R.; Campbell, K. L.; Graetz, D. A.; Portier, K. M. *J. Environ. Qual.* **1982,** *11,* 591-595.

21.   Rogers, H.H.; Davis, D. E. *Weed Sci.* **1972,** *20,* 423-428.

22.   Boyd, C.E. *Econ. Botany* **1976,** *30,* 51-56.

23.   Wooten, J.W.; Dodd, J. D. *Econ. Botany* **1976,** *30,* 29-37.

24.   Brix, H. *Water Sci. Tech.* **1987,** *19,* 107-118.

25.   Gersberg, R.M.; Elkins, B. V.; Lyon, S. R.; Goldman, C. R. *Water Res.* **1986,** *20,* 363-368.

26.   Nichols, D.S. *J. Water Pollut. Control Fed.* **1983,** *55,* 495-505.

27.   Wolverton, B.C.; McDonald, R. C. *J. Water Pollut. Control Fed.* **1979,** *51,* 305-313.

28.   Oron, G.; Porath, D.; Wildschut, L. R. *J. Environ. Engineering* **1986,** *112,* 247-263.

29.   Federle, T.W.; Schwab, B. S. *Appl. Environ. Microbiol.* **1989,** *55,* 2092-2094.

30.   Mo, S.C.; Choi, D. S.; Robinson, J. W. *J. Environ. Sci. Health* **1989,** A*24,* 135-146.

31.   Charpentier, S.; Garnier, J.; Flaugnatti, R. *Bull. Environ. contamin. Toxicol.* **1987,** *38,* 1055-1061.

32.   Pignatello, J.J.; Johnson, L. K.; Martinson, M. M.; Carlson, R. E.; Crawford, R. L. *Appl. Environ. Microbiol.* **1985,** *50,* 127-132.

33.   *Herbicide Handbook of the Weed Science Society of America;* Humburg, N.E.; Colby, S. R.; Hill, E. R.; Kitchen, L. M.; Lym, R. G.; McAvoy, W. J.; Prasad, R., Eds.; Weed Science Society of America: Champaign, IL, **1989.**

34.   Liu, S.-Y.; Freyer, A. J.; Bollag, J.-M. *J. Agric. Food Chem.* **1991,** *39,* 631-636.

35.   Kirby, M. F.; Sheahan, D. A. *Bull. Environ. Contam. Toxicol.* **1994,** *53,* 120-126.

36.   Wang, W. *Environ. Research* **1990,** *52,* 7-22.

37.   Sandermann, H., Jr. *Trends in Biochem. Sci.* **1992,** *17,* 82-84.

38.   Sandermann, H., Jr.; Schmitt, R.; Eckey, H.; Bauknecht, T. *Archives of Biochem. and Biophysics* **1991,** *287,* 341-350.

39.   Wetzel, A.; Sandermann, H. Jr. *Archives of Biochem. and Biophysics* **1994,** *314,* 323-328.

40.   Schmitt, R.; Sandermann, H. Jr. *Z. Naturforsch* **1982,** *37*C, 772-777.

41.   Nichols, D.S. *J. Water Pollut. Control Fed.* **1983,** *55,* 495-505.

42.   Rimes, C.A.; Goulder, R. *J. Appl. Bact.* **1985**, *59*, 389-392.
43.   Goulder, R.; Baker, J. H. In *Microbial Ecology of Leaves;* Andrews, J. H.; Hirano, S. S., Eds.; Springer-Verlag: New York, NY, 1991, pp 60-86.
44.   Fokkema, N.J.; Schippers, B. In *Microbiology of the Phyllosphere;* Fokkema, N. J.; Van Den Heuvel, J., Eds.;. Cambridge University Press: Cambridge, England, **1986**, pp 137-159.
45.   Brix, H. *Water Sci. Tech.* **1987**, *19*, 107-118.
46.   Reddy, K. R.; D'Angelo, E. M.; DeBusk, T. Z. *J. Environ. Qual.* **1989**, *19*, 261-267.
47.   Reilley, K.A.; Banks, M. K.; Schwab, A. P. *J. Environ. Qual.* **1996**, *25*, 212-219.
48.   Krause, A.; Hancock, W. G.; Minard, R. D.; Freyer, A. J.; Honeycutt, R. C. ; LeBaron, H. M.; Paulson, D. L.; Liu, S.-Y., Bollag, J.-M. *J. Agric. Food Chem.* **1985**, *33*, 581-589.
49.   Saxena, A.; Zhang, R.; Bollag, J.-M. *Applied and Environ. Microbiol.* **1987**, *53*, 390-396.
50.   Dries, D.; De Corte, B.; Liessens, J.; Steurbaut, W.; Dejonckheere W.; Verstraete, W. *Biotechnol. Lett.* **1987**, *9*, 811-816.
51.   *Handbook of Environmental Fate and Exposure Data for Organic Chemicals*; Howard, P. H., Eds.; Lewis Publishers: Chelsea, MI, 1991; Vol. 3.

Chapter 11

The Metabolism of Exogenously Provided Atrazine by the Ectomycorrhizal Fungus *Hebeloma crustuliniforme* and the Host Plant *Pinus ponderosa*

J. L. Gaskin and J. Fletcher

Department of Botany and Microbiology, University of Oklahoma, Norman, OK 73019

The mineralization of atrazine by mycorrhizal fungi was examined. The percent mineralization was determined by comparing the amount of $^{14}CO_2$ evolved from the plant and plant-fungus systems when measured amounts of ^{14}C-labeled atrazine were provided under axenic conditions in specially designed exposure chambers. The percent mineralization was 0.1 and 0.3, respectively, for the plant and plant-fungus systems. The study demonstrated that an ectomycorrhizal fungus in association with its host plant increased the metabolism of exogenously provided atrazine above that of the plant by itself, thereby supporting the proposed use of plant-fungal systems in bioremediation of contaminated soils.

The ability of ectomycorrhizae to degrade xenobiotic compounds has been studied with pure cultures examined under laboratory conditions. Donnelly et al. *(1)* screened several mycorrhizal fungi for their ability to degrade two aromatic herbicides, 2,4-D and atrazine. In a 30-day study with ^{14}C labeled atrazine and 2,4-D, it was shown that the rates of degradation as measured by CO_2 evolution varied depending on the species of fungus and herbicide tested *(1)*. In general atrazine was degraded at a higher rate than 2,4-D by the fungi tested. *Gautieria crispa,* was the most active degrader of atrazine; whereas *Oidiodendron griseum,* an ericoid mycorrhizal fungus, was most active against 2,4-D. Maximum mineralization of atrazine occurred between the third and fifth day, while it occurred between the fifth and tenth day for 2,4-D. Robideaux et al. *(2)* found that 10 of 11 ectomycorrhizal fungi tested removed the herbicide hexazinone from liquid medium. The amount of hexazinone removed depended on the species tested and the C/N ratio in the medium. In a study by Donnelly and Fletcher *(3),* examination of the ability of 21 different pure cultures of ectomycorrhizal fungi to metabolize 19 different polychlorinated biphenyl (PCB) congeners, showed that 14 of the fungi metabolized PCB's. These findings with pure cultures suggest that ectomycorrhizae are capable of degrading organic pollutants, but it is not certain that they retain this property when in association with roots of a host plant. The objective of this study was to examine the ability of roots colonized with ectomycorrhizal fungi to metabolize atrazine.

Methods

Ponderosa pine seeds were sterilized in 30% hydrogen peroxide for 55 min., rinsed with sterile water, and placed on solid nutrient medium. When seeds had germinated and grown to approximately 3 cm in length they were examined for microbial contamination, and uncontaminated seedlings were transferred to sterile pouches (Figure 1). The pouches used were plant culture made by Agristar Inc. (Sealy, TX) The bags were modified by inserting paper pads removed from seed germination pouches available from Vaughn's Seed Company (Downers Grove, IL). Although Vaughn plastic polyester pouches have been used previously in ectomycorrhizal work *(4)* we found the Agristar bags to be preferred because they had much higher rates of gas exchange, and could also be autoclaved. Pouches were prepared as shown in Figure 1. Ten ml of Hoagland's solution were added to each pouch. A glass disposable pipette with a foil cap was clipped on the inside of each pouch for future watering. Two fiberglass pads were used as spacers to prevent the sides of the bag from collapsing and thereby permit air circulation. The assembled pouches were autoclaved, permitted to cool to room temperature, and sterile seedlings were carefully inserted to insure that the root and pouch remained axenic.

The plant/pouches were placed in a 22 °C growth chamber and received 14 h of 300-400 µEinsteins of illumination every 24 hr. Sterile water (approx. 5 ml) was added aseptically every three days through the pipette attached to each pouch (Figure 1). Hoaglands solution (10%) was added every two weeks instead of water.

Hebeloma crustuliniforme obtained from Paul Rygiewicz, (Environmental Research Laboratory, Corvallis, Oregon) was grown in petri dishes on mycorrhizal solid medium, which contained in one liter of media: glucose 10 g, $NH_4H_2PO_4$ 0.7 mg, KH_2PO_4 0.4 g, $MgSO_4$ 0.5 g, $CaCl_2$ 100 mg, Fe-citrate 5 mg, $MnSO_4$ 5 mg, $ZnSO_4$ 4.4 mg, yeast extract 50 mg, agar 18 g *(1)*. Ponderosa pines were inoculated with fungal plugs after short roots appeared from lateral roots, the stage of root development considered to be most receptive to colonization *(4,5)*. Inoculum plugs were prepared by removing 4 mm cork borings from the outer edge of 4 week old solid cultures and incubating them in liquid medium until the fungal plug was engulfed in new mycelium. The plugs were then used to inoculate the pines by placing a plug next to a group of short roots, typically comprised of 3-4 roots (Figure 1). Following inoculation, the plant roots and fungi were examined every three days for fungal growth and developmental changes, such as growth toward roots, formation of the mantle, and the appearance of extramatrical hyphae.

When a root had been colonized, the inoculum plug was removed and a filter paper disc laden with ^{14}C-atrazine was carefully placed in individual pouches so that it was in contact (covered) with a dense region of extramatrical hyphae and a few branch roots. The ^{14}C-laden discs were prepared by asceptically transferring 0.5 ml of ^{14}C-atrazine stock solutions (2 µCi·ml[-1]) with a specific activity of 19.1 µCi·µmole[-1] onto autoclaved filter paper discs. The transfers were made with a syringe fitted with a 0.2 µm Acrodisc filter. The discs were left in the transfer hood until the solvent carrier evaporated from the discs. Following addition of the ^{14}C-discs, each pouch was wrapped around a small piece of PVC pipe cylinder (4x10 cm), and a rubber band was used to fit and hold the pouch to the cylinder thereby securing the position of the ^{14}C-disc inside the pouch. Pouches prepared in this manner were then placed into incubation chambers (Figure 2). The ^{14}C laden discs placed over roots in the pouches imbibed approximately 0.5ml of H_2O giving an exposure conentration of 100µM atrazine.

The metabolism of ^{14}C-atrazine was determined by comparing the release of [14]CO_2 from pouches containing: plant, plant-fungus (mycorrhizae), and no living

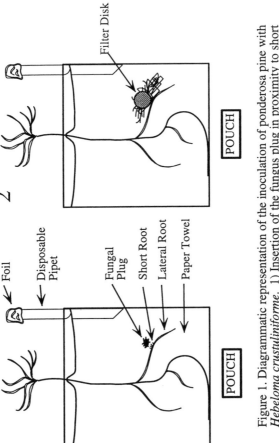

Figure 1. Diagrammatic representation of the inoculation of ponderosa pine with *Hebeloma crustuliniforme.* 1) Insertion of the fungus plug in proximity to short roots. 2) The inoculum plug is removed when extramatrical hyphae appear and the disc is added with radiolabeled compound.

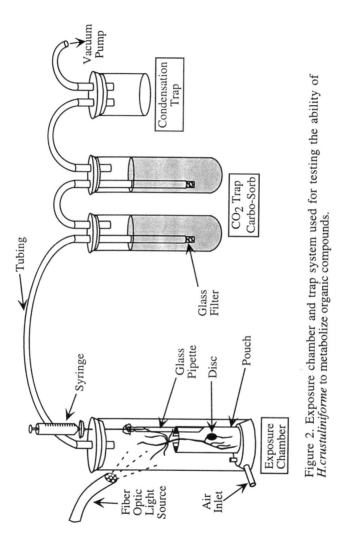

Figure 2. Exposure chamber and trap system used for testing the ability of *H.crustuliniforme* to metabolize organic compounds.

organisms (control). The plant and plant-fungus tests were run in triplicate and the control in duplicate. Specially designed exposure chambers were used to measure the metabolism of ^{14}C-atrazine to $^{14}CO_2$ (Figure 2). Individual chambers holding a single pouch were comprised of a glass cylinder (30 cm tall, 90 cm diameter) with two rubber stoppers at each end. The top and bottom stoppers were fitted with glass tubes inserted through them to permit air to enter the bottom and leave from the top. Each chamber was connected in series by Tygon tubing with two CO_2 traps, and an empty, dry catch-tube. The air entering the CO_2 traps (35x200 mm test-tube) was drawn through a fritted glass bubbler tube submerged in 40 ml of full-strength Carbo-Sorb a CO_2 trapping agent purchased from Packard Instruments (Downers Grove, IL). The second CO_2 trap was used to monitor the possible saturation of the first trap that was intended to retain all evolved CO_2. A steady flow of air was pulled through the chamber/trap system by a vacuum pump.

During the course of a 32 day incubation period water was added to individual pouches through a sterile needle, which was inserted through the top stopper and into a sterile pipette held in each pouch (Figure 2). Water was added as needed from a syringe fitted to an Acrodisc sterilizing filter (Fisher) attached to the needle. Control pouches possessing no plants or fungi also received water.

All chambers were illuminated 14 h a day with a beam of light transmitted through fiber optics. The beam was focused on the needles of the ponderosa pine in chambers holding plants. Each chamber received approximately 500-600 µEinsteins of illumination from 17 fiber-optic fibers whose ends were positioned approximately 3 cm from the glass chamber. The bundle of 17 fibers ran to a trunk line connected to a Fiberstars Inc. (Fremont, CA) illuminator model FS-FIB-401S equipped with a Fiberstar silver halide bulb. This novel method of illumination eliminated the temperature problems normally associated with conventional lighting of small enclosed chambers.

Radioactivity (cpm) evolved from each incubation chamber was determined at regular intervals by collecting two 2 ml samples from each trap (Figure 2). Prior to each sampling the absorbent volume was adjusted to the starting volume (40 ml) by adding additional Carbo-Sorb. Each 2 ml Carbo-Sorb sample was combined with 2 ml of Permafluor E+ scintillation fluid purchased from Packard Instruments. The mixed samples were left in the dark for 24 h and then the radioactivity was determined with a Beckman scintillation counter set for 10 min assays.

The total cumulative amount of radiation (cpm) evolved over time was determined for each incubation chamber. At each sampling time the cpm values of the two 2 ml aliquots removed from a trap were averaged and then adjusted for background radiation by subtracting the cpm value of the two control chambers that possessed no plant or fungal tissue. The amount of radiation in each 40 ml trap at each sampling time was determined by multiplying the cpm value of the 2 ml aliquot by 20. The cumulative amount of radiation that had been released from each incubation chamber at each successive sampling time was determined by adding the sum of the cpm values of previously removed 2-ml aliquots. The percent mineralization of the starting atrazine-^{14}C to $^{14}CO_2$ at different times during the course of the experiment was determined by dividing the total acumulative amount of ^{14}C released as CO_2 by the amount of radiation provided at the beginning of the experiment as ^{14}C-atrazine

The rate of atrazine degradation to $^{14}CO_2$ was calculated by dividing the µmoles of atrazine metabolized per day by the dry weight of either the root-fungus or the estimated fungus in contact with the exposure disc at the completion of the experiment. The total root-fungus weight was established by removing the paper towel and associated root-fungal tissue from individual pouches. The root and fungal tissue directly beneath the exposure disc were removed and dry weights were determined. The amount of fungal weight was then estimated by considering 25% of the total weight to be fungal weight. This correction was based on the 25-40% root to fungal percent ratios reported by Harley and Smith *(6)* and knowledge that the fungal masses observed in the pouches in this study were considered less developed than hyphae associated with plant roots grown in soil. For comparative purposes, the percent mineralization rates reported by Donnelly et al. *(1)* for several different fungi were converted to μmoles atrazine·day^{-1}·mg dry weight^{-1} (Table I). This was accomplished by multiplying the reported percent mineralization rates by the total μmoles of atrazine provided and dividing these rates by days and fungal weight.

Analysis of variance was used to compare the ability of plant and plant-fungus to mineralize ^{14}C-atrazine. The radioactivity present in the last four samples collected were used for these analyses. Treating these as equivalent samples even though they were not collected on the same day was justified by virtue of the fact that the release of radioactive CO_2 had stopped, thereby suggesting that the substrate had been used up and all of the last four values were equally representative of the terminal state of mineralization.

Results and Discussion

Carbon-14 was released from both the mycorrhizae and plant-only systems throughout the course of the experiment. There was substantial variability among the replicates of both the mycorrhizae and plant systems, but in general the amount of $^{14}CO_2$ produced from the mycorrhizae replicates exceeded that from the plant replicates throughout the experiment (Table II). Statistical analyses of the acumulative $^{14}CO_2$ values at the end of the experiment showed a significant difference between the two systems.

The rate of atrazine mineralization by the mycorrhizae was 2-fold faster than the rate by the plant alone (Figure 3). The enhanced rate may have been due to direct metabolism of the xenobiotic by the fungus or perhaps facilitated movement of the compound into the plant root through the fungal hyphae, followed by plant metabolism of atrazine. In either event the ectomycorrhizal system enhanced the degradation of atrazine. Previous studies with pure cultures of ectomycorrhizal fungi have shown that this group of fungi have the genetic capability to degrade xenobiotics, but the degradation of atrazine by the mycorrhizal system used in this study is the first evidence of enhanced xenobiotic degradation by roots colonized with mycorrhizae. The rate of atrazine mineralization by the plant-fungus system was compared to rates reported in the literature for pure cultures of ectomycorrhizae by converting the cpm values in Table I to μmoles atrazine·day^{-1}·mg dry weight^{-1}. This conversion was made by using the specific activity of the parent material to convert the cpm of $^{14}CO_2$ released to μmoles of atrazine metabolized. Rates of atrazine degradation to $^{14}CO_2$ were calculated by dividing the μmoles of atrazine metabolized per day by the dry weight of tissue in contact with the exposure disc at the completion of the experiment. Data from the work of Donnelly et al. *(1)* were recalculated in a similar fashion as explained in the materials and methods. The rate of atrazine degradation calculated for *Hebeloma crustuliniforme* and ponderosa pine roots was 9.0×10^{-7} μmoles atrazine·day^{-1}·mg dry weight^{-1}. Assuming that the fungal tissue was responsible for all of the degradation and that the fungus accounted for 25% of the total root-fungal weight *(6)* then it was estimated that the degradation rate of *Hebeloma crustuliniforme* by itself was 3.6×10^{-6} μmoles atrazine·day^{-1}·mg dry weight $^{-1}$. This value is approximately 10-fold lower than values calculated from the pure culture data reported for several different species of ectomycorrhizae by Donnelly et al *(1)*. The difference in rate between the two studies involves several possible explanations. The lower rate

Table I. Mineralization of Atrazine (μmoles atrazine·day^{-1}·mg dry weight^{-1}) by four different ectomycorrhizal fungi in pure culture as compared to roots of *Pinus ponderosa* with colonized *Hebeloma crustuliniforme.*

Pure Culture[1]				Colonized Plant	
Rhizopogon vulgaris	*Gauteria crispa*	*Gauteria othii*	*Radiigera atragleba*	Root and Fungus	Fungus[2]
2.3x10^{-5}	3.3x10^{-5}	8.3x10^{-5}	9.9x10^{-5}	9.0x10^{-7}	3.6x10^{-6}

[1]Data taken from Donnelly et al *(1)*.
[2]Weight of fungus was considered to be 25% of the total weight of the colonized root *(6)*

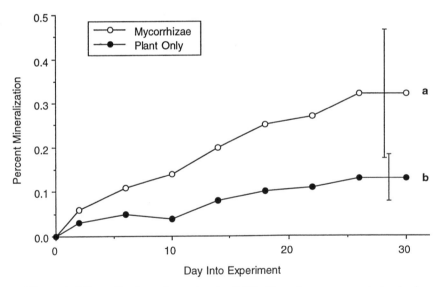

Figure 3. Mineralization of atrazine (p=0.13). Error bars represent standard deviation around the mean of the final two samples pooled together.

Table II. Amount of radioactivity (cpm) recovered in CO_2 when ^{14}C-atrazine was provided to different systems

Day		PLANT-FUNGUS			PLANT		
		A[a]	B	C	A	B	C
2	Sample[b]	172	204	196	166	173	170
	Corrected Sample[c]	22	54	46	16	23	20
	Total Count[d]	440	1080	920	320	460	400
	Cumulative Count[e]	440	1080	920	320	460	400
6	Sample	213	234	228	170	187	197
	Corrected Sample	63	84	79	20	37	48
	Total Count	1260	1680	1580	400	740	960
	Cumulative Count	1304	1788	1672	432	786	1000
10	Sample	209	239	265	177	182	173
	Corrected Sample	59	89	115	27	32	23
	Total Count	1080	1780	2300	540	640	460
	Cumulative Count	1250	2056	2550	612	760	596
14	Sample	258	239	312	169	191	239
	Corrected Sample	104	89	162	19	41	89
	Total Count	2080	1780	3240	380	820	1780
	Cumulative Count	2368	2234	3720	506	1004	1962
18	Sample	272	253	349	182	203	238
	Corrected Sample	122	103	199	32	53	88
	Total Count	2440	2060	3980	640	1060	1760
	Cumulative Count	2936	2692	4784	804	1326	2120
22	Sample	272	230	382	181	212	243
	Corrected Sample	122	80	232	31	62	93
	Total Count	2440	1600	4640	620	1240	1860
	Cumulative Count	3180	2438	5842	848	1612	2396
26	Sample	286	248	403	174	179	249
	Corrected Sample	136	98	253	24	29	99
	Total Count	2720	1960	5060	480	580	1980
	Cumulative Count	3704	2958	6726	770	1076	2702
30	Sample	270	216	405	173	220	239
	Corrected Sample	120	66	255	23	70	89
	Total Count	2400	1320	5100	460	1400	1780
	Cumulative Count	3656	2513	7272	798	1954	2700

[a] A, B, and C represent triplicate exposure chambers
[b] The cpm of each sample taken.
[c] The subtraction of background radiation (average pouch cpm 150).
[d] The correction of volume by multiplying the cpm of a 2ml sample by 20, which brings the volume up to 40ml.
[e] The addition of all previous 2ml samples removed.

may be a feature of species difference since unfortunately *Hebeloma crustuliniforme* was not a part of the Donnelly study and no comparable pure culture data on xenobiotics metabolism is available from another source. Another possibility is that ectomycorrhizal fungi on colonized roots do not metabolize xenobiotics as they do in pure culture, and the degradation observed in our study was actually plant degradation facilitated by hyphal uptake of atrazine. A third explanation is that the fungal tissue was the primary contributer to atrazine degradation, but we overestimated the amount of fungal tissue by assuming that 25% of the fungal root weight was due to the fungus. Each of these possibilities or some combination of them serve as explanations for the 120-fold difference in reported rates (Table I), and emphasizes the need for additional research.

Summary

The mycorrhizal system of ponderosa pine and *Hebeloma crustuliniforme* was shown to mineralize atrazine to a greater extent than the plant by itself. The ability of ectomycorrhizae to enhance the metabolism of atrazine when the fungus is associated with the host plant suggests that ectomycorrhizae could play an active role in remediating organic, soil pollutants.

The advantage of ectomycorrhizae systems in bioremediation over that of other systems such as free-living bacteria or fungi is that degrading strains of these organisms introduced at toxic waste sites are often difficult to maintain for extended periods of time because of competition from wild types whereas ectomycorrhizae introduced along with their host plant theoretically have a survival advantage over competing organisms since the host plant provides an ecological niche for the ectomycorrhizal fungus. The general ecology and structure of ectomycorrhizae are also favorable for soil bioremediation since the long hyphae of the fungus, increase the surface area already explored by roots, and thereby greatly increase the contact between potentially bioreactive organisms and soil contaminants. Although ectomycorrhizae hold great promise in bioremediation, the slow rate of mineralization observed with our test system (ponderosa pine/*Hebeloma crustuliniforme*) clearly shows that the maximum potential of ectomycorrhizal systems in bioremediation can only be realized through more intensive efforts to identify those plant-fungal systems with the greatest genetic potential to degrade a broad spectrum of xenobiotic compounds.

Literature Cited

1. Donnelly, P.K.; Entry A.E.; Crawford, D.L. *Appl. Environ. Microbiol.* **1993**, 59, pp. 2642- 2647
2. Robideaux, M.L.; Fowles, N.L.; King,R.J.; Spriggs, J.W.; Rygiewicz P.T. Abst.12th Annual Meeting. *Society of Environmental Toxicology and Chemistry,* Seattle, WA; **1991**, p.175.
3. Donnelly, P.K.; Fletcher J.S. *Bull. Environ. Contam. Toxicol.* **1995**, 54, pp. 507-513.
4. Fortin, J.A.; Piché, Y; Godbout, C. *Plant and Soil.* **1983**, 71, pp. 275-284.
5. Fortin, J.A.; Piché, Y.; LaLonde, M. *Can. J. Bot.* **1980**, 58, pp 361-365.
6. Harley, J.L.; Smith, S.E. *Mycorrhizal Symbiosis.* Academic Press, London. **1983**.

PHYTOREMEDIATION OF INDUSTRIAL CHEMICALS

Chapter 12

Evaluation of the Use of Vegetation for Reducing the Environmental Impact of Deicing Agents

Patricia J. Rice[1], Todd A. Anderson[2], and Joel R. Coats[1]

[1]Pesticide Toxicology Laboratory, Department of Entomology, Iowa State University, 115 Insectary Building, Ames, IA 50011
[2]The Institute of Wildlife and Environmental Toxicology, Department of Environmental Toxicology, Clemson University, Pendleton, SC 29670

This research project was conducted to evaluate the use of plants for reducing the environmental impact of aircraft deicers. Significant quantities of ethylene glycol-based deicing fluids spill to the ground and inadvertently contaminate soil and surface water environments. Comparisons of the biodegradation of ^{14}C-ethylene glycol ([^{14}C]EG) in rhizosphere soils from five different plant species, nonvegetated soils, and autoclaved control soils at various temperatures (-10 °C, 0 °C, 20 °C) indicate enhanced mineralization ($^{14}CO_2$ production) in the rhizosphere soils. After 28 days at 0 °C, 60.4%, 49.6%, and 24.4% of applied [^{14}C]EG degraded to $^{14}CO_2$ in the alfalfa (*Medicago sativa*), Kentucky bluegrass (*Poa pratensis*) and nonvegetated soils, respectively. Ethylene glycol mineralization was also enhanced with increased soil temperatures. Our results provide evidence that plants can enhance the degradation of ethylene glycol in soil. Vegetation may be a method for reducing the volume of aircraft deicers in the environment and minimizing offsite movement to surface waters.

Under FAA regulation, deicing agents must be used to remove and prevent ice and frost from accumulating on aircraft and airfield runways. Aviation deicing-fluids used in North America primarily consist of ethylene glycol (EG) and/or propylene glycol (PG) with a minimal amount of additives (*1*). Vast quantities of glycols enter the environment through deicing of aircraft, spills, and improper disposal of used antifreeze. Approximately 43 million L/yr of aircraft deicing products are used nationwide. During severe storms, large planes may require thousands of gallons of deicing-fluid per deicing event (*1*). An estimated 80% of the fluids spill onto the ground, which may lead to the contamination of soil, surface water, and groundwater (*1-3*). Runoff may also be collected in airport storm-sewer systems and directly released (untreated) into streams, rivers, or on-site retention basins (*1,2,4,5*). Airport runoff and storm-sewer discharge have been found to contain concentrations of EG ranging from 70 mg/L to > 5,000 mg/L (*1*). Hartwell et al. (*3*) reported 4,800 mg/L EG in a creek that had received drainage from an airport storage basin. Ethylene glycol has been detected in groundwater at 415 mg/L (*1*) and 2,100 mg/L (*6*). Surface waters contaminated with airport runoff have been shown to be harmful to aquatic communities (*1,3,7*). Fisher and co-workers (*8*) studied the acute impact of airport storm-water discharge on aquatic life and reported a 48-h LC50 of 34.3 and 69.3% effluent for *Pimephales promelas* and *Daphnia magna*, respectively. The primary concern of untreated runoff released into surface waters is the high biological oxygen demand produced by the rapid biodegradation of EG and PG. Even dilute levels of contamination may deplete the available dissolved oxygen, resulting in asphyxiation (*1,2,4,7*). Fish kills have been observed in waters with direct discharge of airport runoff and waste (*1*).

Vegetation can enhance the removal of human-made organic compounds and pollutants in soil environments by microbial degradation in the rhizosphere and plant uptake (*9,10*). The rhizosphere is the region of soil influenced by the roots. Plant roots secrete energy rich exudates and mucilages, which support large and diverse populations of microorganisms (*11-14*). Increased diversity and biomass of microbial communities in the rhizosphere render this zone better for degradation of organic pollutants. Previous research has shown enhanced degradation of industrial chemicals such as trichloroethylene (*15,16*), polycyclic aromatic hydrocarbons (*17*), and petroleum (*18*) in rhizosphere soil compared with root-free soil. In addition to enhanced degradation in the rhizosphere, plants may take up contaminants as part of their transpiration stream (*9*). Vegetation may play a vital role in remediating polluted ecosystems and preventing further contamination by enhancing degradation and uptake into tissues, thereby reducing migration to surface waters and groundwater aquifers.

Previous research has revealed that microbial degradation of EG can occur in both aerobic and anaerobic environments. Several genera of bacteria that utilize EG as a carbon and energy source have been isolated (*19-21*). Only recently has the fate of EG been studied in the soil, despite the widespread use of this compound (*5,22*). McGahey and Bouwer (*22*) studied the biodegradation of EG in simulated subsurface environments, utilizing inocula from soil, groundwater, and wastewater. They concluded that naturally occurring microorganisms were capable of degrading EG and that substrate concentration, soil type, temperature, and quantity of oxygen affect the rate of biodegradation. In addition, Klecka and co-workers (*5*) measured the biodegradation rates of five different aircraft deicing-fluids in soil collected near an airport runway. Rates of degradation for the deicers ranged from 2.3 to 4.5 mg/kg soil per day and 66.3 to 93.3 mg/kg soil per day for samples at -2 °C and 25 °C, respectively.

Recently, there has been interest in reducing the contamination of glycol-based deicing agents in the environment, because of their widespread use and adverse effects on aquatic ecosystems. The purpose of our research was to evaluate the use of plants to enhance the biodegradation of glycols in soil. In addition, we observed the influence of two potential rate-limiting factors (soil temperature and substrate concentration) on the mineralization rate of EG in the rhizosphere and nonvegetated soils.

Materials and Methods

Chemicals. Ethylene glycol (EG) and ethylene glycol-1,2-^{14}C ([^{14}C]EG) were purchased from Fisher Scientific (Fair Lawn, NJ) and Aldrich Chemical Company (Milwaukee, WI).

Upon receipt, the [^{14}C]EG was diluted with ethylene glycol to yield a stock solution of 0.277 μCi/μl.

Soil Collection. Pesticide-free soil was collected from the Iowa State University Agronomy and Agricultural Engineering Farm near Ames, (Boone County) Iowa. Ten golf-cup cutter (10.5 cm x 10 cm, Paraide Products Co.) soil samples were randomly removed from the field and combined for each replicate. Samples were sieved (2.0 mm), placed in polyethylene bags, and stored in the dark at 4 °C until needed. Soils were analyzed by A & L Mid West Laboratories (Omaha, NE) to determine physical and chemical properties. The sandy loam soil had a measured pH of 6.6 and consisted of 54% sand, 29% silt, 17% clay, 3.1% organic matter.

Rhizosphere soils from several different grass and legume plant species were used in this study. Plants were grown from seed for 6 to 8 weeks in pesticide-free soil under the same environmental conditions (25 °C, 14:10 light:dark cycle). The different plant species consisted of tall fescue (*Festuca arundinacea*), perennial rye grass (*Lolium perenne L.*), Kentucky blue grass (*Poa pratensis L.*), alfalfa (*Medicago sativa*), and birdsfoot trefoil (*Lotus corniculatus*). These plants were chosen to represent vegetation that may be found adjacent to airport deicing areas, airport runways, and leguminous plants capable of fixing atmospheric nitrogen. Six different rhizosphere soils were studied. They included rhizosphere soil from each plant species and a mixed rhizosphere from soil that contained the cool season grasses (*F. arundinacea*, *P. pratensis*), a legume (*M. sativa*), and *L. perenne*. Rhizosphere soils were carefully collected from the roots. Soils were sieved (2 mm), placed in a polyethylene bag, and stored in the dark at 4 °C for less than 48 h before they were used in the degradation studies.

Degradation Study: Treatment and Incubation. Portions of the [^{14}C]EG stock solution were diluted with acetone and ethylene glycol to make a 100 μg/g (0.5 μCi/0.004 g), 1,000 μg/g (0.5 μCi/0.04 g), and 10,000 μg/g (0.5 μCi/0.4 g) treating solutions. [^{14}C]EG was applied at a rate of 1,000 μg/g to the rhizosphere, nonvegetated, and autoclaved (autoclaved 3 consecutive d for 1 h) control soils and also a rate of 100 μg/g and 10,000 μg/g to the *M. sativa* rhizosphere and nonvegetated soils. After the acetone evaporated from the soil, four 10- or 20-g (dry weight) subsamples of the treated soils were transferred to individual incubation jars and the soil moistures were adjusted to 1/3 bar (-33 kPa). One sample from each soil treatment was extracted three times with either 30 ml 9:1 (v/v) CH$_3$OH:H$_2$O or 30 ml CH$_3$OH to determine the actual quantity of ^{14}C applied to the soil. The extraction efficiencies ranged from 95% to 103%. The three remaining samples were the three replicates for each soil treatment. A vial containing 3 ml 2.77 M NaOH was suspended in the headspace of each incubation jar to trap ^{14}CO$_2$ evolved from the mineralization of [^{14}C]EG. These traps were replaced every 24 h for the first 3 d, and every 48 h thereafter for the remainder of the study. The quantity of [^{14}C]EG mineralized to ^{14}CO$_2$ was determined by radioassaying subsamples of the NaOH on a RackBeta model 1217 liquid scintillation counter (Pharmacia LKB Biotechnology, Inc., Gaithersburg, MD). Soils were incubated at -10 °C, 0 °C, and 20 °C for 30 d (28 to 30 d).

Mineralization is considered the ultimate degradation of an organic compound. The ^{14}CO$_2$ produced during the mineralization of a radiolabeled substrate can be used to determine the degradation rates of that compound (*23*). Therefore we calculated the mineralization time 50% (MT50), the estimated time required for 50% of the applied

[14C]EG to mineralize, by using formulas previously used for determining degradation rate constants and half-lives (*24,25*). Calculations of MT50s were based on the assumption that the dissipation of ethylene glycol from the soil by mineralization followed first-order kinetics. Linear regressions of the natural log of percentage $^{14}CO_2$ (100% of applied ^{14}C - % $^{14}CO_2$ evolved) vs. time were used to determine the MT50 and coefficients of determination (r^2). Data points used to calculate these values include the quantity of $^{14}CO_2$ produced from the initial treatment of the soil through the log or exponential phase of the mineralization curve (Figure 1). The lag phase was accounted for in the calculations as described by Larson (*25*). Lag time in this study was defined as the number of days before $^{14}CO_2$ exceeded 2% of the applied radiocarbon. The MT50 values compared well with the actual time required for 50% of the applied ^{14}C to mineralize (further discussed in the results). These calculated MT50s were only used to compare the differences between the different soil types at -10 °C, 0 °C, and 20 °C, because oversimplification of the actual mineralization rates may have occurred. Analysis of variance and the least squared means were used to test for significant differences between the different soils at the $p \leq 0.05$ level of significance (*26*).

Soil Extraction and Analyses. At the completion of the study, soils were extracted three times with either 30 ml 9:1 (v/v) $CH_3OH:H_2O$ or 30 ml CH_3OH. The extractable ^{14}C was analyzed on a liquid scintillation counter (Pharmacia LKB Biotechnology, Inc., Gaithersburg, MD). The extracted soils were air dried then crushed and homogenized in a plastic bag. Subsamples of the soils were made into pellets (0.5 g soil and 0.1 g hydrolyzed starch) and combusted in a Packard sample oxidizer (Packard Instrument Co.). The $^{14}CO_2$ produced from the soil combustion was trapped in Permafluor V and Carbo-Sorb E. Spec-Chec ^{14}C standard (9.12 x 10^5 dpm/ml) was used to determine the trapping efficiency. Three to six soil pellets were combusted for each replicate. The soil-bound radiocarbon was quantified by liquid scintillation. The data were statistically analyzed by analysis of variance and least significant differences at 5% (*26*).

Results

Mineralization of [14C]EG in Rhizosphere and Nonvegetated Soils. The mineralization rates of different [14C]EG concentrations in nonvegetated and *M. sativa* rhizosphere soil, incubated at 0 °C, is shown in Figure 2 and Figure 3. An inverse relationship was evident between the concentration of [14C]EG applied to the soils and the percentage of radiocarbon mineralized. Significantly ($p \leq 0.05$) smaller percentages of the applied [14C]EG was transformed to $^{14}CO_2$ as the substrate concentration increased. After 28 days, 55.2%, 20.5%, and 7.14% of applied ^{14}C evolved as $^{14}CO_2$ in the nonvegetated soils treated with 100 μg/g, 1,000 μg/g, and 10,000 μg/g [14C]EG, respectively. Comparison of the data in the nonvegetated soil (Figure 2) and the *M. sativa* rhizosphere soil (Figure 3) indicated significantly ($p \leq 0.05$) enhanced mineralization in the rhizosphere soil. Within 8 days after treatment, the production of $^{14}CO_2$ in the 100 μg/g [14C]EG *M. sativa* rhizosphere soils was elevated by 26% compared with the nonvegetated sample at the same concentration. After 28 days, 62.2%, 49.7% and 21.2% of the added ^{14}C was liberated as $^{14}CO_2$ in the 100 μg/g, 1,000 μg/g, and 10,000 μg/g rhizosphere soils, respectively. Overall, *M. sativa* rhizosphere soils significantly enhanced the mineralization of ethylene glycol by 7% to 29% as compared with the nonvegetated soils with similar [14C]EG

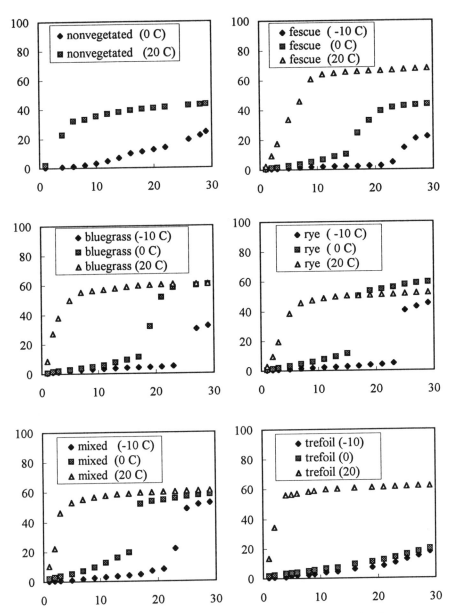

Figure 1.   Mineralization of [^{14}C]ethylene glycol in nonvegetated soils and bluegrass (*P. pratensis*), fescue (*F. arundinacea*), rye (*L. perenne*), trefoil (*L. corniculatus*), and mixed rhizosphere soils at -10 °C, 0 °C, and 20 °C. Mixed rhizosphere soils were collected from soil that contained *M. sativa*, *F. arundinacea*, *L. perenne*, and *P. pratensis*.

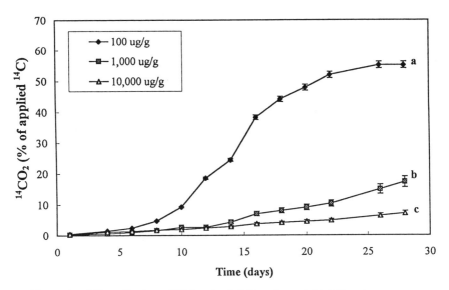

Figure 2. Mineralization of 100 μg/g, 1,000 μg/g, and 10,000 μg/g [^{14}C]ethylene glycol in nonvegetated soil incubated at 0 °C. Data points are the mean of three replicates ± one standard deviation.

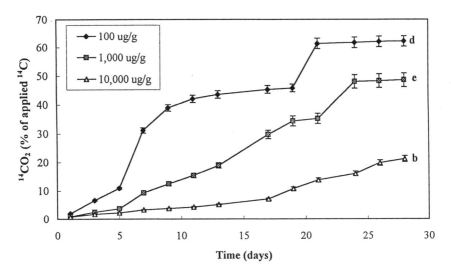

Figure 3. Mineralization of 100 μg/g, 1,000 μg/g, and 10,000 μg/g [^{14}C]ethylene glycol in *M. sativa* rhizosphere soil incubated at 0 °C. Data points are the mean of three replicates ± one standard deviation.

concentrations. Furthermore, the total percentage of applied radiocarbon that evolved as $^{14}CO_2$ from the 1,000 μg/g nonvegetated soils and the 10,000 μg/g $M.$ $sativa$ rhizosphere soils was not significantly different.

The effect of vegetation and temperature on the degradation of [^{14}C]EG in the soil was studied by comparing the mineralization of 1,000 μg/g EG in several rhizosphere soils, nonvegetated soils, and sterile soils, incubated at -10 °C, 0 °C, and 20 °C. Examination of $^{14}CO_2$ produced after 15 days showed significantly greater ($p \leq 0.05$) mineralization of [^{14}C]EG as the temperature increased, except for the sterile soils (Figure 4). A average of 2.7%, 12.2%, and 50.3% of applied radiocarbon was evolved as $^{14}CO_2$ in the $L.$ $perenne$ rhizosphere soils incubated at -10 °C, 0 °C, and 20 °C, respectively. $L.$ $corniculatus$ rhizosphere soil produced the greatest quantity of $^{14}CO_2$ within the initial 15-day incubation period at -10 °C. No significant differences were observed between the $F.$ $arundinacea,$ $L.$ $perenne,$ and $P.$ $pratensis$ and the mixed rhizosphere soils. A comparison of the rhizosphere soils, nonvegetated soils, and sterile soils at 0 °C and 20 °C indicated that the rhizosphere soils significantly enhanced the mineralization of ethylene glycol. After 15 days, the greatest quantity of $^{14}CO_2$ produced at 0 °C occurred in the mixed and $M.$ $sativa$ rhizosphere soils. Over 17.3% and 19.3% of the applied radiocarbon was mineralized in the mixed and $M.$ $sativa$ rhizosphere soils compared with 6.73% in the nonvegetated soils. Significant differences were observed between all the soils studied at 20 °C. The transformation of [^{14}C]EG to $^{14}CO_2$ in descending order was $F.$ $arundinacea$ rhizosphere $> M.$ $sativa$ rhizosphere $>L.$ $corniculatus$ rhizosphere $>P.$ $pratensis$ rhizosphere $> L.$ $perenne$ rhizosphere $>$ mixture rhizosphere $>$nonvegetated $>$sterile soils. After 15 days, 65.5%, 50.3%, 37.9%, and 0.27% of the applied radiocarbon mineralized in the $F.$ $arundinacea,$ $L.$ $perenne,$ nonvegetated, and sterile soils, respectively.

One month (28 d to 30 d) after the application of EG, the different rhizosphere soils continued to enhance the mineralization of [^{14}C]EG by 1.7 to 2.4 times and 1.2 to 1.6 times greater than the nonvegetated soils at 0 °C and 20 °C, respectively (Table I). Our results showed significantly ($p \leq 0.05$) greater quantities of $^{14}CO_2$ evolved in the soils tested at 20 °C compared with -10 °C, with the exception of the mixed rhizosphere soils. A measured 52.9%, 56.8%, and 53.9% of the applied parent compound was mineralized in the -10 °C, 0 °C, and 20 °C mixed rhizosphere soils, respectively. Further examination of the data at 0 °C and 20 °C (Table I) revealed no significant differences between the production of CO_2 at 30 days in the $L.$ $perenne,$ $P.$ $pratensis,$ and mixed rhizosphere soils. After 30 days, the largest quantity of $^{14}CO_2$ that evolved at -10 °C, 0 °C, and 20 °C occurred in the mixed rhizosphere soil, $P.$ $pratensis$ and mixed rhizosphere soils, and the $M.$ $sativa$ and $F.$ $arundinacea$ rhizosphere soils, respectively.

At the completion of the degradation study, the percentage of extractable radiocarbon ranged from 2.40 % to 95.6% (Table I). Significantly greater quantities of extractable ^{14}C was detected in the sterile soil samples compared with the nonvegetated and rhizosphere soils. Over 93% of the applied radiocarbon was detected in the soil extracts of the autoclaved soils incubated at -10 °C and 0 °C. In addition, extractable ^{14}C was significantly ($p \leq 0.05$) more abundant in the nonvegetated soils incubated at 0 °C than the rhizosphere soils. With the exception of $L.$ $perenne$ rhizosphere soil, significantly greater quantities of extractable radiocarbon were detected in the -10 °C soils compared with the 20 °C soils. The extractable radiocarbon was not significantly different between the nonvegetated and rhizosphere soils at 20 °C.

Table I. The effect of vegetation and soil temperature on the degradation of [^{14}C]EG after a 30 d incubation period (reported as percentage of applied ^{14}C)

Soil sample	Temperature (°C)	CO_2[a]	Extractable[a]	Soil-bound residues[a]	Mass balance
Sterile	-10	0.03 A	95.6 A	3.2 AB	98.8
Sterile	0	0.03 A	93.6 A	2.7 A	96.3
Sterile	20	1.7 AB	78.1 B	4.7 B	84.5
Nonvegetated	0	24.4 C	62.8 C	17.5 CD	105
Nonvegetated	20	42.6 D	5.2 D	29.2 E	77.0
M. sativa rhizosphere	0	49.6 EF	3.9 D	34.0 F	87.5
M. sativa rhizosphere	20	71.9 G	4.8 D	26.8 E	104
F. arundinacea rhizosphere	-10	22.2 C	24.8 E	23.3 G	70.3
F. arundinacea rhizosphere	0	43.6 D	5.6 D	22.1 G	71.3
F. arundinacea rhizosphere	20	67.8 G	3.5 D	23.0 G	94.3
L. perenne rhizosphere	-10	45.2 DF	3.8 D	23.5 G	72.5
L. perenne rhizosphere	0	47.1 DFH	3.9 D	17.5 CD	68.5
L. perenne rhizosphere	20	52.4 EHI	3.3 D	18.7 C	74.4
P. pratensis rhizosphere	-10	32.2 J	26.7 E	24.6 G	83.5
P. pratensis rhizosphere	0	60.4 K	4.2 D	23.4 G	88.0
P. pratensis rhizosphere	20	60.7 K	7.5 D	23.1 G	91.3
L. corniculatus rhizosphere	-10	19.5 C	50.2 F	15.5 D	85.2
L. corniculatus rhizosphere	0	20.1 C	42.8 G	12.7 H	75.6
L. corniculatus rhizosphere	20	62.0 K	2.4 D	11.9 H	76.3
mixed rhizosphere[b]	10	52.9 EI	4.0 H	23.3 G	80.2
mixed rhizosphere[b]	0	56.8 IK	3.7 D	19.3 C	79.8
mixed rhizosphere[b]	20	53.9 I	3.0 D	18.0 C	74.9

[a]Means in each column followed by the same letter are not significantly different ($p = 0.05$).
[b]Samples were collected from soil planted with a mixture of *M. sativa, F. arundinacea, L. perenne,* and *P. pratensis.*

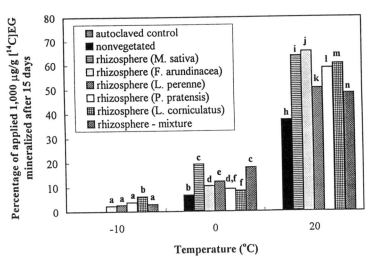

Figure 4. The effects of vegetation and soil temperature on the mineralization of [^{14}C]ethylene glycol after a 15 d incubation period. Each bar is the mean of three replicates. Bars followed by the same letter are not significantly different (p=0.05).

The quantity of soil-bound residues detected in the soil samples, ranged from 2.7% to 34.0% of the applied radiocarbon (Table I). Examination of the data in Table I indicated that the rhizosphere and nonvegetated soils had significantly ($p \leq 0.05$) greater quantities of bound residues than sterile soils.

Calculated MT50 and Mineralization Rate of [^{14}C]EG Mineralization. Ethylene glycol was mineralized at a faster rate in rhizosphere soils than nonvegetated or sterile soils. The MT50s were determined for all the different soil types studied at the various temperatures (Table II). Smaller MT50 values represent faster mineralization rates. The MT50 for [^{14}C]EG in the sterile soils, nonvegetated soils, and *F. arundinacea* rhizosphere soils incubated at 20 °C was 1523 d, 43 d, and 7 d, respectively. Calculated MT50 values compared well with the actual time required for 50% of ethylene glycol to be mineralize in the soil. Approximately 50 % of the ethylene glycol applied to *P. pratensis* and *F. arundinacea* rhizosphere soils at 0 °C and 20 °C was mineralized in 20 d to 21 d and 7 d to 8 d compared with 20 d and 7 d for the calculated MT50s, respectively. Among the soils evaluated at -10 °C, the rate of ethylene glycol mineralization was greatest to least for mixed rhizosphere > *L. perenne* rhizosphere > *P. pratensis* rhizosphere > *L. corniculatus* rhizosphere> *F. arundinacea* rhizosphere > sterile soils. Except for the *L. corniculatus* rhizosphere soils, the MT50s were not significantly different between the rhizosphere soils incubated at 0 °C. Based on the MT50s (Table II), mixed rhizosphere soils mineralized ethylene glycol approximately 1.5 times to 19.7 times faster than the other rhizosphere soils at the same temperature and 1.6 times faster than the nonvegetated soils at 20 °C.

Furthermore, the data (Table II) indicate the MT50s significantly ($p \leq 0.05$) decreased with increased temperatures. The MT50 for *F. arundinacea* rhizosphere soil at -10 °C, 0 °C, and 20 °C were 533 d, 28 d, and 7 d, respectively. Increasing the temperature from -10 °C to 20 °C for *F. arundinacea* rhizosphere soils enhanced the mineralization rate by a factor of 76. Generally, a 15 d to 21 d and a 21 d to 27 d lag phase was observed in the 0 °C and -10 °C soil samples, respectively (Figure 1). Low quantities of $^{14}CO_2$ (<6% of applied ^{14}C) were produced during the lag phase. Nonvegetated and rhizosphere soils incubated at 20 °C showed no lag phase and consistently mineralized >45% of the applied radiocarbon within 9 d after treatment.

Discussion

Results obtained from our investigation indicate that vegetation can enhance the mineralization rate of [^{14}C]EG in the soil. Significantly ($p \leq 0.05$) greater quantities of $^{14}CO_2$ were consistently produced in the *M. sativa, F. arundinacea, L. perenne, P. pratensis, L. corniculatus*, and mixed rhizosphere soils than the amount of $^{14}CO_2$ produced in both the sterile and nonvegetated soils. A comparison of rhizosphere soils and nonvegetated soils showed a two- to four-fold increase in the transformation of [^{14}C]EG to $^{14}CO_2$. The accelerated mineralization rate observed in these soils may be a result of greater microbial biomass and activity generally found in rhizosphere soils (*11-14*). Previous research has shown enhanced biodegradation of industrial chemicals (*16,17*) and pesticides (*27-31*) in rhizosphere soils compared with nonvegetated soils. In addition, microorganisms that utilize ethylene glycol as a carbon and energy source have been previously isolated (*20,21*).

Table II. Calculated MT50s for [^{14}C]ethylene glycol

Soil sample	Temperature (°C)	MT50 (r^2)[a]
Sterile	-10	>10,000 (r^2=0.81) A
Sterile	0	>10,000 (r^2=0.81) A
Sterile	20	1,523 (r^2=0.99) B
Nonvegetated	0	73 (r^2=0.93) C
Nonvegetated	20	43 (r^2=0.70) D
M. sativa rhizosphere	0	26 (r^2=0.96)E
M. sativa rhizosphere	20	6 (r^2=0.91) F
F. arundinacea rhizosphere	-10	533 (r^2=0.50) G
F. arundinacea rhizosphere	0	28 (r^2=0.69) E
F. arundinacea rhizosphere	20	7 (r^2=0.92) F
L. perenne rhizosphere	-10	40 (r^2=0.56) D
L. perenne rhizosphere	0	20 (r^2=0.83) E,H
L. perenne rhizosphere	20	10 (r^2=0.92) F,H
P. pratensis rhizosphere	-10	59 (r^2=0.56) I
P. pratensis rhizosphere	0	20 (r^2=0.80) E,H
P. pratensis rhizosphere	20	9 (r^2=0.96) F
L. corniculatus rhizosphere	-10	107 (r^2=0.91) J
L. corniculatus rhizosphere	0	103 (r^2=0.95) J
L. corniculatus rhizosphere	20	3 (r^2=0.97) F
mixed rhizosphere[b]	-10	27 (r^2=0.71) E
mixed rhizosphere[b]	0	20 (r^2=0.86) E,H
mixed rhizosphere[b]	20	5 (r^2=0.91) F

[a]Means followed by the same letter are not significantly different ($p = 0.05$).
[b]Samples collected from soil planted with a mixture of *M. sativa, F. arundinacea, L. perenne,* and *P. pratensis.*

Results from this study provide strong evidence that mineralization was the predominant factor involved in the dissipation and reduction of ethylene glycol in the soil. Within 30 days, 42.6% to 71.9% of the applied radiocarbon evolved as $^{14}CO_2$ from the biologically active soils (nonvegetated and rhizosphere soils) at 20 °C. Ethylene glycol mineralization at 0 °C in the sterile soils was minimal (0.03%) compared with the nonvegetated (24.4%) and rhizosphere soils ($\geq$43.6%) indicating that transformation of this aircraft deicer was a microbiological process. Several genera of bacteria have been shown to utilize ethylene glycol as a source of carbon and energy for growth (*20,21*). Our results indicate significantly ($p \leq 0.05$) greater quantities of radiocarbon were detected in the soil-bound residues of the biologically active soils compared with the sterile soils. Previous research has shown ethylene glycol does not adsorb to soil (*32*). Lokke (*32*) observed the mobility of ethylene glycol through an anaerobic soil column and reported that very little to no ethylene glycol adsorbed onto the subhorizon of melt water sand, sandy till, and clayey soils. Therefore, we conclude that the increased quantity of ^{14}C soil-bound residues in the biologically active soil was a result of [^{14}C]EG mineralization and, thus, portions of the radiocarbon were incorporated into the cell constituents.

Substrate concentration significantly influenced the mineralization of ethylene glycol in the soil. Our results showed an increase in [^{14}C]EG concentration significantly reduced the percentage of applied radiocarbon that evolved as $^{14}CO_2$ in both the nonvegetated and rhizosphere soils. McGahey and Bouwer (*22*) noted an increase in the time required for 95% of the applied ethylene glycol to be removed from the samples with increased substrate concentrations. Comparisons of the various [^{14}C]EG concentrations in nonvegetated and *M. sativa* rhizosphere soils clearly indicate that the rhizosphere soil significantly enhanced the mineralization of EG compared with the nonvegetated soils.

A positive relationship occurred between the soil temperature and the ethylene glycol mineralization rate. Increasing the temperature from -10 °C to 20 °C in the biologically active soils resulted in enhanced mineralization rates that were approximately 6 to 7 times faster than the rates noted in the -10 °C soils. Klecka et al. (*5*) also noted an increase in the biodegradation rate of ethylene glycol from the soil with increased temperatures. Temperature has been shown to greatly effect the enzyme activity and the growth rate of microorganisms (*33,34*). Generally a 10 °C increase approximately doubles the rate of biological reactions (*13,14,33,34*). Examination of our data also indicates that the biologically active soils had significantly ($p \leq 0.05$) greater mineralization rates at -10 °C than the sterile soils at 20 °C. These results indicate the microbial communities were able to survive and mineralize ethylene glycol at this cold temperature. Microorganisms are capable of growing and metabolizing organic compounds at low temperatures as long as water continues to exist as a liquid (*13,14,23*). The presence of ethylene glycol contamination in the soil may have reduced the freezing point of the water within the soil. Thus, psychrophilic bacteria may have been able to metabolize ethylene glycol at the subzero temperature. Lag phases were observed in the soils incubated at the two cooler temperatures (-10 °C and 0 °C). This may be due to lower enzyme and biological activity at the cooler temperature and, therefore, acclimation time was needed. No lag phase was observed in the soils incubated at 20 °C. Rather a large evolution of $^{14}CO_2$ occurred within the first few days after [^{14}C]EG application. Comparisons of the rhizosphere soils and nonvegetated soils at various temperatures indicate that rhizosphere soils significantly ($p \leq 0.05$) enhanced the mineralization of ethylene glycol in the soil.

Rhizosphere soils of different plant species were studied to determine their effect on the mineralization rate of ethylene glycol. Soils were collected from the root zone of various grasses (*F. arundinacea, L. perenne, P. pratensis*), legumes (*M. sativa, L. corniculatus*), and a mixture of these plant species. The mixed rhizosphere soil had the shortest MT50 of the soils incubated at -10 °C. A comparison of the MT50s in the soils incubated at 0 °C indicates the mixed, *P. pratensis*, and *L. perenne* rhizosphere soils had significantly faster mineralization rates than the *M. sativa* and *F. arundinacea* rhizosphere soils, but they were not significantly different from each other. No particular rhizosphere soil collected from an individual plant species was predominately the most efficient at mineralizing ethylene glycol at all three temperatures. The rate of [^{14}C]EG transformation to $^{14}CO_2$ in the mixed rhizosphere soils was unsurpassed at the cooler temperatures (-10 °C and 0 °C) with the most significant difference noted at -10 °C. Approximately, 7% and 30% more $^{14}CO_2$ was produced at -10 °C in mixed rhizosphere soils compared with other rhizosphere soils from individual plant species. In addition, the mineralization rate of [^{14}C]EG in the mixed rhizosphere soils incubated at -10 °C was 1.6 times faster than the mineralization rate in the nonvegetated soils incubated at 20 °C. These results suggest that a mixed culture of plant species would enhance the degradation of aircraft deicers more than a monoculture. Bachmann and Kinzed (*35*) studied the rhizosphere soils of six different plant species and noted the metabolic activity of the soils were variable depending on the species. The mixed rhizospheres in our study probably had more diverse exudates secreted into the soil from the mixed plant culture than the monocultures. The mixed rhizosphere soils may have contained more diverse and abundant microbial communities that resulted in greater degradation of ethylene glycol at -10 °C.

The enhanced mineralization of ethylene glycol observed in the rhizosphere soils from these studies may be underestimated in comparison with rhizosphere soils in the natural environment. Plant-soil interactions are responsible for maintaining the increased microbial biomass and activity in the rhizosphere soil. Therefore, by removing the soil from the roots, we may have lost some of the beneficial rhizosphere properties by the end of the experiment (*36*). Additional studies are needed that include the intact plant.

Conclusion

Our results provide evidence that vegetation may be an effective method for remediating soils contaminated with aircraft deicing fluids. Rhizosphere soils consistently enhanced the degradation of ethylene glycol compared with the nonvegetated soils, regardless of changes in the soil temperature and substrate concentration. In addition, mixed rhizosphere soils were the most prominent ($p \leq 0.05$) soil type for mineralizing ethylene glycol at subzero temperatures. Therefore, a mixed culture of cold-tolerant plant species could be planted alongside airport deicing areas and runways to help enhance the biodegradation of glycol-based deicers that inadvertently contaminate the soil. Facilitating the biodegradation of these deicers in the soil will reduce the offsite migration and minimize the concentration of glycol-based deicers that reach the surface waters, thus reducing their environmental impact.

Acknowledgments

This research was supported by a grant from the U.S. Air Force Office of Scientific Research. The authors would like to thank Jennifer Anhalt, Karin Tollefson, Brett Nelson, John Ramsey, and Piset Khuon for their technical support. Journal paper J-17175 of the Iowa Agriculture and Home Economics Experiment Station, Ames, Iowa, Project No. 3187, and supported by Hatch Act and State of Iowa funds.

Literature Cited

1. Sills, R. D.; Blakeslee, P. A. In *Chemical Deicers and the Environmen;* D'Itri, F. M., Ed.; Lewis Publishers: Chelsea, MI, 1992, pp 323-340.
2. Evans, W. H.; David, E. J. *Water Res.* 1974, *8*, 97-100.
3. Hartwell, S. I.; Jordahl, D. M.; Evans, J. E.; May, E. B. *Environ. Toxicol. Chem.* 1995, *14*, 1375-1386.
4. Jank, R. D.; Cairns, V. W. *Water Res.* 1974, *8*, 875-880.
5. Klecka, B. M.; Carpenter, C. L.; Landenberger, B. D. *Ecotoxicol. Environ. Saf.* 1993, *25*, 280-295.
6. Flathman, P. E.; Jergers, D. E.; Bottomley, L. S. *Ground Water Monit. Rev.* 1989, winter, 105-119.
7. Pillard, D. A. *Environ. Toxicol. Chem.* 1995, *14*, 311-315.
8. Fisher, D. J.; Knott, M. H.; Turley, S. D.; Turley, B. S.; L. T. Yonkos,; Ziegler, G.P. *Environ. Toxicol. Chem.* 1995, *14*, 1103-1111.
9. Shimp, J. F.; Tracy, J. C; Davis, L. C.; Lee, E.; Huang, W.; Erickson, L. E.; Schnoor, J. L. *Crit. Rev. Environ. Sci. Tech.* 1993, *23*, 41-77.
10. Davis, L. C.; Erickson, L. E.; Lee, E.; Shimp, J. F.; Tracy, J. C. *Environ. Prog.* 1993, *12*, 67-75.
11. Anderson, T. A.; Guthrie, E. A.; Walton, B. T. *Environ. Sci. Technol.* 1993, *27*, 2630-2636.
12. Foster, R. C.; Rovira, A. D.; Cook, T. W. *Ultrastructure of the Root-Soil Interface;* The American Phytopathological Society: St. Paul, MN, 1983; 1-10.
13. Pelczar, M. J.; Chan, E. C. S. Jr.,; Krieg, N. R. *Microbiology;* McGraw-Hill Book Company: New York, NY, 1986; pp 107-549.
14. Paul, E. A.; Clark, F. E. *Soil Microbiology and Biochemistry;* Academic Press, Inc.: San Diego, CA, 1989; pp 26-31.
15. Walton, B. T.; Anderson, T. A. Anderson. *Appl. Environ. Microbiol.* 1990, *56*, 1012-1016.
16. Anderson, T. A.; Walton, B. T. *Environ. Toxicol. Chem.* 1995, *14*, 2041-2047.
17. Aprill, W.; Sims, R. C. *Chemosphere.* 1990, *20*, 253-265.
18. Rasolomanana, J. L.; Baladreau, J. *Rev. Ecol. Biol. Sol.*, *24*, 443-457.
19. Child, J.; Willetts, A. *Biochem. Biophys. Acta* 1977, *538*, 316-327.
20. Wiegant, W. M.; De Bont, J. A. M. *J. Gen. Microbiol.* 1980, *120*, 325-331.
21. Gaston, L. W.; Stadtman, E. R. *J. Bacteriol.* 1963, *8*, 356-362.
22. McGahey, C; Bouwer, E. J. *Water Sci. Technol.* 1992, *26*, 41-49.
23. Atlas, R. M.; Bartha, R. *Microbial Ecology: Fundamentals and Applications;* The Benjamin/Cummings Publishing Co., Inc.: Menlo Park, CA, 1987; pp 233-259.
24. Walker, A. *Rev. Weed. Sci.* 1987, *3*, 1-17.

25. Larson, R. J. In *Environmental Extrapolation of Biotransformation Data: Role of Biodegradation Kinetics in Predicting Environmental Fate;* Maki, K. L.; Dickson, K.L.; Cairns, J., Eds.; Biotransformation and Rate of Chemicals in the Aquatic Environment; American Society for Microbiology: Washington, DC, 1980; pp 67-86.

26. Steel, R. G. D.; Torrie, J. H. *Principles and Procedures of Statistics, A Biometrical Approach;* McGraw-Hill Book Company: New York, NY, 1980; pp 137-191.

27. Reddy, B. R.; Sethunathan, N. *Appl. Environ. Microbiol.* **1983,** *45,* 826-829.

28. Lee, J. K.; Kyung, K. S. *J. Agric. Food. Chem.* **1991,** *39,* 588-591.

29. Hsu, T. S.; Bartha, R. *Appl. Environ. Microbiol.* **1979,** *37,* 36-41.

30. Anderson, T. A.; Kruger, E. L.; Coats, J. R. *Chemosphere.* **1994,** *28,* 1551-1557.

31. Anderson, T. A.; Kruger, E. L.; Coats, J. R. In *Bioremediation of Pollutants in Soil and Water;* Schepart, B. S., Ed.; ASTM: Philadelphia, PA, pp 149-157.

32. Lokke, H. *Water Air Soil Pollut.* **1984,** *22,* 373-387.

33. Alexander, M. *Science.* **1981,** *211,* 132-138.

34. Manahan, S. E. 1992. *Toxicological Chemistry.* Lewis Publishers: Chelsea, MI, 1992; pp 128-168.

35. Bachmann, G.; Kinzel, H. and H. Kinzel. *Soil Biol. Biochem.* **1992,** *24,* 543-552.

36. Walton, B. T.; Guthrie, E. A.; Hoylman, A. M. In *Bioremediation Through Rhizosphere Technolog;* Anderson, T. A.; Coats, J. R., Eds.; ACS Symposium Series; American Chemical Society: Washington DC, 1994, Vol. 563; pp 11-26.

Chapter 13

Phytoremediation of Trichloroethylene with Hybrid Poplars

Milton Gordon[1], Nami Choe[1], Jim Duffy[3], Gorden Ekuan[2], Paul Heilman[2], Indulis Muiznieks[2], Lee Newman[1], Marty Ruszaj[3], B. Brook Shurtleff [1], Stuart Strand[1], and Jodi Wilmoth[1]

[1]Department of Biochemistry, University of Washington, Box 357350, Seattle, WA 98195–7350
[2]Washington State University at Puyallup, Natural Resource Sciences, 7612 Pioneer Way E, Puyallup, WA 98371
[3]Occidental Chemical Corporation, 2801 Long Road, Grand Island, NY 14072

We tested the ability of hybrid poplar to absorb trichloroethylene (TCE) from groundwater. Initial studies used axenic tumor cultures of H11-11 grown in the presence of ^{14}C-TCE. These cells metabolized the TCE to produce trichloroethanol, di- and trichloroacetic acid. Some of the TCE was incorporated into insoluble, non-extractable cell residue, and small amounts were mineralized to ^{14}C-CO_2. Rooted poplar cuttings grown in PVC pipes produced the same metabolites when exposed to TCE. Mass balance studies indicate that the poplars also transpire TCE. In addition we are conducting one of the first controlled field trials for this technology. Trees were planted in cells lined with high density polyethylene and dosed with TCE via an underground water stream during the growing season. Cells containing trees had significantly reduced TCE levels in the effluent water stream compared to control cells containing only soil. These results show that significant TCE uptake and degradation occur in poplars, which bodes well for the future use of poplars for *in situ* remediation of TCE.

Trichloroethylene (TCE) has been widely used since its discovery in the mid-nineteenth century *(1)* for a variety of purposes: as an anesthetic, a dry cleaning agent and, until recently, in enormous quantities as a degreasing agent. While most TCE is of anthropomorphic origin, TCE has recently been found to be produced by marine algae *(2)*. It has contaminated ground water at many localities, and is now the most commonly found volatile organic contaminant at many of these sites *(3)*.

The fact that TCE is a suspected carcinogen *(4)* has lent urgency to clean-up efforts. The stability, persistence and accumulation of TCE in non-miscible underground pools, "DNAPLs", that are not easily found or accessible render attempts at remediation difficult. The principle methods for remediating groundwater contaminated with TCE are pumping water from the aquifer and stripping the TCE by aeration or by charcoal absorption. These procedures can take years or decades and can be very expensive *(5)*. Other techniques utilize bacteria to degrade TCE, although this often requires inducers such as toluene or phenol for degradation to occur *(6)*.

Recent research investigating the use of plants to remediate environmental pollution is yielding exciting results. With some species of plants, such as *Lespedeza cuneata* (Dumont), a legume; *Paspalum notatum* var. *saurae* Parodi, a grass; *Solidago* sp., a composite; and *Pinus taeda* (L.), loblolly pine; the major site of TCE degradation appears to be in the rhizosphere *(7,8)*.

In our laboratory, we have investigated the use of poplar hybrids, especially a clone of *Populus trichocarpa* x *Populus deltoides*, designated H11-11, as a means to remediate TCE pollution. We used poplar for a variety of reasons. The tree has a very wide geographical distribution and can be grown from southern Alaska into Central America. The members of the species can be easily crossed sexually. Propagation by cuttings is simple and, of course, results in clones of a given individual. The species can be grown axenically in culture and exogenous genes can be inserted into the poplar genome by agrobacterial transformation. The College of Forest resources at the University of Washington has worked with poplars for over twenty years, and has accumulated an enormous amount of background data that we can draw upon. The absorption surface of roots in a stand of poplars is enormous and can approach 300,000 km/ha. The water usage under the warm, arid conditions of eastern Washington State is about 140 cm per year in a stand of 5 year old trees at a density of 1,750 trees/ha.

The investigations discussed in this chapter first utilized axenic cultures of H11-11 tumor cells, then small potted plants treated with water containing TCE, and, finally, a number of experimental cells which mimicked field trials. We also determined the fate of TCE in a bioreactor that enabled us to account for the distribution of most of the TCE metabolized by small H11-11 rooted cuttings. We will elaborate upon our results using these procedures.

Axenic tumor cell experiments

The axenic H11-11 tumor cells were produced by transforming shoots of H11-11 with *Agrobacterium tumefaciens* A281. Tumor cells are used because they can be easily grown on simple minimal medium, Murashige and Skoog basal salt medium, in which the only organic components are glycine, sucrose and vitamin B_1 *(9)*. The cells were grown on the above medium at various degrees of saturation of TCE with shaking and illumination for 3 or 5 days at 22°C. The toxicity of TCE towards the H11-11 cells was determined by the vital stain, trypan blue (Figure 1). The results served as a guide for the metabolic studies. The level of TCE used was not toxic to the cells. The cells were incubated with 0.080 gm TCE per Liter. The suspension was then centrifuged and, if needed, the pellet and supernatant stored at -20°C. A sample of the cells was extracted at 20°C with 1N H_2SO_4/10% NaCl, then with methanol and then methyl tertiary butyl ether. In order to test for the presence of di- and trichloroacetic acid, a second batch of cells was extracted with sodium hydroxide, the extracts acidified, back extracted with methyl tertiary butyl ether, and the organic acids in the ether extract esterified with diazomethane. The methyl esters were then analyzed.

The cell cultures were also tested for the formation of $^{14}C\text{-}CO_2$ from TCE. The gases evolved during four or ten day incubations with $^{14}C\text{-}$ TCE were trapped in NaOH and confirmed to be CO_2 by precipitation as $BaCO_3$ and re-emission of CO_2 by acid. The latter procedure ensures that any traces of TCE carried over into the NaOH traps, where it probably forms glyoxylic acid, is not quantified as CO_2. The cells and media were separated by centrifugation and the amount of counts remains in the media determined by direct counting. The cells were extracted with H_2SO_4 and methanol extractable counts were determined. The insoluble cell residue was combusted and analyzed for radioactivity. The results of these analyses are given in the accompanying table (Table I). The products from the TCE metabolism, trichloroethanol, di- and trichloroacetic acid and CO_2, are also among the products produced by the activity of rat and mouse hepatic cytochrome P450 *(10)*. These findings suggest that the oxidative metabolism of TCE in poplars is similar to the processes in mammals.

Whole plant experiments

We next utilized whole plants by growing rooted poplar cuttings in PVC pipes, 20.5 cm diameter, which contained a 30 cm bottom layer of sand with 60 cm of soil overlay. The plants were dosed with water containing 50 ppm TCE or water alone by addition directly to the sand layer through an inner watering tube. A total of .8gm of TCE was added over a period of 8 months. During the period of growth, the plants were assayed to determine whether any TCE was transpired by enclosing the leaves in plastic bags and pulling the atmosphere through a charcoal filter for 0.5 hours. The leaves transpired approximately 1.0µg of TCE/leaf/hour. The results were highly variable and indicated transpiration values from undetectable to about 1.6 micrograms TCE per leaf/hour. After eight months, the plants in PVC pipes were harvested and various morphological measurements were taken (Table II). The major difference noted was that the density of the fine roots in the sand layer of TCE exposed plants was visually less than that of the controls. The plant tissues were analyzed for metabolites of TCE to give the results noted in Table III. The nature and levels of the chlorinated metabolites suggest that the TCE is oxidized as it moves from bottom roots to the upper sections of the crown of the plant. These results, together with the products derived from the axenically cultured poplar cells, strongly argue for a role of the plant in the metabolism of TCE, in addition to the well known degradation of TCE by soil microorganisms.

Mass balance studies. The mass balance of TCE in hybrid poplar trees was determined in a bioreactor. The crown, stem and roots of the plant had to be kept separate so that the transpiration of the compound by each organ could be determined without interference by volatiles from other plant tissues or the soil. The chamber is shown in the accompanying diagram (Figure 2) and is constructed of glass, aluminum foil and inert inorganic materials. Note that the separate chambers for roots, stems and crown are independently aspirated to prevent volatile transpirates from leaking into other chambers. The TCE is added to the bottom chamber to simulate contaminated soil; controls show only, 0.03% leakage of the TCE into the crown chamber.

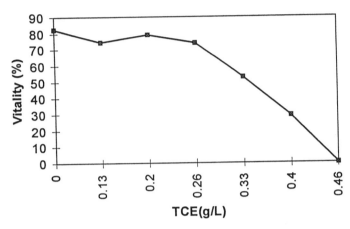

Figure 1. Toxicity of TCE on H11-11 tumor cells. Viability was determined by trypan blue exclusion as only dead cells are stained by this dye. One hundred cells per point are counted. The error per point is about ± 2%

Table I. Concentrations of TCE and metabolites found in supernatant and axenic cells exposed to TCE. Concentrations given are nanograms of TCE or metabolite per gram of sample. ND = not detected at stated limit. Errors are ± 5%.

	TCE	Chloral hydrate	Trichloro-ethanol	Dichloro-acetic acid	Trichloro-acetic acid
Control 1					
pellet	ND40	ND40	ND40	ND10	ND10
supernatant	ND40	ND40	ND40	ND10	ND10
Control 2					
pellet	ND40	ND40	ND40	ND10	ND10
supernatant	ND40	ND40	ND40	ND10	ND10
Exposed - batch 1					
pellet	ND40	ND40	60	12000	ND10
supernatant	2000	ND40	760	1600	ND10
Exposed - batch 2					
pellet	ND40	ND40	80	39000	130
supernatant	ND40	ND40	110	3800	30

Table II. Comparison of TCE treated and control plants. Morphological measurements of plants exposed to TCE were compared to control plants grown under the same greenhouse conditions for 8 months. Except for the clear differences in root distribution, the outward appearance of the treated and control plants was similar. Results given are from duplicate experiments.

Plant Parameter	% of control
Height	80-92
Stem weight	70-72
Leaf area	75-78
Number of leaves	50-80
Root weight	52-70
Length of fine roots	22-32

Table III. Concentration of TCE and metabolites found in tissues of plants exposed to TCE under greenhouse conditions. Concentrations are nanogram of TCE or metabolite per gram of sample. ND = not detected at stated limit. Errors are ± 5% of stated value.

	tissue	clone*	TCE	Chloral hydrate	Trichlo-ethanol	Dichloro acetic acid	Trichloro acetic acid
control 1	leaves	A	ND40	ND40	ND40	ND10	ND10
	stems		ND40	ND40	ND40	ND10	ND10
control 2	leaves	B	ND40	ND40	ND40	ND10	25
	stems		ND40	ND40	ND40	ND10	ND10
control 3	leaves	B	ND40	ND40	ND40	ND10	ND10
	stems		15	ND40	ND40	ND10	ND10
TCE 1	leaves	A	13	ND40	180	ND10	1100
	stems		770	ND40	140	ND10	31
TCE 2	leaves	B	49	ND40	19	180	7200
	stems		1900	ND40	170	ND10	22
TCE 3	leaves	B	27	ND40	24	ND10	2100
	stems		1300	ND40	125	ND10	100
TCE 4		A					
	Roots: upper		13	ND40	200	320	44
	middle		150	ND40	110	25	21
	lower		640	ND40	31	270	44

* clone designations
  A *Populus trichocarpa* x *P. deltoides*, hybrid number H11-11
  B *Populus trichocarpa* x *P. deltoides*, hybrid number 50-189

Small rooted poplar cuttings (ca. 20 cm tall) were planted in a peat moss/ vermiculite mixture in the root chamber and ^{14}C-TCE added to the roots at 5 ppm. After 7 days approximately 0.8% of the added ^{14}C was detected in the transpirate, and a questionable trace was converted to CO_2.

Several conclusions can be drawn from these three laboratory experiments. First, that TCE is absorbed by poplars when present in low concentrations in soil. Second, some of the material is transpired by the plants, and some is metabolized to known extractable metabolites. Finally, some non-extractable material is fixed in tissue; the nature of incorporation in tissues is under investigation. We would predict that the contribution of each of these pathways towards the removal of TCE would depend upon variables such as concentration of TCE, size of tree, type of soil, temperature, light intensity, relative humidity, wind velocity, etc.

Controlled field trials

A controlled field trial was set up in collaboration with the Occidental Chemical Corporation in which we constructed an artificial aquifer. Double walled cells, 3.7 m x 6.1 m x 1.5 meters deep, were constructed of heavy gauge plastic sheeting. The bottoms had about 0.3 m of sand overlaid with 1.1 m of Sultan silt loam. In order to ensure a uniform flow of input water, a "T" shaped input pipe was utilized at the bottom and a 1/40 slope was oriented toward the effluent well. Four cells were planted with 30 cm plants of *Populus trichocarpa* x *P. deltoides*, H11-11. Two of these cells were dosed with water containing 50 ppm TCE while the other two cells received only water. The fifth cell, which had no vegetation, also received water with 50 ppm TCE. Each of the cells received the same volume of liquid; however, variable amounts of liquid were pumped out of the cells to maintain a level of about 20 cm of liquid in the bottom of the cells.

About 9 weeks after injection of TCE into the cells, breakthrough of TCE and related metabolites, particularly *cis*-1,2-dichloroethylene, occurred in the cells which did not contain trees. The occurrence of this compound is indicative of anaerobic dehalogenation of TCE. In cells with trees, very little TCE or metabolites could be detected in the effluent until after leaf drop (Figure 3). The stems, roots and leaves of the trees exposed to TCE showed the above products of oxidative metabolism, which were also seen in the greenhouse studies and tissue culture experiments. The one-year-old trees (ca. 3.5 meters tall) were very effective in removing TCE from the input water. These experiments are being continued.

Conclusion

The results of the above experiments show that TCE has multiple fates in poplars. The trees transpire unaltered TCE. The compound also can undergo oxidation to chloromethyl derivatives, trichloroethanol, dichloroacetic acid, trichloroacetic acid, or complete mineralization to CO_2 which are oxidative metabolites found in mammalian livers. Variable amounts of TCE are found fixed in insoluble, nonextractable residues. Plant uptake coupled with TCE metabolism serves to remove TCE from groundwater and from the soil environment. The details of these reactions and methods for increasing the mass flow through these pathways are subjects of current investigations.

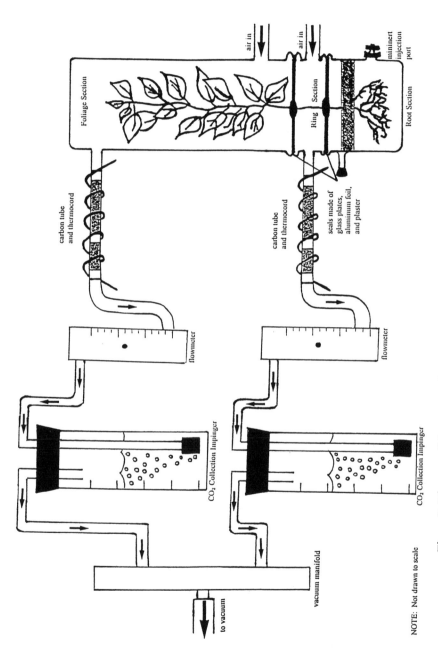

Figure 2. Diagram of bioreactor. The temperature of the carbon tubes was left at 40°C to prevent condensation of water.

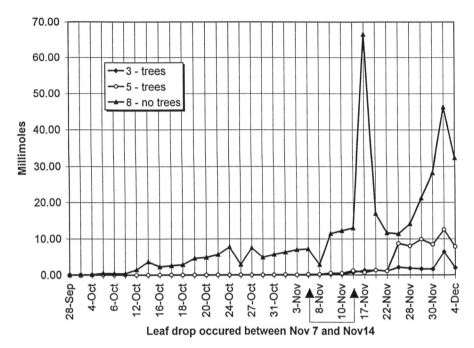

Leaf drop occured between Nov 7 and Nov14

Figure 3. Micromoles of TCE and metabolites recovered in effluent water 28 September through 4 December 1995. Cells 3 and 5 contained trees. Cell 8 is the non-vegetation control. The spikes on Nov. 17 and Dec. 1 are due to flooding during periods of very heavy rainfall.

These experiments show that actively metabolizing poplars are able to intercept a moving plume of TCE contaminated water and reduce the levels of this compound significantly. The system shows promise where there is sufficient space to plant trees and when the roots can reach the contaminated region.

Acknowledgments

This work was generously supported by grants from Occidental Chemical Corporation, the U. S. Environmental Protection Agency 10#R822329-01-3 and the U. S. National Institute of Environmental Health and Safety Superfund Grant # 2P42 ES04696-09

References

1. Defalque, R. J. *Clin. Pharmacol. Ther.*, **1961**, *2*, 665-688.
2. Abrahamson, K.; Ekdahl, A.; Collén, J.; Pedersén, M. *Limnol Oceanographic*, **1995**, *40*, 1321-1326.
3. Rajagopal, R. *Environ. Prof.*, **1986**, *8*, 244-264.
4. National Cancer Institute. Carcinogenesis bioassay of trichloroethylene., **1976**, CAS no. 76-01-6. U. S. Department of Health, Education and Welfare publication (NIH) 76-802.
5. Travis, C. C.; Doty, C. B. *Environ. Sci. and Technol.*, **1990**, *24*, 1464-1466.

6.   Hopkins, D.; Munakan, J.; Semprini, L.; McCarty, P. L. *Environ. Sci Technol.*, **1993**, *27*, 2542-2547.
7.   Anderson, T. A.; Walton, B. T. *Environ. Toxicol. and Chem.* **1995**, *14:12*, 2041-2047.
8.   Anderson, T. A.; Guthrie, E. A.; Walton, B. T. *Environ. Sci. and Technol.* **1993**, *27:13*, 2630-2636.
9.   Murashige, T.; Skoog. F. *Physiol. Plant.*, **1962**, *15*, 473-497.
10. Dekant, W. *New Concepts and Developments in Toxicology.*; Elsevier Science Publishers B. V. Biomedical Division.: Amsterdam, The Netherlands, **1986**, P. L. Chambes, P. Gehring and F. Sakai, eds. Metabolic conversion of tri- and tetrachloroethylene: formation and deactivation of genotoxic intermediates, pp. 211-221.

Chapter 14

Field Study: Grass Remediation for Clay Soil Contaminated with Polycyclic Aromatic Hydrocarbons

Xiujin Qiu[1], Thomas W. Leland[1], Sunil I. Shah[1], Darwin L. Sorensen[2], and Ernest W. Kendall[3]

[1]Central Research and Engineering Technology, Union Carbide Corporation, P.O. Box 8361, South Charleston, WV 25303
[2]Utah Water Research Laboratory, Utah State University, Logan, UT 84322–8200
[3]Union Carbide Remediation, Union Carbide Corporation, Houston, TX 77058

A three-year field-pilot study has demonstrated that Prairie Buffalo-grass (*Buchloe dactyloides* var. 'Prairie') has accelerated naphthalene concentration reduction in clay soil. Extensive soil sampling and analyses have been performed to evaluate statistically significant differences between soil PAH concentrations with and without grasses. Comparative performance evaluation for other PAH compounds was restricted by analytical variability. A parallel experiment to assess the performances of twelve, warm season grass species from different genetic origins has shown that Kleingrass (*Panicum coloratum* var. 'Verde') has a greater potential to remove PAHs from rhizosphere soil than other tested grasses at the site soil. Preliminary data indicated that Kleingrass root zone soil concentrations of both low and high molecular weight PAHs were approximately one order of magnitude lower than those with Prairie Buffalograss. Grass tissue analysis found no evidence of PAH bioconcentration. Food chain effects should not be a concern for phytoremediation of aged PAH-contaminated soils. Additional soil sampling is planned in the future to confirm the findings.

PAH-contaminated clay soils are frequently observed at petrochemical manufacturing sites. Engineered remedial technologies based on excavation and subsequently thermal or chemical treatment often entail high cost and a greater health risk by transferring contaminants to atmosphere. Applications of in-situ biotechnologies to date are mostly limited to the treatment of water-soluble contaminants associated with high permeability soils. Hydrophobic PAHs sorbed onto clay soils are difficult to treat, because of mass transfer restrictions. Union Carbide intends to develop a low cost and low risk technology using vegetation. Several laboratory studies have been conducted since 1989 in collaboration with Utah State University. The results were promising (*1,2,3,4*). A grass remediation field-pilot study was started in 1992 to evaluate grass-rhizosphere effects on PAH degradation under field conditions and to identify competent grass species. A previous publication in 1993 presented field study methodology and preliminary results (*5*). This paper updates the results after three years of monitoring.

Background

PAH degradation in soil and sediment is slow, although a wide variety of pure cultures of bacteria, fungi and algae and their purified enzymes have the ability to metabolize PAHs (6,7,8). Microbial activity is constrained in the natural environment by the availability of nutrients and electron acceptors. (9). PAH degradation rates are further restricted by the desorption kinetics, the population of PAH-degrading microorganisms, and the competing reaction for electron acceptor utilization (10,11,12). Microorganisms typically degrade PAHs aerobically by incorporating oxygen atoms into the ring structure generating dihydrodiols via oxygenases. The derivative is further mineralized through aromatic ring cleavage and subsequent oxidation (6). The biodegradation rates of PAHs are significantly higher under oxic conditions than those under anoxic ones. Low-molecular-weight PAHs (with two or three condensed benzene rings) are amenable to microbial degradation under aerobic and denitrifying environment. However, negligible biodegradation occurs under the sulfate-reducing or methanogenic environments (13,14,15). According to thermodynamic calculations high-molecular-weight PAHs (with four or more condensed benzene rings) are reported theoretically nondegradable at low redox potentials (16). Moreover, high-molecular-weight PAHs do not serve as substrates for microbial growth, though they may be subject to cometabolic transformation (17,18,19).

Heterotrophic microorganisms are responsible for the catabolic degradation of organic chemicals in nature, however, many are unable to degrade PAHs. Total heterotrophic population does not indicate the potential for PAH degradation by microorganisms. Prolonged exposure to chemical toxicants can cause adaptations in microbial populations that result in greater resistance to toxicity or enhanced ability to utilize toxicants as substrates for metabolism or co-metabolism. Studies have shown microbial adaptations and increased PAH degradation in response to chronic PAH exposure, but the actual mechanisms of these adaptations are not known (19,20).

In situ bioremediation of PAH-contaminated clay soil is a challenge. The high adsorption capacity of clay soils may limit the amount of PAHs available to microorganisms. The low flux of nutrients and electron acceptors through low permeability clay soil may also restrict microbial activities. An engineered process may accelerate biodegradation, however a system of delivering electron acceptors, substrates, nutrients, and enzymes to numerous microsites would be technically and economically unfeasible. Nevertheless, plant roots are known to perform suchlike functions.

The rhizosphere provides a complex and dynamic microenvironment, where microbial communities associated with plant roots, have great potential for detoxification of organic contaminants. Deep fibrous root system of certain grasses may improve aeration in soil by removing water through transpiration and by altering clay soil structure through agglomeration. Plant release photosynthate to soil through exudation and sloughing of dead root cells increases soil organic matter. Root exudates (carbohydrates, amino acids, etc.) sustain a dense microbial community in the rhizosphere, which may enhance degradation, mineralization, and/or polymerization of organic toxicants (21). The consortium of bacteria and fungi associated with the rhizosphere may possess highly versatile metabolic capabilities (21). Certain root exudates may support microbial cometabolism of high molecular weight PAHs, which bacteria and fungi cannot use as a sole carbon source (1). In addition, the greater soil organic content in rhizosphere soil may alter toxicant sorption, bioavailability, and leachability (22).

Plants use a variety of reactions to degrade complex aromatic structures to more simple derivatives. Aromatic ring-cleavage reactions in plant tissues, which lead to complete catabolism of the aromatic nuclei to carbon dioxide, have been reported (23). Benzo[a]pyrene, a five-ring PAH, can be metabolized to oxygenated derivatives in plant tissues (24). Although some of these derivatives are known to be more toxic than the original compounds, they appear to be polymerized into the insoluble plant lignin fraction, which may be an important mechanism for their detoxification. With plant seedlings, Benzo[a]pyrene was assimilated into organic acids including amino acids (25). Complete degradation of Benzo[a]pyrene to carbon dioxide was also observed with a wide range of plants (25).

A well-established grass cover conserves soil and substantially reduces the risk of exposure to soil contaminants. PAHs associated with soil generally do not pose a risk, unless soil particles are transported to a receptor via direct ingestion, inhalation, or dermal contact. Owing to their hydrophobic nature, PAHs have poor mobility. High-molecular-weight PAHs are virtually immobile with water flux. Although PAH degradation, mineralization, and polymerization in soil may take considerable time, grasses reduce the risk to human health and the environment immediately after a dense grass ground cover is developed. Grasses also significantly reduce potential contaminant migration via surface runoff and any infiltration to groundwater.

Risk may arise from plant bioconcentration of carcenogenic PAHs via food chains due to animals grazing grasses. Many plant roots and shoots are capable of bioconcentrating soil-borne chemicals such as pesticides and petroleum compounds. Plant concentration of PAHs has been reported (25,26,27,28,29). However, PAHs are mostly accumulated in the roots and not translocated to plant shoots because of their hydrophobic nature (26,29).

In summary, PAH degradation in clay soil may be limited by low redox potential, desorption kinetics, lack of responsible microbial population, and lack of nutrients. A plant root zone system may facilitate PAH biodegradation. Planting grasses substantially reduces the risk of exposure to contaminated soils and may ultimately transform PAHs to innocuous products (CO_2 and H_2O). PAHs may be bioconcentrated to plant roots, however, are not translocated to plant shoots. Therefore, animal grazing grasses should not cause subsequent risk through food webs.

Methods

Several laboratory studies had been conducted prior to this field-pilot study. PAH disappearance was enhanced in the presence of prairie grasses and cow manure in a screening study(1). Subsequent soil microcosm studies using radiolabeled PAH compounds showed that a variety of native grasses enhanced PAH mineralization and incorporation into soil organic fraction (2,3,4).

To evaluate the grass effects under field conditions we initiated a field study at Union Carbide Seadrift Plant in Texas. Three plots were constructed at the olefins' production area between the old foundations of a demolished natural gas compressor building. Plot 1 (10 ft x 20 ft) was an unvegetated control. Plot 2 (10 ft x 20 ft) was sodded with Prairie Buffalograss. These two main plots were designed to evaluate the grass rhizosphere effects on PAH degradation in clay soil by periodic sampling and analysis. Plot 3 (10 ft x 90 ft) was designed to screen the performance of a variety of grasses from different genetic origin mainly by visual observation.

Unlike the laboratory studies, PAH concentrations in field were highly variable. Sufficient number of samples must be taken to accurately estimate the statistical significance of vegetation effects. A systematic randomized sampling strategy was used to reduce sample variance (5). Field PAH concentrations displayed a lognormal distribution. Geometric mean with its corresponding confidence limit is the appropriate estimate of central tendency. Log concentration data were used throughout the statistical analysis using JMP® computer software (SAS Institute, Inc,). Detailed field-study methodology can be found in Qiu, 1994 (5).

Experimental design. A three-way factorial design model had been proposed in our earlier publication to test the statistically significant differences in soil PAH concentrations for the effects of vegetation treatment, soil depth, and treatment time. This test model was rather inappropriate, because of serious systematic variability in the field experiments. The experiments were not exclusively comparative with regard to sampling depths, plots, and sampling times. First, PAH concentrations were correlated with soil depths before treatment. Concentrations in surface soil (0 - 1 ft) were significantly higher than those in subsurface soil (>1 ft) (*31*). Because of the nonrandom vertical concentration distribution, the surface and subsurface soil should be considered as subgroups in a stratified population. In addition, the vegetation treatment effects on surface soil were essentially different from those on the subsurface soil. During the three years of treatment, the grass roots penetrated less than one foot deep. No rhizosphere had developed in the subsurface soil, but the grass root exudates leaching down to the subsurface soil may have altered PAH sorption, bioavailability, and leachability. Secondly, systematic variation among the six sampling and analysis events was possible due to differences in analytical procedures, sampling temperatures, and sample holding time.

To eliminate the known sources of discrepancies, a nested design model was used to separate the total variation into parts assignable to various sources, i. e., vegetated versus control plots, soil depths, and treatment times. Table I shows a 2 x 7 x 2 nested design with the three sources of variation and ten replicates. The test soil was divided into two depths, surface (0 - 1 ft) and subsurface (1 - 4 ft). Each depth is to be independently sampled seven times (six rounds completed so far). For each sampling time, ten replicate samples have been taken from each of the two plots, vegetated and unvegetated control at each depth, yielding a total of 40 samples per event. Analysis of variance (ANOVA) estimated three separate component variances: the variance from surface to subsurface soils, the variance from between treatment periods, and the variance from vegetated to unvegetated control plots. Statistical significance was tested for the effects of vegetation treatment, treatment time, and soil depth.

Table I.   A 2 x 7 x 2 Nested Experimental Design with Ten Replicates

Source of Variation																												
Depth	Surface Soil (0 - 1 ft)												Subsurface Sol(1 - 4 ft)															
Treatment Time (days)	0		162		301		462		757		1045			0		162		301		462		757		1045		
Plot	C	V	C	V	C	V	C	V	C	V	C	V	C	V	C	V	C	V	C	V	C	V	C	V	C	V	C	V
Number of Replicate	10	10	10	10	10	10	10	10	10	10	10	10	10	10	10	10	10	10	10	10	10	10	10	10	10	10	10	10

(where, C = unvegetated control plot and V = Buffalograss-vegetated plot)

Sample Analysis. The field soil samples were extracted by Soxhlet extraction and the extracts were analyzed for PAHs by GC/MS (gas chromatography/mass spectrophotometer) for the four earlier sample sets. We were unable to detect high molecular weight PAHs (HMW PAHs) and several low molecular weight PAHs (LMW PAHs) by GC/MS. The machine used had relatively high quantitation limits (QL), ranging from 5 to 7 mg/kg-soil. The HMW PAH and several LMW PAH concentrations in the site soils ranged from approximately 0.1 to 5 mg/kg-soil. To obtain data necessary we implemented an analytical method to lower QL for PAHs to below the PAH cleanup standards. The method uses HPLC with UV and fluorescence detector in series (a modification of EPA SW846 method 8310) achieving a QL of <0.5 mg/kg-soil.

Using HPLC/UV/FD all PAHs were detected in the later two rounds of samples (9/14/94 and 6/30/95).

Analytical errors are often less significant than sampling errors. However, the former can be serious when the measurements are close to the QL. Quality control analysis indicated that naphthalene data measured by HPLC were consistent with those measured by GC/MS, while the other PAH data (acenaphthene, acenaphthylene, phenanthrene, and anthracene) were not. Naphthalene concentration range (>100 mg/kg) was significantly higher than those of other PAHs (<5-10 mg/kg), resulting in relatively low analytical errors. In this report, data analysis for PAHs other than naphthalene are limited to the last two sample sets, measured by HPLC.

Soil texture, physicochemical properties, and nutrient status were analyzed before treatment. The soil has a high clay content (51% - 61%) and contains 2% of soil organic matter. Its nutrient status and pH were suitable for vegetation (5). Common agricultural fertilizer (13-13-13) was applied once a year at a rate of 1 lb-N/1000 ft^2. Grass plots (except tall grasses) were mowed as needed. Regular irrigation was applied during the first month after planting to ensure root establishment. Later, the test plots were watered only during the extremely dry season. Relatively dry soil is desired for deep rooting.

Results and Discussion

Rhizosphere Effects on Naphthalene. Prairie Buffalograss accelerated naphthalene removal in soil. Naphthalene concentration in vegetated soil was significantly lower than that in the unvegetated soil after two years of treatment. Figures 1 and 2 contain the mean log naphthalene concentrations over time in surface and subsurface soils for both vegetated and unvegetated plots, respectively. Each data point is the mean log concentration of ten discrete soil samples. The vertical error bars are the 95% confidence intervals. ANOVA showed that for surface soil, the naphthalene concentrations in the vegetated plot was significantly lower than that in the unvegetated control plot on days 301, 757, and 1045 at >95% confidence level, while they were insignificant on Days 162 and 462 (*31*). For subsurface soil, there were no significant differences in the naphthalene concentrations at 95% confidence level between vegetated and control plots during the course of treatment.

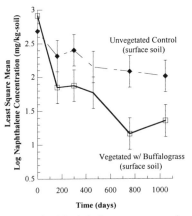

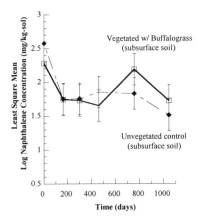

Figure 1. Naphthalene concentrations in surface soil over time

Figure 2. Naphthalene concentrations in subsurface soil over time

The statistical analysis excludes time zero concentration data, because we did not have good quality control in chemical analysis at time zero. Statistical analysis indicated an enhanced naphthalene concentration reduction in rhizosphere soil. However, overlying vegetation had no significant effects on the subsurface soil where grass roots had not yet developed, perhaps due to extremely wet condition.

The field naphthalene concentration data did not fit a typical first order degradation rate model, therefore the correspondent degradation rate can not be derived. However, ANOVA showed that naphthalene concentration in surface soil significantly decreased over time for both vegetated and unvegetated control plots (*31*). A slow dissipation of naphthalene in the control soil indicated a potential intrinsic degradation of naphthalene in the site soil.

Rhizosphere Effects on Other PAH Compounds. Comparative performance evaluation for other PAH compounds was restricted by analytical variability. Soil PAH data (last two sets analyzed by HPLC) show no significant changes in PAH concentrations between Days 757 (Sept. 1994) and 1045 (Jun. 1995). Unfortunately, the GC/MS used in analyzing the earlier four sets of samples failed to accurately quantify PAHs other than naphthalene, because the concentration levels were either below or marginally higher than the quantitation limit.

Figures 3 and 4 show LMW and HMW PAH concentrations in soil for the Buffalograss plot and the control plot, respectively. Table II presents whether the differences in PAH concentrations between Buffalograss and control plots were statistically significant or not. Interestingly, all PAHs concentrations in subsurface soil were higher in vegetated plot than those in the control plot. So were most PAHs in surface soil, except for four LMW PAHs, despite that the differences were mostly insignificant at 95% confidence level in the surface soil, except for the seven heavier HMW PAHs in 1994. Conversely, the differences were mostly significant in the subsurface soil, except for the five LMW PAHs in year 1995. Most LMW PAH (except fluoranthene) concentrations in vegetated surface soil were lower than those in control surface soil, although this difference was only significant for naphthalene. We do not know whether these results were simply due to sampling and analysis variability or caused by complex physicochemical and biological processes in the test plots. However, similar pattern was also observed in the surface soil for several other grasses as discussed later.

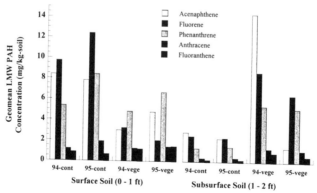

Figure 3. Comparison of low-molecular-weight PAH concentrations in soil (94 = samples taken on day 757 in 1994; 95 = samples taken on day 1045 in 1995; cont = unvegetated control plot; and vege = Buffalograss-vegetated plot)

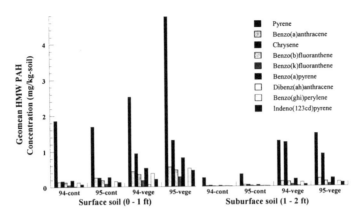

Figure 4. Comparison of high-molecular-weight PAH concentrations in soil (94 = samples taken on day 757 in 1994; 95 = samples taken on day 1045 in 1995; cont = unvegetated control plot; and vege = Buffalograss-vegetated plot.)

TABLE II.   Statistical Significant Differences in Soil PAH Concentrations

Sample Depth	Surface (0 - 1 ft)		Subsurface (1 - 2 ft)	
Sample Time	1994	1995	1994	1995
Compound	Soil concentrations significantly different between Buffalograss-vegetated and unvegetated control plots?			
	Control	Vege	Control	Vege	Control	Vege	Control	Vege
Naphthalene	High	Low	High	Low	No	No	No	No
Acenaphthene	No	No	No	No	Low	High	No	No
Fluorene	No	No	No	No	Low	High	No	No
Phenanthrene	No	No	No	No	Low	High	No	No
Anthracene	No	No	No	No	Low	High	No	No
Fluoranthene	No	No	No	No	Low	High	No	No
Pyrene	No	No	No	No	Low	High	Low	High
Benzo[a]anthracene	No	No	No	No	Low	High	Low	High
Chrysene	Low	High	No	No	Low	High	Low	High
Benzo[b]fluoranthene	Low	High	No	No	Low	High	Low	High
Benzo[k]fluoranthene	Low	High	No	No	Low	High	Low	High
Benzo[a]pyrene	Low	High	No	No	Low	High	Low	High
Dibenz[a,h]anthracene	Low	High	No	No	Low	High	Low	High
Benzo[g,h,i]perylene	Low	High	No	No	Low	High	Low	High
Indeno[1,2,3,cd]pyrene	Low	High	No	No	Low	High	Low	High

Where, control = unvegetated control plot; vege = Buffalograss-vegetated plot; No = statistically insignificant between unvegetated-control and Buffalograss-vegetated plots; Low = significantly low; and High = significantly high.

     If the results were not simply owing to the sampling and analysis variability, then why were soil PAH concentrations in the vegetated plot in most cases higher than those in unvegetated control plot? Could it be caused by the differences in solvent extractability of PAHs incorporated in soil matrix? The presence of organic acids and phenols released from grass roots may increase the apparent solubility of hydrophobic compounds. Over time hydrophobic PAHs may be more extractable and bioavailable. While the LMW PAHs could have been degraded, HMW PAH degradation could be more slowly or could have been recalcitrant under the likely anoxic conditions.

Mihelcic et al. reported that acenaphthene and naphthalene in soil slurry biodegraded under denitrifying conditions (*10*). Leduc et al. reported that biodegradation of acenaphthene, acenaphthylene, fluorene, and anthracene occurred under aerobic and denitrifying environment, however, it was insignificant under the sulfate-reducing or methanogenic environments. Degradation of HMW PAHs under anoxic conditions has not been reported. Soil redox potentials were not measured during the study, however, near saturated conditions were known to exist very often as a result of poor site drainage and intermittent flooding. While the surface soil may be aerated occasionally, the subsurface soil may have been constantly under anaerobic conditions. The latter was exasperated by arising capillary fringe of the shallow ground water table. Even LMW PAHs would be recalcitrant under sulfate reducing or methanogenic environment.

Annual fertilization has supplied a large amount of nitrate, which can be used as electron acceptor under denitrifying conditions. Fertilization could have stimulated LMW PAH degradation in the test plot. Mihelcic et al. reported that the significance of PAH degradation under denitrifying condition in the field depends on the availability of nitrate, as well as on the ratio of PAH compound to the mineralizable organic carbon content, typically less than 1.4% of the total natural soil organic carbon (*32*). The latter competes for nitrate utilization. Therefore, the effects of root exudation may be complicated.

Furthermore, the increases of soil organic content in vegetated plot may cause more PAHs partitioning onto soil organic matter and reduce their water solubility. Manilal and Alexander found that the rate of phenanthrene mineralization was slower in an organic soil than that in a mineral soils with lower organic content (*9*). Mihelcic et al. and Al-Bashir et al. reported that the sorption-desorption interactions with soil may limit the amount of total PAH compound available to the microorganisms and control the rate of mineralization (*10,12,13*).

Further study is necessary to determine the vegetation effects on PAH-contaminated soils, especially for carcinogenic HMW PAHs. The rhizosphere effects on PAH removal, including complex interactions among sorption, degradation, hydrolysis, mineralization, and humification, are not well understood. Other research needed includes establishing environmentally acceptable cleanup standards for phytoremediation, based on potential risks to human health and the environment. This study suggested that PAHs incorporated in aged contaminated soil may not be completely extractable by organic solvents, meanwhile only a fraction of the organic solvent extractable PAHs is water soluble, presumably bioavailable. The fraction may be a function of soil physicochemical and biological characteristics. The organic solvent-extractable PAH concentrations in aged contaminated soil do not measure a real risk level to human health and the environment. Analyzing site-specific soil characteristics is particularly important, when conducting phytoremediation, risk assessment, and determining cleanup goals.

Buffalograss Root Development. High soil moisture content suppressed Buffalograss root development. Buffalograss is drought tolerant, but suffers in wet soil. Figure 5 shows the Buffalograss root zone depth over time. Grass roots developed mostly during the second year, when the weather was unusually dry. During the first and the third years, repeated abnormal heavy precipitation and the poor drainage condition caused intermittent water logging in the test plot, which severely inhibited grass root development. The site is surrounded by above ground cooling water basins, resulting in high groundwater table. The local Lake Charles clay soil has very low permeability. Soil moisture content analysis indicated that the test plot soil may have been seasonally or permanently saturated, excluding air from the soil pores and resulting in anaerobic conditions. Grass roots were reluctant to extend downward into anaerobic soil environment. A parallel green house study demonstrated that PAH removal was enhance due to small decreases in moisture content (small increase in air-filled porosity) in soil near saturation (*33*).

Managing soil moisture content at optimal levels is important for biodegradation and plant growth. A drainage system controlling soil saturation level would be desirable where economically feasible. Selectively growing water-tolerant plants, capable of removing excessive soil moisture and maintaining appropriate soil redox potential may be a low-cost alternative.

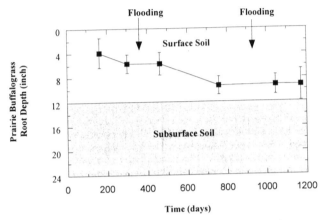

Figure 5. Prairie Buffalograss root depths over time

Identifying Competent Grass Species.

Twelve warm season plant species, naturally adapted to the coastal areas of Texas, from different genetic origins are being tested at Plot 3. A competent grass species must possess anatomical, biochemical, and physiological adaptations to the site-specific environmental condition. The important characteristic is a deep, dense root zone system to improve soil aeration condition and selectively foster PAH-degrading bacteria growth. Table III lists their names and application forms. Most of the species selected are short grasses along with three tall grasses. Also we included two herbaceous plants, natural flowers, to examine the potential of plants other than grasses. Switchgrass was planted in the second study year replacing Bluegramagrass. The latter did not grow. The two natural flowers had germinated, but died soon after, perhaps the soil was too wet.

Experimental Design and Methods. A randomized complete block experiment was designed to assess the performance of the twelve grass species. Plot 3 is divided into three blocks. Each block contains twelve experimental units. Each unit has a size of 5 ft x 5 ft. The twelve species were randomly located on the twelve units within a block. Each grass has triplicate experimental units located in the three blocks, respectively. Periodic visual observations have been conducted unit by unit to evaluate: 1) rate of seed germination; 2) percentage survival of seedlings or solid sodding; 3) grass density and height; and 4) extent of root zone establishment. Appraisal is based on a nine-grade grading criteria. Root zone soils were sampled and analyzed for PAHs after three years.

TABLE III. Twelve Plant Species Tested on Plot 3

No	Name	Type	Application Form
1	Seashore Paspalum *Paspalum vaginatum* var. 'Adalayd'	Short grass	Sod
2	Buffalograss *Buchloe dactyloides* var. 'Prairie'	Short grass	Sod
3	Bermudagrass *Cynodon dactylon* var. 'Texturf 10'	Short grass	Sod
4	Zoysiagrass *Zoysia japonica* var. 'Meyer'	Short grass	Sod
5	St. Augustinegrass *Stenotaphrum secundatum* var. 'Raleigh'	Short grass	Sod
6	K.R. (King Ranch) Bluestem *Bothriochioa ischaemum*	Short grass	Seed
7	Common Buffalograss *Buchloe dactyloides*	Short grass	Seed
8	Weeping Lovegrass *Eragrostis curvula*	Tall grass	Seed
9	Kleingrass *Panicum coloratum* var. 'Verde'	Tall grass	Seed
10	Switchgrass *Panicum virgatum* var. 'Alamo'	Tall grass	Seed
11	Texas Bluebonnet *Lupinus texensis*	Herbaceous	Seed
12	Winecup *Callirhoe involucrata*	Herbaceous	Seed

Performance Evaluation. Performances of the twelve grasses were evaluated based on visual appearance, root zone development, and root zone soil PAH concentration level.

Visual Appearance. Statistical analysis of the visual appearance grading data showed that the top five grasses above average are (1) Verde Klein, (2) Zoysia (3) Bermuda, (4) Prairie Buffalo, and (5) Common Buffalo. Verde Kleingrass was significantly better than all but Zoyzia and Bermuda, while the latter two were not significantly better than Praire Buffalo or Common Buffalo. There were no significant differences in visual appearances between tall and short grasses, between seed and sod forming grasses, and among the three blocks. Verde Kleingrass appears the most healthy, dense, and the best ground coverage.

Root zone development. After 3-year growth the Verde Kleingrass and the Common Buffalograss developed deeper roots than the others (Figure 6). Zoysiagrass and Bermudagrass displayed relatively dense but short roots. The vertical error bars are standard deviations among the triplicate experimental units. The root zones of other grasses were either poorly developed or never developed at all.

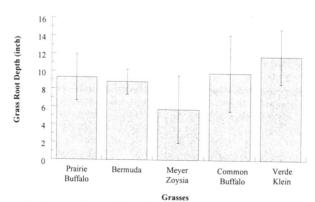

Figure 6. Comparison of root depths among different grass species

Soil PAH Concentrations Associated with Different Grass Species Two of the top five grasses, Common Buffalo and Zoysia had similar PAH concentration levels in root zone soil to Prairie Buffalograss, but Verde Kleingrass. Root zone soils were collected from each of the triplicate experimental units of three top-grading grasses, Verde Kleingrass, Meyer Zoysiagrass, and Common Buffalograss, after three years of growth. Bermudagrass plots were not sampled. For each grass two composite soil samples were analyzed for PAHs. Figure 7 shows that the mean soil concentrations of LMW PAHs, associated with the four grasses, including Prairie Buffalograss (plot 2) were generally lower than those with the control (plot 1). The concentrations associated with Verde Kleingrasses were either nondetectable or approximately one to two orders of magnitude lower than those with the other grasses and the control. The mean soil concentrations of HMW PAHs, associated with three grasses, except Kleingrass, were close to or higher than those with the control (Figure 8). Still, the soil HMW PAH concentrations in Verde Kleingrasses root zone soil were much lower than those in the control soil. To verify the differences between grass species, more soil samples were collected from each of the triplicate experimental units of six grasses, including Klein, Bermuda, Switch, Common Bufflo, Zoysia, and St. Augustine, in May 1996. Analytical results indicated that root zone soil PAH concentrations of Kleingrass and Bermudagrass were significantly lower than those of the other four grasses. Additional soil sampling and analysis are planned in 1997 or later to obtain sufficient data for statistical analysis.

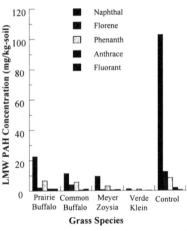

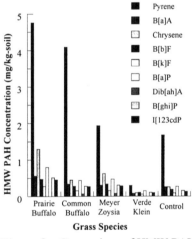

Figure 7. Comparison of LMW PAH concentrations in root zone soils of the four top-grading grasses against unvegetated control

Figure 8. Comparison of HMW PAH concentrations in root zone soils of the four top-grading grasses against unvegetated control

Kleingrass appeared to have a greater potential to remove PAH compounds from rhizosphere soil than other tested grasses. It withstood sustained water-logging and exceptional seasonal drought and developed healthier, deeper, and dense roots in the clay soil. This bunch-type tall grass with deep fibrous roots may be naturally fit for the field soil. According to USDA's soil survey, the local Lake Charles clay soil formed under a dense stand of tall grasses. We do not know whether the possible enhanced removal of PAHs was simply due to the improved soil aeration condition or a series of complex chemical and/or enzymatic reactions in the presence of a unique substrate secreted from Kleingrass roots. Researchers reported that many heterotrophic bacteria in natural soil are unable to degrade PAHs. Only a small adapted population respond to PAH degradation (*17,18*). Despite the many uncertainties, we believe that certain plants have the potential of remediating PAH-contaminated soils by influencing soil chemistry and biochemistry, improving oxygen transport for microbial activities, and perhaps selectively fostering PAH-degrading bacteria growth.

Bioconcentration

Risk arising from plant bioconcentration of PAHs has been studied, because most HMW PAHs are known to be carcinogenic (*35*). Animals feeding on the plant might receive high doses of the PAHs and cause subsequent risks via food chain. We examined potential bioconcentration of PAHs to grass tissues at our test plots.

Grass shoot and root tissues were collected from all the experimental units of Verde Kleingrass, Meyer Zoysiagrass, Prairie Buffalograss, and Common Buffalograss. Soil particles clinging on the roots were removed by ultrasonic water bath. Duplicate composite tissue samples were soxhlet extracted and analyzed for PAH concentrations. No PAHs were detected from either shoots or roots. Figures 9 and 10 shows the detection limits of the LMW and HMW PAHs in grass tissue versus the associated soil concentrations, respectively.

Contrary to several published reports (*26,27,28,29*), our study showed no PAH bioconcentration in the plant shoots or roots. Previously published data are either from municipal wastewater sludge tests or from laboratory experiments, in which PAH compounds were freshly spiked onto the soil. Those experiments may not reflect field application of phytoremediation, where plants grow on aged-contaminated soils. Freshly supplied chemicals in laboratory soils or municipal sludges remaining on the exterior surface of soil/sludge particles could have a chance to partition onto plant root surfaces. However, as a contaminated site aged, the PAHs at the exterior soil aggregate surfaced would have already degraded or been transported away. The remaining portion, mainly trapped in or adsorbed to the interior surfaces of soil

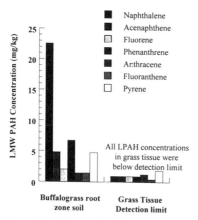

Figure 9. Comparing the detection limits of LMW PAHs in grass tissue to the associated soil concentrations

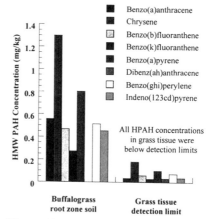

Figure 10. Comparing the detection limits of HMW PAHs in grass tissue to the associated soil concentrations

micropores, would have little chance to redistribute to plant roots. According to the theoretical calculation of bioconcentration factors, PAH concentrations in the grass shoots and roots would be less than the detection limits of this study (*33*). Neither food chain effects nor volatilization through leaves should be a concern for phytoremediation of aged PAH-contaminated soils.

Conclusions

Prairie Buffalograss accelerated naphthalene concentration reduction in soil, despite the fact that high soil moisture content has suppressed grass root development. The field soil PAH concentration reduction did not fit the first order reaction kinetics and the corresponding rate can not be derived. Comparative performance evaluation for other PAH compounds was restricted by analytical variability. PAH degradation in soil could have been limited by low redox potential and desorption kinetics.

A drainage system controlling soil saturation level would be desirable where economically feasible. Selectively growing water-tolerant plants, capable of removing excessive soil moisture and maintaining appropriate soil aeration condition may be a low-cost alternative.

Verde Kleingrass demonstrated a greater potential to remove PAHs from wet clay soil than the other tested grasses. The Kleingrass roots were well established in the wet clay soil. After three years of growth, it's root zone soil concentrations for both low and high molecular weight PAHs were approximately one order of magnitude lower than those associated with Buffalograss.

Contrary to several published reports, no bioconcentration of PAHs in the grass shoots or roots was observed. Previously published data from experiments with freshly supplied chemicals may not reflect actual field conditions, where plants grow on aged-contaminated soils. Neither food chain effects nor volatilization through leaves should be a concern for phytoremediation of aged PAH-contaminated soils.

Acknowledgments

Union Carbide Seadrift Plant EP Department performed the pilot study. Field personnel included Mr. Joe Padilla, Mr. Trinidad Gomez, Mr. Marcus May, and Mr. Joe Stubbs. The authors thank Mr. Steve Galemore, Mr. Craig Falcon, Mr. David McMahon, Mr. Jack Batterola, and Mr. Ron Pace for their administrative support.

At Utah State University, Dr. Ronald C. Sims is the principal investigator of the program. Mr. Venubabu Epuri, Mr. Jeff Watkins, Dr. Ari Ferro, Mr. Wyne April, and Mr. James Herrick contributed to laboratory studies and/or field sample analysis.

Special thanks go to Dr. Milton C. Engelke, Texas A&M University, for his valuable input to this study. Thanks are due to Mr. John Matthews and Mr. Scott G. Huling, USEPA Kerr Laboratory, and Dr. Daniel F. Pope, Dynamac Corporation, for partially funding the studies at Utah State University. The author sincerely thanks Mr. Richard A. Conway and Dr. James L. Hansen for their helpful discussion.

Literature Cited

1. April, W; Sims, R.C. *Chemosphere.* **1990**, *20(1-2)*:253-265.
2. Ferro, A.M.; Sims, R.C; Bugbee, B.G. *J. Envir. Quality.* **1994**, 23(2):272-279.
3. Watkins, J.W.; Sorensen, D.L.; Sims, R.C. In *Bioremediation through Rhizosphere Technology;* Anderson T. A.; Coats, J.R. Eds.; ACS Symposium Series 563; American Chemical Society: Washington, DC, **1994**, pp 123-131.
4. Epuri,V.; and Sorensen, D.L. In *Phytoremediation of Soil and Water Contaminants*; Kruger, E.L., Anderson T.A. and Coats, J.R., Eds.; ACS Symposium Series; (American Chemical Society: Washington, DC. in press).

5. Qiu, X.; Shah, S.I.; Kendall, E.W.; Sorensen, D.L.; Sims, R.C.; Engelke, M.C. In *Bioremediation through Rhizosphere Technology;* Anderson T.A.; Coats, J.R. Eds.; ACS Symposium Series 563; American Chemical Society: Washington, DC, **1994**, pp 142-159.
6. Cerniglia, C.E. *Adv. Appl. Microbiol.* **1984**, 30:31-71.
7. Kihohar, H.; Nagao, K. *J. Gen. Microbiol.* **1978**, 105:69-75.
8. Schocken, M.J.; Gibson, D.T. *Appl. Environ. Microbiol.* **1984**, 48:10-16.
9. Manilal, V.B.; Alexander, M. *Appl. Microbial. Biotechnol.* **1991**, 35:401-405.
10. Mihelcic, J.R.; Luthy, R.G. *Appl. Envir. Microbiol.,* **1988**, 54(5):1188-1198.
11. Raddy, K.R.; Rao, P.S.C.; Jessup, R.E. *Soil Sci. Soc. Am. J.* **1982**, 46:62-68.
12. Al-Bashir, B.; Cseh, T.; Lecuc, R.; Samson, R. *Appl. Miocrobiol. Biotechnol.* **1990**, 34:414-419.
13. Mihelcic, J.R.; Luthy, R.G. *Appl. Envir. Microbiol.,* **1988**, 54(5):1182-1187.
14. Leduc,R.; Samson,R.; Al-Bashir,B.; Al-Hawari, J.; Cseh, T. *Wat. Sci. Tech.* **1992**, 26(1-2):51-60.
15. Bauer, J.E.; Capoint, D.G. *Appl. Envir. Microbiol.,* **1985**, 50:81-90.
16. McFarland, M.J.; Sims, R.C. *Journal of Groundwater,* **1991**, 29(6):885-698.
17. Keck, J.; Sims, R.C.; Coover, M.; Park, K.; Symons, B. *Water Res.,* **1989**, 23(12):1467-1476.
18. Bulman, T.; Lesage, S.; Fowlie P.J.A.; Weber, M.D. *PACE Report;* Petroleum Association for Conservation of the Canadian Environment, Ottawa, Canada, **1985**, 85-2.
19. Cernignia, C.E.; Heitkamp, M.A. In *Metabolism of Polycyclic Aromatic Hydro-Carbons in the Aquatic Environment.* Varanasi, U. Ed.; CRC Press, Inc.: Boca Raton, FL, **1989**, pp 41-68.
20. Grosser, R.J.; Warshawsky, D.; Vestal, J.R. *Envir. Toxicol. Chem.* **1995**, 14(3):375-382.
21. Fitter, A.H.; Hay, R.K.M. *Envir. Physiology, 2nd Ed.* Academic Press, London, UK, **1987**, pp 423.
22. Walton, B.T.; Guthrie, E.A.; Hoylman, A.M. In *Bioremediation through Rhizosphere Technology;* Anderson T. A.; Coats, J.R. Eds.; ACS Symposium Series 563; American Chemical Society: Washington, DC, **1994**, pp 11-27.
23. Ellis, B.E. *Lloydia.* **1974**, 37(2):168-184.
24. Harms, H.; Dehren, W.; Monch, W. *Z. Naturforsch.* **1977**, 32:321-326.
25. Sims, R.C.; Overcash, M.R. *Residue Revs.* **1983**, 88:2-68.
26. Ellwardt, P. *Proc. Series - Soil Organic Matter Studies.* **1977**, 2:291-298.
27. Muller, V.H. *Z. Pflanzenernaehr. Bodenkd.* **1976**, 6:685-695.
28. Topp, E.; Scheunert, I.; Attar,A.; Korte, F. *Ecotoxicol. Environ. Safety.* **1986**, 11:219-228.
29. Harms, H.; Sauerbeck, D. *Angew. Botanik.* **1984**, 58:97-108.
30. Qiu, X. *Field Scale Pilot Study: Vegetation Enhanced Soil Bioremediation, 1993-1994 Annual Progress Report;* Union Carbide Corporation: Tech. Center, South Charleston, WV, **1995.**
31. Qiu, X. *Field Scale Pilot Study: Vegetation Enhanced Soil Bioremediation, 1994-1995 Annual Progress Report;* Union Carbide Corporation: Tech Center, South Charleston, WV, **1996.**
32. Raddy, K.R.; Rao, P.S,C.; Jessup, R.E. *Soil Sci. Soc. Am. J.* **1982**, 46:62-68.
33. Leland, T.W. *Matric Potential, Vegetation, and Mycorrizae Effects on PAH Degradation in Clay Soil.* MS Thesis. Dept. of CEE, Utah State University: Logan, UT, **1996.**
34. Warshawsky, G.D.; Vestal, J.R. *Envir. Toxicol. Chem.* **1995**, 14:3:375-382.
35. Bell, R.M. *Higher Plant Accumulation of Organic Pollutants from Soils;* EPA/600/R-92/138; PB92-209378; NTIS: Springfield; VA, **1992**, pp 127.

Chapter 15

Benzo(a)pyrene and Hexachlorobiphenyl Contaminated Soil: Phytoremediation Potential

V. Epuri and Darwin L. Sorensen[1]

Utah Water Research Laboratory, Utah State University, Logan, UT 84322–8200

Benzo(a)pyrene (B[a]P) is a carcinogenic polynuclear aromatic hydrocarbon and hexachlorobiphenyl (HCB) is a polychlorinated biphenyl (PCB) congener. Spiked, radiolabeled B[a]P and HCB mineralization, volatilization, solvent extractability, soil binding and plant accumulation were measured in soil microcosms for differences between unvegetated and vegetated treatments. Aroclor 1260 and PAH contaminated loamy sand from a New Jersey plastics plant was used. Soil was planted with Tall Fescue (*Festuca arundinacea* Screb.) or was left unplanted. Incubation under artificial lighting for 180 days with vegetation resulted in decreased B[a]P volatilization, increased mineralization, and increased solvent extractability but had no detectable effect on soil binding. Vegetation had no effect on HCB volatilization or soil binding but enhanced its mineralization and decreased its extracability.

Benzo(a)pyrene (B[a]P) belongs to a class of compounds known as polycyclic aromatic hydrocarbons (PAHs). It has five fused benzene rings. Hexachlorobiphenyl (HCB) belongs to class of compounds known as polychlorinated biphenyls (PCBs). It has six chlorines attached to the biphenyl. B[a]P is a carcinogen and PCBs are toxic and are possible carcinogens (*1*). Since both of these compounds are toxic, there is a need to minimize their concentrations in the environment.

Both B[a]P and HCB have low aqueous solubilities and are relatively non volatile (*2, 3*). They are persistent in the environment and not readily biodegradable. Biodegradation of HCB has been reported (*4*). There was a significant reduction in the concentrations of hexachlorobiphenyls in liquid media, when treated with a strain of *Alcaligenes eutrophus*, isolated from a PCB contaminated soil. Biodegradation of chlorobiphenyls (mono to penta) has been reported in other studies (*5-15*). Biodegradation of B[a]P has been reported both in liquid media and soil (*16-22*). In

---

[1]Corresponding author

most of the studies, specific microorganisms had been used (*16-20*). Mineralization of B[a]P by indigenous microorganisms has been reported in two studies (*21, 22*).

Bioremediation is the technique of converting hazardous organic chemicals into harmless compounds using biological processes (usually microbial metabolism). Bioremediation may be enhanced by the use of vegetation, a process known as phytoremediation. The rhizosphere soil (soil adhering to the root) has higher microbial numbers, biomass and activity than the surrounding root-free soil (*23*) and hence enhanced degradation may be possible in the rhizosphere (*24*). Phytoremediation has been shown to increase the mineralization of a few pesticides (*25, 26, 27*). It has also enhanced the mineralization of industrial chemicals such as trichloroethylene and pentachlorophenol (*28, 29*). In the case of recalcitrant compounds such as PAHs having four rings or more, vegetation has enhanced the disappearance of extractable chemicals from soil (*30*). There has been only one study published that investigated the phytoremediation of B[a]P (*30*) and none for HCB. There is a need to learn more about the phytoremediation potential of carcinogenic PAHs and PCBs. We performed a laboratory scale "treatability study" of B[a]P and HCB to evaluate the potential for enhanced phytoremediation of soil contaminated with these kinds of compounds.

Materials and Methods

Experimental Design. The experimental design used in this study was a randomized complete block design with factorial treatments (*31*). The two factors were vegetation and time. Unvegetated and vegetated treatments were used to assess the effect of vegetation and time was evaluated by sacrificing three sets of microcosms after 12, 102 and 180 days of incubation. The experimental design of the study is shown in Table 1.

At day 0, when the radiolabeled compound was spiked into the soil, the compound would have been in a state of flux among phases and among various associations with the solid phase. It was anticipated that relatively rapid physical and chemical sorption of the compound would have occurred during the first few days. The situation after this "equilibration" period would be more representative of the field situation. For this reason, the first set of microcosms was sacrificed after 12 days. The decision to sacrifice the third set after 180 days was based on the economics of conducting the experiment. The second set was sacrificed after 102 days to learn about the distribution of the compound approximately midway through the experiment.

Soil Sampling and Characterization. Soil was collected from CAMU 2, at the Union Carbide Corporation plant in Boundbrook, New Jersey. In 1987, a few buildings were demolished at the plant as a part of a renovation plan. During the process, a transformer valve was accidentally opened and transformer oil ran into the soil. The transformer oil contained PAHs (byproducts of petroleum refining) and PCBs (added to improve the thermal stability and electric resistivity) and thus the soil was contaminated with PAHs and PCBs.

Table 1. Experimental design for phytoremediation treatability of B[a]P and HCB

Factor	Number	Description
Compound	2	Benzo[a]pyrene; Hexachlorobiphenyl
Treatment	2	Vegetated; Unvegetated
Time of sacrifice	3	12, 102, 180 days
Replicates	3	
Total Number of Microcosms	36	

Soils were collected from two different locations at CAMU 2. The soil was sieved at the site through a 6.4 mm sieve, to remove large stones and building debris. The soil was then shipped to Utah State University. In the laboratory, the soil was sieved to pass a 2 mm screen and mixed thoroughly to minimize heterogeneities. It was stored at approximately 5° C until use. The results of soil characterization analyses are given in Table 2.

Table 2. Characteristics of the CAMU 2 soil samples

Characteristic	Value
Soil Texture	Loamy Sand
% Sand	78
% Silt	15
% Clay	7
Field capacity (moisture content; %)	17.4
Cation exchange capacity (cmol(+)/kg)	10.7 ± 0.4*
pH	7.7 ± 0.4
Organic carbon (%)	1.46 ± 0.01
Electrical conductivity (mmhos/cm)	0.30 ± 0.04
Total Kjeldahl nitrogen (%)	0.037 ± 0.003
Calcium (mg/kg)	32 ± 11.9
Magnesium (mg/kg)	2.6 ± 0.97
Sodium (mg/kg)	5 ± 1.35
Potassium (mg/kg)	3.0 ± 0.76
Nitrate-nitrogen (mg/kg)	<1.03
Sulphate (mg/kg)	51 ± 28.1
Chloride (mg/kg)	800 ± 56
Ammonia-nitrogen (mg/kg)	2.0 ± 0.25
Phosphate-phosphorus (mg/kg)	<1.8

*Mean ± Standard Deviation (Each analysis was performed in triplicate)

Analysis of Organic Contaminants. A soil sub-sample was analyzed for polychlorinated biphenyls using ultrasonic extraction (USEPA Method 3550A) and gas chromatography with electron capture detection (USEPA Method 8080) (*32*). The results of the analysis are given in Table 3. Only Aroclor 1260 was detectable. Approximately 38% of Aroclor 1260 (by weight) is hexachlorobiphenyls (33). Three soil sub-samples were analyzed for 10 PAHs. The samples were soxhlet extracted using USEPA Method 3540. The extracts were analyzed by USEPA Method 8310, which uses high performance liquid chromatography with ultraviolet and fluorescence detection (34). The results of the analysis are given in Table 4.

Table 3. Concentrations of PCBs in soil

Compound	Concentration (mg/kg)
Aroclor 1016	< 0.03
Aroclor 1221	< 0.06
Aroclor 1232	< 0.03
Aroclor 1242	< 0.03
Aroclor 1248	< 0.03
Aroclor 1254	< 0.03
Aroclor 1260	26

Table 4. Concentrations of selected PAHs in soil

Compound	Concentration (mg/kg)
Naphthalene	$8 \pm 38^*$
Phenanthrene	54 ± 8
Anthracene	10 ± 2
Pyrene	48 ± 5
Benzo(a)anthracene	19 ± 2
Chrysene	19 ± 3
Benzo(a)fluoranthene	18 ± 3
Benzo(k)fluoranthene	10 ± 1
Benzo(a)pyrene	20 ± 3
Benzo(g,h,i)perylene	13 ± 3

*Mean ± Standard Deviation

Grass Selection and Seed Disinfection. Seed germination tests were conducted on soil to be used in the study. Four species of grass seeds were selected for the study based on the characteristics of soil and climatalogical conditions at the site. They were Tall Fescue (*Festuca arundinacea* Screb.), Perennial Rye grass (*Lolium perenne* L.), Reliant Hard Fescue (*Festuca longifolia* Thuill.) and Nassau Kentucky Blue grass (*Poa pratensif* L.).

In the germination test, 40 g of soil was added to petri dishes. Ten seeds were added to the soil in each petri dish in a circular pattern. Five grams of water was added to each petri dish. The petri dishes were observed for germination each day until the germination percentage remained the same for 7 days.

Tall Fescue had the highest rate of germination and it was selected for use in the experiment. It can produce a voluminous, space filling root system to depths of 250 cm (35).

Tall Fescue seeds used in the experiment were disinfected for two reasons. First, to prevent damage to plant growth by pathogens present on the seed coat. Second, to ensure that the biodegradation measured was caused by microorganisms present in the soil and not the microorganisms present on the seed coat.

Disinfection was done by soaking the seeds in 5% Clorox solution for 2 minutes and then rinsing them in sterile deionized water for 2 minutes. This procedure was repeated twice. To check for the effectiveness of disinfection, the seeds were placed on tryptic soy agar plates, and incubated at 28° C for 7 days to detect culturable heterotrophic bacteria. Bacterial growth was not detected.

Experimental Setup and Microcosm Initiation. The schematic of the experimental setup is shown in Figure 1. Each microcosm consisted of a 1,000 mL vacuum trap flask (Kontes) connected to a drying tube (Fisher) which contained two polyurethane plugs (S/P Brand diSPo plugs, Baxter) to trap volatile organics (36, 37). This was connected to two glass culture tubes, in series, each of which contained 20 mL of 0.1 N KOH to trap carbon dioxide (38). The second carbon dioxide trap was connected through a manifold to a vacuum source controlled by a vacuum regulator (Vacu/Trol, Spectrum). An acetal screw clamp (Fisher) was clamped to the tubing between the second carbon dioxide trap and the vacuum source to regulate air flow.

Plastic tubing (Formula R-3603 vacuum tubing, Tygon) was used to connect various components of this setup. Plastic tubing may absorb volatile organic compounds. To minimize errors due to compounds absorbing into the tubing, the connection between the flask and the polyurethane plug trap was kept less than 1 cm. Polypropylene and polyethylene connectors (Fisher) were used wherever necessary.

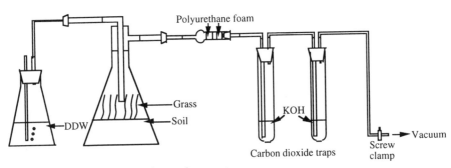

Figure 1. Schematic of experimental apparatus

A 500 mL Erlenmeyer flask containing deionized water was originally connected to the microcosm to provide for the flow of humidified air through the microcosm to minimize evaporation of soil water. The humidifier was later deleted from the apparatus.

Four hundred grams of soil were added to each 1000 mL flask. To minimize the potential for nutrient limitation of plant growth and microbial activity, 50 mg/kg of NO_3-N and 5 mg/kg of PO_4-P were added to the soil as $NaNO_3$ and KH_2PO_4. The soil was spiked with either B[a]P-7-[14]C (826,600 DPM) or uniformly labeled [14]C 2,2´,4,4´,5,5´ HCB (830,000 DPM). The total amount of radioactive material to be added to each microcosm was made up to 4 mL with methanol. This was distributed uniformly over the soil in small drops. Then the soil was mixed with a glass rod to distribute the radioactive material as homogeneously as possible. Less than 10 µg/kg of either radioactive B[a]P or HCB was added to the soil.

Ten Tall Fescue grass seeds (Steve Reagan Company) were added to microcosms belonging to the vegetated treatment. After addition of seeds, the microcosms were watered with deionized water to bring the soil to approximately 80 percent of field capacity (14% moisture content). Unvegetated microcosms were prepared similarly without the addition of grass seed.

Microcosm Operation. The microcosms were placed in a growth chamber constructed from an uninsulated, unsealed steel storage shed assembled inside the Utah Water Research Laboratory. Lighting was provided by two 1000 watt metal halide lamps (Venture, Energy Technics). After 102 days, one 1000 watt metal halide lamp was turned off. Additional lighting was provided by two fluorescent light fixtures, each containing four tubes designed to support plant growth (Grolux, Sylvania; Plantlight, General Electric). The light intensity was measured using a quantum sensor (LI-COR, Inc). The photosynthetic photon flux (PPF) at the surface of the soil, with two metal halide lamps, ranged from 8.1 mol/m^2 day to 13.3 mol/m^2 day. The PPF at the surface of the soil, with one metal halide lamp, ranged from 5.2 mol/m^2 day to 8.9 mol/m^2 day. A PPF value of 11.5 mol/m^2 day is typical of winter sunlight on a sunny day and 5.6 mol/m^2 day is typical of winter sunlight on a cloudy day (*39*). A 16 hour photoperiod was maintained and controlled by automatic timers.

Initially, air was drawn through the microcosms at a flow rate of approximately 25 mL/min. The flow rate was checked using a soap bubble electronic flowmeter (Tekmar). An air conditioner and an exhaust fan were operated to dissipate the heat generated by the metal halide lamps. The temperature in the growth chamber varied from 23.2 °C to 28.2 °C through the experimental period of 180 days. The average temperature progressively increased with time during the experimental period.

The moisture content of soil in the microcosms was maintained at approximately 80 percent of its field capacity. In order to make correction for the weight of plants, tall fescue grass was grown in the greenhouse simultaneously. Each week, one plant was removed (including roots) and weighed. This weight was multiplied by the number of plants in each microcosm to estimate additional weight gained by microcosms due to plant growth.

Radiolabeled Compound Analysis. Analysis for radioactivity in the trapping materials was conducted approximately weekly during the experiment, except for the period between 45 and 84 days. During this period, analysis was suspended, pending evaluation of the design of the experiment.

The polyurethane foam plugs were extracted with methylene chloride:acetone (1:1 v/v) using sonication (*34*). The extraction and analysis of the second polyurethane plug from each microcosm was used as a quality control measure to check the assumption that all the volatile organics were absorbed in the first polyurethane plug. After the extraction, an aliquot of the extract was analyzed for radioactivity using a scintillation counter (LS 6000SE, Beckman).

The two carbon dioxide traps (0.1 N KOH) from each microcosm were analyzed weekly for radioactivity. The analysis of the second carbon dioxide trap was used as a quality control measure to check the assumption that all the carbon dioxide is absorbed in the first trap. An aliquot of KOH solution was analyzed for radioactivity using the scintillation counter.

Extractability of the radiolabeled compound with methylene chloride:acetone (1:1 v/v) was evaluated by analyzing the extract for radioactivity using the scintillation counter. Soil incorporation of the radiolabeled compound was determined by combusting samples of extracted soil in a biological oxidizer (OX600, R.J. Harvey Instrument Corporation). The CO_2 evolved from the oxidizer was trapped and analyzed for radioactivity using a scintillation counter.

To measure the plant accumulated radiolabeled compound, the entire plant material, including roots, was dried at a relatively low temperature (70° C) and ground using a mortar and pestle. The ground plant material was combusted in a biological oxidizer and the CO_2 evolved was trapped and analyzed for radioactivity.

Microbial Enumeration. The number of total viable aerobic heterotrophic bacteria and total viable fungal propagules in soil samples were estimated using the spread plate technique (*40*). Plate counts were performed on untreated soil and soils from unvegetated and vegetated treatments. For the unvegetated and vegetated treatments, plate counts were performed on all the microcosms that remained at the end of the study (180 days). Microbial enumeration in untreated soil stored at 5° C in the dark was also performed after 180 days. It was kept at room temperature for 2 days before the counts were made.

Decimal dilutions of soil were made in phosphate buffered saline. The plates were incubated in the dark at 28° C under humid conditions. Heterotrophic bacteria were grown on tryptic soy agar medium and colonies were counted after 7 days. Martins rose bengal medium was used for enumeration of fungal propagules and the colonies were counted after 3 days (*40*).

Progress of the Experiment and Changes Made. All the seeds planted in vegetated microcosms germinated within 6-10 days. One set of microcosms was sacrificed after 12 days and analyzed. The microcosms that were sacrificed were selected randomly using a random digit table (*41*). This procedure was also followed when the second set of microcosms was sacrificed. During the first 51 days of the experiment, the rate of loss of water from the microcosms was very slow and water was added once a week to maintain the microcosms at 80% of field capacity. There was little radioactivity in the first traps and practically no radioactivity in the second traps from each microcosm analysed during the first three weeks. It was decided to not analyze the second trap until there was a significant amount of radioactivity in the first trap.

After 39 days, grass in a few microcosms began to turn yellow. By 45 days, all of the grass in a few microcosms was yellow. A few grass plants which had become completely yellow were removed from the microcosms and were examined for pathology. It was suspected that the humid conditions in the microcosms may have encouraged the growth of pathogenic fungi on the plants. To remedy the problem, all the humidifiers were removed after 51 days. Dead plants were also removed that day. The poor growth and death of plants may also have been influenced by the low flow rate of air and the low supply of carbon dioxide. To avoid this, various flow rates were tried and trapping efficiencies were calculated. The flow rate of air through the microcosms was increased to 250 mL per minute and the volume of KOH in the trap was increased to 250 mL after 75 days. The microcosms in which the plants died were reseeded with the number of seeds added equaling the number of dead plants. After these changes, the yellowing of plants stopped, all the seeds germinated, and the plants were healthy through the end of the study. After the humidifiers were removed, the rate of water loss in microcosms increased and water was added daily to maintain the moisture content. When the grass grew to the top in a particular microcosm, it was harvested and stored in plastic bags at -20° C untill the end of the study and then analyzed.

Data Analysis, Statistics, and Quality Control. Two-way analysis of variance (ANOVA) was used to determine the significance of time and vegetation on volatilization, mineralization, solvent extractability and soil binding of each of the compounds. Effects were considered statistically significant only if the level of confidence was above 95%. The Least Significant Difference (LSD) procedure (*41*) was used to identify significant differences between specific pairs of means when significant effects were indicated by the ANOVA.

A mass balance approach was used as a quality control procedure to account for the amount of radiolabeled compound added to the soil in each microcosm and to help assure the accuracy of each analysis. Extraction efficiencies were measured for all extraction procedures. To measure the extraction efficiency of sonication, three polyurethane plugs were spiked with a known amount of radioactive naphthalene and analyzed with the regular samples. To measure the extraction efficiency of soil sample sonication, three 40 g soil samples were spiked with a known amount of radioactive B[a]P and three with radioactive HCB. Each sample was extracted and analyzed with the regular samples. To measure the efficiency of combustion of organic matter in soil samples by the biological oxidizer, three 1 g soil samples were spiked with a known amount of radioactive B[a]P and three with radioactive HCB. Each sample was combusted and analyzed with the regular samples. To measure the efficiency of combustion of organic matter in plant material by the biological oxidizer, three 0.25 g plant material samples (ground) were spiked with a known amount of radioactive B[a]P and three with HCB. Each sample was combusted and analyzed with the regular samples. Percent recoveries of radioactive compounds were calculated. Reagent blanks and method blanks were used in all analyses.

Results And Discussion

The extraction efficiencies for sonication of the polyurethane plugs and soil were above 85% and the combustion efficiencies of soil and plant material were above 90% indicating that the extraction and combustion procedures were adequate. The percent mass balance for each of the microcosms spiked with radiolabeled B[a]P or HCB is given in Tables 5 and 6, respectively. Mass balances varied between 84% and 107%. Since more than 84% of the radiolabeled carbon was accounted for, we are confident that the data are reliable.

Volatilization. The total ^{14}C in the volatile organics trap of each microcosm was determined. Volatilization in each microcosm was expressed as a percent of total DPM spiked into the soil. Cumulative percent volatilization at each time interval was calculated as the sum of percent volatilization up to the time of measurement. The cumulative volatilization of radiolabeled B[a]P and/or its degradation product(s) in unvegetated and vegetated treatments during the experimental period is shown in Figure 2. The cumulative volatilization of radiolabeled HCB and/or its degradation product(s) in unvegetated and vegetated treatments is shown in Figure 3. The Least Significant Differences (LSDs) were very small and would not be visible in the figures.

Total volatilization of B[a]P and/or its degradation product(s) in both treatments was below 1%. The total volatilization of HCB and/or its degradation product(s) in both treatments was also below 1%. Both B[a]P and HCB have relatively low vapor pressures and their volatilization potential is low. The low volatilization of B[a]P observed in this study agrees with the study done by Park, Sims and Dupont (*42*), where volatilization of B[a]P was undetectable. There are two possible reasons for the low volatilization observed. Either these compounds were not transformed into volatile intermediate(s) during the course of the study or the intermediate(s) that were formed were not volatile. It is also important to note that the purity of radioactive compounds used was 99% or more and the volatilization observed could have been the volatilization of low molecular weight contaminants of the reagent.

Time and vegetation had a significant effect on the volatilization of radiolabeled B[a]P and/or its degradation product(s). The influence of vegetation had no significant influence on volatilization over time. After 31 days, volatilization in both treatments was low but volatilization in the unvegetated treatment was marginally, but significantly higher.

Volatilization of radiolabeled HCB and/or its degradation product(s) increased over time, but vegetation did not significantly affect this process.

Mineralization. Total ^{14}C in the carbon dioxide trap of each microcosm was determined. Mineralization in each microcosm was expressed as a percent of total DPM spiked into the soil. Cumulative percent mineralization at each time interval was calculated as the sum of percent mineralization up to the time of measurement. The cumulative mineralization of radiolabeled B[a]P in unvegetated and vegetated treatments during the experimental period is shown in Figure 4. The cumulative mineralization of radiolabeled HCB in unvegetated and vegetated treatments is shown

Table 5. Mass balance of radiolabeled carbon from benzo(a)pyrene and/or its degradation product(s)

Treatment	Volatil-ization (%)	Mineral-ization (%)	Solvent extractable ^{14}C (%)	Soil bound ^{14}C (%)	Plant incorp-orated ^{14}C (%)	Total (%)
12 Days						
Unvegetated	0.02	0.05	47.74	36.26		84.07
	0.07	0.04	55.86	43.70		99.67
	0.07	0.02	48.48	47.49		96.06
Vegetated	0.04	0.03	56.24	49.97		106.28
	0.03	0.04	55.01	43.19		98.27
	0.06	0.03	56.19	38.64		94.93
102 Days						
Unvegetated	0.51	0.70	38.38	53.35		92.94
	0.41	0.72	40.01	56.72		97.86
	0.35	0.58	34.57	49.68		85.18
Vegetated	0.35	0.58	40.83	50.60	0.02	92.38
	0.53	0.45	40.56	44.80	0.04	86.37
	0.46	0.60	40.83	58.04	0.01	99.93
180 Days						
Unvegetated	1.02	3.81	23.95	58.90		87.68
	0.91	3.78	28.31	69.20		102.20
	0.90	4.54	23.95	70.02		99.41
Vegetated	0.91	4.30	30.21	49.99	0.24	85.66
	0.86	4.07	28.85	53.23	0.41	87.42
	0.91	3.95	26.40	61.35	0.19	92.80

Table 6. Mass balance of radiolabeled carbon from hexachlorobiphenyl and/or its degradation product(s)

Treatment	Volatil-ization (%)	Mineral-ization (%)	Solvent extractable ^{14}C (%)	Soil bound ^{14}C (%)	Plant incorp-orated ^{14}C (%)	Total (%)
12 Day						
Unvegetated	0.05	0.02	55.75	33.80		89.62
	0.08	0.03	53.01	44.73		97.85
	0.07	0.02	58.93	42.32		101.34
Vegetated	0.06	0.02	55.41	49.96		105.44
	0.07	0.02	55.68	46.86		102.62
	0.10	0.02	52.96	37.06		90.14
102 Days						
Unvegetated	0.43	0.06	49.87	40.72		91.08
	0.46	0.06	49.33	54.97		104.81
	0.37	0.06	45.53	47.23		93.19
Vegetated	0.45	0.09	42.55	46.78	0.01	89.89
	0.31	0.07	44.18	43.24	0.01	87.80
	0.41	0.10	46.62	52.31	0.01	99.45
180 Days						
Unvegetated	0.89	0.44	32.79	57.51		91.64
	0.82	0.56	34.15	51.58		87.11
	0.88	0.40	33.61	63.18		98.07
Vegetated	0.91	0.65	27.10	62.54	0.14	91.35
	0.77	0.72	30.63	62.05	0.17	94.33
	0.97	0.56	26.83	71.76	0.10	100.22

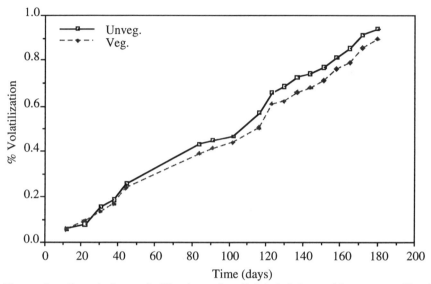

Figure 2.   Cumulative volatilization of radiolabeled benzo(a)pyrene and/or its degradation products.

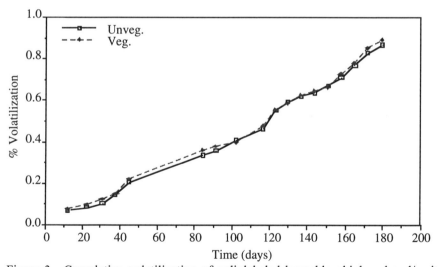

Figure 3.   Cumulative volatilization of radiolabeled hexachlorobiphenyl and/or its degradation product(s).

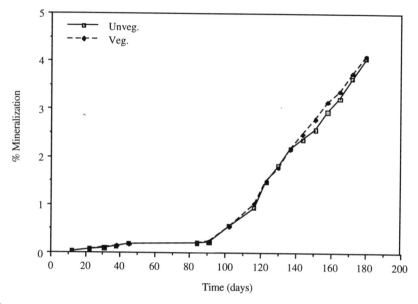

Figure 4. Cumulative mineralization of radiolabeled benzo(a)pyrene.

in Figure 5. The Least Significant Differences (LSDs) are very small and hence not shown in the figures.

Total mineralization of B[a]P was 4.0% and 4.1% in unvegetated and vegetated treatments, respectively. Mineralization of B[a]P depends on the degradative capabilities of the species of microorganism(s) in soil. Mineralization of B[a]P by indigenous microorganisms in soil has been reported in two studies conducted by Grosser, Warshawsky and Vestal (*21, 22*). In their 1991 study, 2.8%, 3.4% and 25.3% of radiolabeled B[a]P was mineralized in three soils from abandoned coal gasification plants over a period of 220 days. In their 1994 study, mineralization of radiolabeled B[a]P in soils from an abandoned coal tar refinery was less than 8% over a period of 160 days. The mineralization rate of B[a]P observed in the present study appears to be similar to those reported in these studies.

Time and vegetation had a significant effect on the mineralization of radiolabeled benzo(a)pyrene but there was no significant change in the effect of vegetation on mineralization over the incubation period. Vegetation significantly enhanced the mineralization of B[a]P after 144 days but the increase was small. While statistically significant, the difference between the treatments observed between 144 and 180 days would not be important in soil remediation practice. A longer incubation time may have helped to determine if the effect of vegetation would prove to be of any practical significance.

Cumulative mineralization of HCB in both treatments was below 0.8%. This is the first study where mineralization of HCB has been observed. Apparent

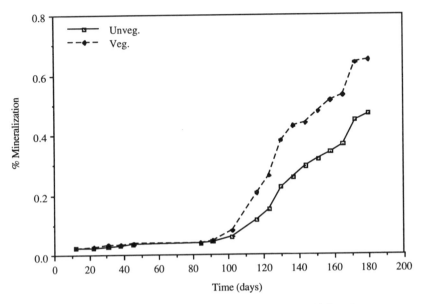

Figure 5. Cumulative mineralization of radiolabeled hexachlorobiphenyl.

mineralization of HCB increased steadily through 180 days. However, it is important to note that the purity of the radioactive compound used was 99% or more and the mineralization observed could have been the mineralization of contaminants in the reagent.

The effect of vegetation on hexachlorobiphenyl mineralization changed significantly during the study. The first statistically significant increase in mineralization in a vegetated microcosms was observed at 22 days of incubation. After 91 days, the difference between treatments increased consistently until the end of the study.

Vegetation significantly enhanced the mineralization of B[a]P and HCB. Earlier studies (25, 26, 28, 29, 44) have also found enhanced mineralization of organic contaminants. The increase may be a result of higher microbial community density and/or higher activity of microorganisms in the vegetated treatment soil. The rate of mineralization of B[a]P and HCB and/or their derivative(s) in both treatments was comparatively low until about 100 days and then it increased substantially and remained steady until the end of the study. The reasons for the 100 day acclimation period for mineralization are not known. It seems reasonable to assume that since the amount of radiolabeled compound added was very small compared to the original concentrations in soil, the increased toxicity due to the addition of radiolabeled compound would be negligible. An increase in toxicity from the addition of methanol (4 mL; 6% w/w) as the solvent for the radiolabeled compound may not be ruled out.

Microorganisms could also have taken some time for recovery from soil disturbances and to adapt to the laboratory microcosm environment.

Solvent extractable and soil-bound radiolabeled carbon. Radiolabeled compound (B[a]P or HCB) and/or its degradation product(s), which were extractable from soil by solvent or that remained on the soil after extraction (soil-bound) were expressed as a percent of total DPM spiked in soil.

The behavior of the compounds in the soil after several days should be more representative of behavior of the compound in soil in a field situation. The relatively rapid physical processes (e.g., large gradient driven mass transport) effecting the distribution of radiolabeled compound would have been essentially complete after several days. During this time, some of the compound may have been bound to soil by rapid sorption processes. Hence day 12 was chosen for convenience as the starting point and solvent extractable and soil-bound radiolabeled carbon were measured from this day.

Time and vegetation had a significant effect on the percent radiolabeled B[a]P and/or its degradation product(s) in soil that were solvent extractable, but the effect of vegetation did not change significantly over the period of the study. Figure 6 shows the apparently linear decrease in percent solvent extractable radiolabeled B[a]P and/or its degradation product(s) in both vegetated and unvegetated soil with time.

Figure 7 shows that, averaged over the period of the experiment, there was more solvent extractable radiolabeled B[a]P and/or its degradation product(s) in vegetated soil. In contrast, in a study done by Aprill and Sims (*30*), vegetation decreased the solvent extractability of B[a]P from soil. The reasons for the higher extractability in

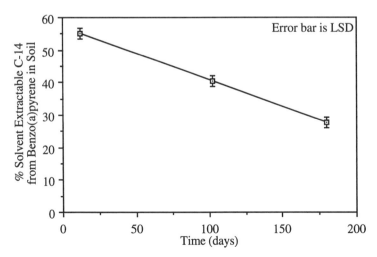

Figure 6. Decrease in percent solvent extractable radiolabeled benzo(a)pyrene and/or its degradation product(s) in soil with time.

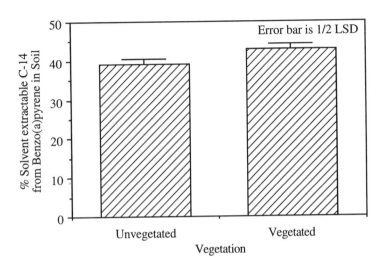

Figure 7. Increase in percent solvent extractable radiolabeled benzo(a)pyrene and/or its degradation product(s) in soil due to vegetation.

the vegetated treatment are not known. The increase is small (4%) and a longer study is needed to determine if the increase would be of any practical significance.

The percent soil-bound radiolabeled B[a]P and/or its degradation product(s) increased significantly over time (Figure 8). Vegetation did not significantly influence the binding of B[a]P and/or its degradation product(s) to soil.

The extractability of HCB and/or its degradation products decreased significantly with time (Figure 9). The effect of vegetation on extractability was significant and the effect did not change significantly during the study. Figure 10 shows the time-averaged decrease in percent solvent extractable radiolabeled HCB and/or its degradation product(s) in soil due to vegetation. The decrease in solvent extractability from soil may be due to increased sorption and/or humification. Again, a longer study could determine if this decrease has more practical significance.

Reflecting the loss of extractability, the amount of soil bound radiolabel from HCB increased over time (Figure 11). The effect of vegetation on the amount of soil binding of HCB was not statistically significant.

Figure 6 shows that the percent solvent extractable radiolabeled carbon from B[a]P and/or its degradation product(s) in soil decreased with time. Figure 8 shows that the percent soil-bound radiolabeled carbon from B[a]P and/or its degradation product(s) increased with time. Similar observations can be made for HCB from figures 9 and 11. There was, apparently, a steep decrease in solvent extractable radiolabeled carbon from 0 to 12 days (from 100% to approximately 53% and 55% for B[a]P and HCB, respectively). By comparison, the decrease from 12 to 180 days is slow. This type of biphasic sorption has been observed for compounds like pyrene (*43*) and

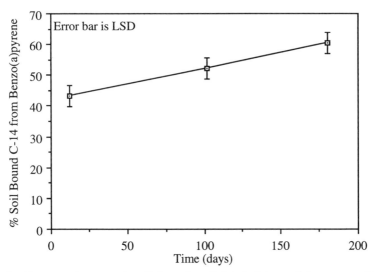

Figure 8. Increase in percent soil bound radiolabeled benzo(a)pyrene and/or its degradation product(s) with time.

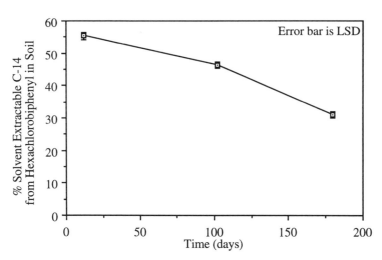

Figure 9. Decrease in percent solvent extractable radiolabeled hexachlorobiphenyl and/or its degradation product(s) in soil with time.

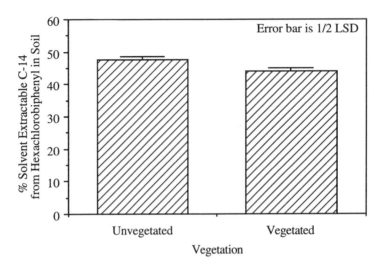

Figure 10. Decrease in percent solvent extractable radiolabeled hexachlorobiphenyl and/or its degradation product(s) in soil due to vegetation.

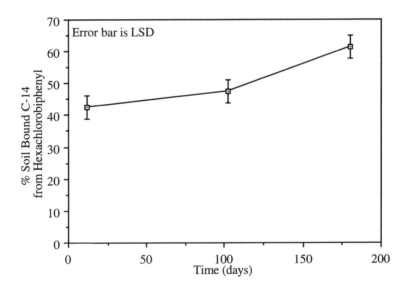

Figure 11. Increase in percent soil bound radiolabeled hexachlorobiphenyl and/or its degradation product(s) with time.

parathion (*44*). It is believed that the rapid phase occurs due to surface adsorption to the soil while the slower phase occurs by the diffusion of compound into the soil material.

As time progressed, lesser amount of radiolabeled carbon was extractable by the solvent. This may be due to processes such as sorption and/or humification. What ever the processes were, either they were irreversible during the duration of the experiment or the forward processes were much faster than reverse processes.

Sorption may be a combination of one or more of the processes such as bonding by van der Waals forces, ligand exchange, covalent bonding and hydrophobic partitioning. Hydrophobic partitioning is generally believed to be the main mechanism for the sorption of nonpolar organic compounds to soils (*45*). Humification may be due to biotic processes or a combination of biotic and abiotic processes (*46*). Nonpolar organic compounds retained in soil by sorption can be degraded biologically by microorganisms or abiotically by functional groups in humic and fulvic acids. The degradation products of these nonpolar organic compounds can undergo polymerization to form humus (*47*). Although monomers are the basic units in humus formation, larger fragments including polycyclic structular units can participate in the reaction (*48*).

The decrease in percent solvent extractable radiolabeled carbon from HCB and/or its degradation product(s) in soil (Figure 9) was nonlinear compared to B[a]P (Figure 6). This pattern is reflected in the increase of percent soil-bound radiolabeled carbon from HCB and/or its degradation product(s) (Figure 11). Through the procedure of contrasts (*41*) it was found that the slope of the line between 102 and 180 days was significantly higher than the slope of the line between 12 and 102 days in Figure 8. Similarly it was found that the slope of the line between 102 and 180 days was significantly higher than the slope of the line between 12 and 102 days in Figure 11. This suggests that there is a change in mechanism(s) after 180 days which was causing increased soil binding.

Plant Growth and Incorporation. Table 7 gives data on plant incorporation and plant growth with the progress of time in soil spiked with B[a]P. Table 8 gives data on plant incorporation and plant growth with progress of time in soil spiked with HCB.

Table 7. Plant incorporation and plant growth with progress of time in soil spiked with B[a]P

Time (days)	Microcosm	Percent ^{14}C associated with plant material	Wet weight of plant material produced and combusted (grams)
102	1	0.02	0.52
	2	0.04	0.68
	3	0.01	0.45
180	1	0.24	2.83
	2	0.41	4.29
	3	0.19	3.11

Table 8. Plant incorporation and plant growth with progress of time in soil spiked with HCB

Time (days)	Microcosm	Percent ^{14}C associated with plant material	Wet weight of plant material produced and combusted (grams)
102	1	0.01	0.34
	2	0.01	0.56
	3	0.01	0.42
180	1	0.14	1.85
	2	0.17	2.61
	3	0.10	1.98

Tables 7 and 8 show that the percent of radiolabeled compound and/or its degradation product(s) associated with plant material increased with time for both B[a]P and HCB. The increase in plant growth (by mass) from 102 days to 180 days is higher than the increase in plant growth up to 102 days in all the microcosms. This is probably due to poorer growth conditions and the death of a few plants in microcosms due to fungal attack before 102 days.

The grass in the microcosms was harvested a total of four times during the study, since the grass had grown to the top of microcosms each time. The harvested grass was stored at -20° C until it was combusted. Roots grew to the bottom in each microcosm and then grew laterally across the bottom of the flask.

The PPF (light intensity) in the chamber varied between typical values for winter sunlight on a sunny day and winter sunlight on a cloudy day. The grasses seemed to be growing well under these conditions.

Microbial Enumeration. The initial number of total viable aerobic heterotrophic bacteria in the soil used in this study was estimated to be 6.9×10^8 colony forming units per gram of soil. The number of total viable fungal propagules in the same soil was estimated to be 2.6×10^5 per gram of soil. The data were log-transformed and analysis of variance was used to determine the effect of vegetation.

Tables 9 and 10 show that the number of total viable microorganisms in vegetated treatment are significantly higher than unvegetated treatment for both B[a]P and HCB. Increases in numbers of microbial populations (as estimated by spread plate technique) due to vegetation have been reported in the literature (27, *49*). It should be noted that plating techniques underestimate the true microbial population density as all growth media are selective. Values estimated by plate counts are approximately 1-10% of the values observed by direct count techniques (*50*).

All of the additional microorganisms in vegetated treatment soil may not be capable of degrading and/or metabolizing B[a]P or HCB. In fact, it is highly probable that most of them are capable only of metabolizing natural organic compounds such as carbohydrates, amino acids, etc., and not recalcitrant compounds like B[a]P and HCB.

Table 9. Density of microorganisms in the soil in microcosms spiked with B[a]P after 180 days

Life Form	Treatment	Mean	95% Confidence interval Lower Limit	95% Confidence interval Upper Limit
Fungi (propagules per gram of soil)	Unvegetated	6.5×10^5	5.1×10^5	8.2×10^5
	Vegetated	1.4×10^6	1.1×10^6	1.8×10^6
Aerobic bacteria (colonies per gram of soil)	Unvegetated	3.6×10^{10}	3.3×10^{10}	3.9×10^{10}
	Vegetated	7.6×10^{10}	7.0×10^{10}	8.2×10^{10}

Table 10. Density of microorganisms in the soil in microcosms spiked with HCB after 180 days

Life Form	Treatment	Mean	95% Confidence interval Lower Limit	95% Confidence interval Upper Limit
Fungi (propagules per gram of soil)	Unvegetated	2.0×10^5	1.2×10^5	3.5×10^5
	Vegetated	1.4×10^6	8.1×10^5	2.4×10^6
Aerobic bacteria (colonies per gram of soil)	Unvegetated	3.7×10^{10}	3.2×10^{10}	4.3×10^{10}
	Vegetated	7.8×10^{10}	6.8×10^{10}	9.1×10^{10}

At the end of the experiment, the number of microorganisms in either the B[a]P or HCB unvegetated soil was higher than in the soil prior to use. The number of viable aerobic heterotrophic bacteria increased 100 fold during the experiment. The addition of nutrients, such as nitrate and phosphate, and maintenance of soil at 80% of field capacity throughout the experimental period apparently enhanced the growth of microorganisms.

Conclusions. Results of the experiment show that there were beneficial effects due to vegetation including increase in mineralization and decrease in solvent extractability.

However, the differences observed within the time frame of the experiment are not environmentally significant enough to be of practical use. Differences between treatments continued to increase throughout the study and may have continued if the study had been extended for a longer period of time. Phytoremediation appears to be a slow process. A period of a several years may be required for treatability studies to fully understand the applicability and effectiveness of phytoremediation.

The half life of B[a]P in microcosms was calculated to be 172 days, based on decrease in solvent extractability from soil. The half life of B[a]P was 1,454 days based on measurement of mineralization. The half life of HCB was 202 days based on decrease in solvent extractability from soil. Regulatory actions relating to cleanup of contaminated soil are often chosen based on solvent extractability of contaminants from soil. Based on mineralization, the half life of HCB in the unvegetated treatment was 13,002 days and in the vegetated treatment it was 10,495 days. All half lives were calculated using a first order model. For calculating half lives based on mineralization, an acclimation period of 102 days was omitted. Soil binding appears to be the major removal mechanism for both B[a]P and HCB. Soil binding may be due to sorption and/or humification.

From the results of this study, it appears that no significant health risks result from the uptake of B[a]P and HCB into Tall Fescue. The plant incorporation of both radiolabeled compounds were less than 0.5%. The risk to health of animals by ingestion seems to be low. Vegetation did not enhance the volatilization of B[a]P and HCB and total volatilization was less than 1% for both. The risk of contamination of the atmosphere using grasses for phytoremediation appears to be low.

Phytoremediation also has application in reducing the leaching of rainfall through soil, reducing the potential for groundwater contamination. Establishment of vegetation on contaminated soil also prevents soil erosion into surface waters including rivers, lakes, etc. and thus prevents their contamination. Vegetation provides a cover on the surface of soil that prevents dust emissions. It is also aesthetically pleasing.

The only undesirable effect caused by vegetation in the study was an increase in solvent extractability of B[a]P accompanied by a decrease in soil binding. Comparing this with the advantages of phytoremediation, field studies and laboratory studies extending for periods of a few years should be conducted.

Acknowledgments

We thank Union Carbide Corporation for financial support for this work. Xiujin Qiu of Union Carbide served as project officer. We also thank Bruce Bugbee, Ronald Sims, and Anne Anderson of Utah State University for their very helpful advice and encouragement.

Literature Cited

1. Safe, S. *Environ. Health Perspect.* 1992, 100, 259-268.
2. *The Merck Index;* Budavari, S., Ed.; Merck, Rahway, NJ, 1989.
3. Mackay, D.; Shiu, W. Y.; Ma, K. C. *Illustrated Handbook of Physical-Chemical Properties and Environmental Fate for Organic Chemicals*, Lewis Publishers: Boca Raton, FL, 1992.

4. Bedard, D. L.; Wagner, R. E.; Brennen, M. J.; Haberl, M. L.; Brown, J. F. *Appl. Environ. Microbiol.* 1987, 5, 1094-1102.
5. Ahmed, M.; Focht, D. D. *Can. J. Microbiol.* 1972, 19, 47-52.
6. Furukuwa, K.; Matsumura, F. *J. Agric. Food Chem.* 1976, 24, 251-256.
7. Focht, D. D.; Brunner, W. *Appl. Environ. Microbiol.* 1985, 50, 1058-1063.
8. Rhee, G. Y.; Bush, B.; Brown, M. P.; Kane, M.; Shane, L. *Water Res.* 1989, 23, 957-964.
9. Adrianes, P.; Focht, D. *Environ. Sci. Technol.* 1990, 24,1042-1049.
10. Viney, I.; Bewley, R. J. F. *Arch. Environ. Contam. Toxicol.* 1990, 19, 789-796.
11. Thomas, D. R.; Carswell, K. S.; Georgiou, G. *Biotechnol. Bioeng.* 1992, 40, 1395-1402.
12. Hickey, W. J.; Searles, D. B.; Focht, D. D. *Appl. Environ. Microbiol.* 1993, 59, 1194-2000.
13. Barriault, D.; Sylvestre, M. *Can. J. Microbiol.* 1993, 39, 594-602.
14. Alder, A. C.; Haggblom, M. M.; Oppenhelmer, S. R.; Young, L. Y. *Environ. Sci. Technol.* 1993, 27, 530-538.
15. Rhee, G. Y.; Sokol, R. C.; Bush, B.; Bethoney, C. M. *Environ. Sci. Technol.* 1993, 27, 714-719.
16. Barnsley, E. A. *Can. J. Microbiol.* 1975, 21, 1004-1008.
17. Cerniglia, C. E.; Gibson, D. T. *J. Biol. Chem.* 1979, 254, 12174-12180.
18. Bumpus, J. A.; Tien, M.; Wright, D. Aust, S. D. *Science*, 1985, 228, 1434-1436.
19. Sanglard, D.; Leisola, M. S. A.; Fiechter, A. *Enzyme Microb. Technol.* 1986, 8, 209-212.
20. Heitkamp, M. A.; Cerniglia, C. E. *Appl. Environ. Microbiol.* 1989, 55, 1968-1973.
21. Grosser, R. J.; Warshawsky, D.; Vestal, J. R. *Appl. Environ. Microbiol.* 1991, 57, 3462-3469.
22. Grosser, R. J.; Warshawsky, D.; Vestal, J. R. *Environ. Toxicol. Chem.* 1995, 14, 375-382.
23. Atlas, R. M.; Bartha, R. *Microbial Ecology: Fundamentals and Applications*; Benjamin/Cummings: Redwood City, CA, 1992.
24. Anderson, T. A.; Guthrie, E. A.; Walton, B. T. *Environ. Sci. Technol.* 1993, 27, 2630-2636.
25. Hsu, T. S.; Bartha, R. *Appl. Environ. Microbiol.* 1979, 37, 36-41.
26. Reddy, B. R.; Sethunathan, N. *Appl. Environ. Microbiol.* 1983, 45, 826-829.
27. Anderson, T. A.; Kruger, E. L.; Coats, J. R. *Chemosphere* 1994, 28, 1551-1557.
28. Walton, B. T.; Anderson, T. A. *Appl. Environ. Microbiol.* 1990, 56, 1012-1016.
29. Ferro, A. M.; Sims, R. C.; Bugbee, B. *J. Environ. Qual.* 1994, 23, 272-279.
30. Aprill, W.; Sims, R. C. *Chemosphere* 1990, 20, 253-265.
31. Ostle, B. *Statistics in Research*, Iowa State University Press: Ames, IA, 1963.
32. USEPA. *Test methods for evaluating solid waste, SW-846*; U. S. Environmental Protection Agency: Washington, DC, 1992.
33. Shiu, W. Y.; McKay, D. *J. Phys. Chem. Ref. Data* 1986, 15, 911-929.
34. USEPA. *Test methods for evaluating solid waste, SW-846*; U. S. Environmental Protection Agency: Washington, DC, 1986.
35. Kutschera, L.; Lichtenegger, E.; Sobotik, M. *Wurzelatlas mitteleuropaischer Grunlandpflazen*; Gustav Fischer Verlag: NewYork, NY, 1982.
36. Marinucci, A. C.; Bartha, R. *Appl. Environ. Microbiol.* 1979, 38, 1020-1022.

37. Badkoubi, A. *Pentachlorophenol mineralization with different amounts of organic matter in liquid and soil slurry by* Phanerochaete chrysosporium, Ph.D. Dissertation, Utah State University, Logan, UT, 1994.

38. Cheng, W.; Coleman, D. C. *Soil Biol. Biochem.* 1989, 21, 385-388.

39. Bugbee, B. *Crop physiology laboratory manual;* Utah State University: Logan, UT; 1995.

40. Wollum, A. G. In *Methods of soil analysis*, Page, A. L. Ed.; Agronomy No. 9, American Society of Agronomy: Madison, WI 1982; Part 2; pp. 781-802.

41. Moore, D. S.; McCabe, G. P. *Introduction to the practice of statistics*, W. H. Freeman: New York, NY, 1989.

42. Park, K. S.; Sims, R. C.; Dupont, R. R. *J. Environ. Eng.* 1990, 116, 632-640.

43. Karickhoff, S. W. In *Contaminants and sediments*, Baker, R. A., Ed. Ann Arbor Science: Ann Arbor, MI, 1980; pp. 193-205.

44. Leenheer, J. A.; Ahlrichs, J. L. *Soil Sci. Soc. Am. Proc.* 1971, 35, 700-705.

45. Karickhoff, S. W.; Brown, D. S.; Scoot, T. A. *Water Res.* 1979, 13, 241-248.

46. Stevenson, F. J. *Humus chemistry: Genesis, composition, reactions*, John Wiley and Sons: New York, NY, 1982.

47. Bollag, J. M. In *Aquatic and terrestrial humic material*, Christman, R. F.; Gjessing, E. T., Ed.; Ann Arbor Science: Ann Arbor, MI, 1983; pp. 127-141.

48. Aiken, G. R.; Mcknight, D.; Wershaw, R.; McCarthy, P. *Humic substances in soil, sediment and water*, John Wiley and Sons: New York, NY, 1985.

49. Lee, E.; Banks, M. M. *J. Environ. Sci. Health, Part A: Environ. Sci. Eng.* 1993, 28, 2187-98.

50. Alexander, M. *Introduction to soil microbiology, second edition*; John Wiley & Sons: New York, NY, 1977.

Chapter 16

Fate of Benzene in Soils Planted with Alfalfa: Uptake, Volatilization, and Degradation

A. Ferro[1], J. Kennedy[1], W. Doucette[2], S. Nelson[3], G. Jauregui[4], B. McFarland[3], and B. Bugbee[5]

[1]Phytokinetics, Inc., 1770 North Research Parkway, Suite 110, North Logan, UT 84341
[2]Department of Civil and Environmental Engineering, Utah State University, Logan, UT 84322–8200
[3]Chevron Research and Technology, 1003 West Cutting Boulevard, Richmond, CA 94804
[4]Chevron USA Products Company, 6001 Bollinge Canyon Road, San Ramon, CA 94583
[5]Department of Plants, Soils and Biometeorology, Utah State University, Logan, UT 84322–8200

The fate of benzene in planted and unplanted soils was investigated using high-flow sealed test systems specifically designed for the recovery of volatile organic compounds (VOCs). Test systems, containing established alfalfa plants or unplanted controls, were subirrigated with aqueous solutions of [^{14}C]benzene to produce soil concentrations of 40 (low dose) or 620 µg/kg soil (high dose). The test systems allowed us to determine separately root uptake and foliar uptake of the radiolabel and to establish a mass balance for the ^{14}C-label. During the 7 to 10 days experiment, the efflux of ^{14}C-labeled VOCs and $^{14}CO_2$ resulting from mineralization were measured at 12h intervals. At the end of each experiment, soils and plant tissues were analyzed for ^{14}C and, in some experiments, benzene was analyzed using gas chromatography. Less than 2% of the recovered ^{14}C was associated with the plant shoots and between 2% to 8% in the root fraction (root tissue plus rhizosphere soil). No benzene was detected in the soils or plant tissue by gas chromatography, even in the high dose experiments. The average total recovery of added radiolabel in all experiments was greater than 90%. Comparisons of planted and unplanted soils indicated that alfalfa did not enhance the degradation of benzene in this experimental system. The amount of benzene that was volatilized directly from soil was far greater than any that might have evolved from the plants themselves.

Understanding the fate of organic chemical contaminants in plant-soil systems is crucial for evaluating the effectiveness of phytoremediation. Several studies have reported that some plants can accelerate the removal of some classes of organic chemical contaminants from soils (*1-7*), but for volatile organic compounds (VOCs), technical complications in analyzing plant-soil systems have made it difficult to distinguish between degradation, volatilization, and plant uptake (*8-10*). Mathematical models have been developed which predict the behavior of VOCs in plants (*11-15*); so far, however, there have been very few empirical studies (*8-11, 16-20*).

Phytoremediation is now being evaluated on soils contaminated with petroleum hydrocarbons. For soils from some sites, preliminary evidence suggests that the technology may be more effective (and inexpensive) than conventional technologies such as *in situ* bioremediation. However, such sites frequently contain benzene and other VOCs in the near-surface soils. The fate of these compounds in specific plant-soil systems is extremely difficult to predict based on the existing literature (*8, 11-20*). With benzene, for example, it is uncertain if plants can take-up and accumulate levels of the compound in foliage which might create an ecotoxicological hazard, if such uptake might be concentration dependent, or if plants might increase volatility. Alternatively, if benzene is rapidly degraded in the rhizosphere soil, plant uptake may be greatly reduced. The objective of this project was to study the fate of benzene in a specific plant-soil system in order to resolve some of these uncertainties.

Materials and Methods

High-Flow Sealed Test System. A diagram of a single test system is shown in Figure 1 and is a modification of the system described previously (*21*). Three modules were run simultaneously in order to generate triplicate sets of data. The bell jar housed glass columns that contained planted or unplanted soils (Figure 2). Plants were grown in the greenhouse in soil packed in glass columns (Figure 2), and the established plant-soil systems transferred into the bell jar systems. At the start of the experiment, [^{14}C]benzene solutions were injected into a single place ~12 cm below the surface of the soil using long syringe needles. A rapid flow of air through the bell jar (1 L/min) was necessary to remove transpirational water vapor and to reduce the accumulation of ^{14}C-labeled VOCs and $^{14}CO_2$ in the chamber headspace. As discussed previously (*21*), the dilution of $^{14}CO_2$ by CO_2 in the air stream was sufficiently high that no measurable radiolabel in the shoots (see below) could have resulted from $^{14}CO_2$ fixation. Leaks were minimized by keeping the pressure in the bell jar close to atmospheric pressure. The low pressure-differential in the bell jar was achieved by pumping air into and out of the jar. Flow-rates were matched using pairs of rotometers, and pressure in the bell jar was monitored using manometers. Transpirational water vapor in the air effluent from the bell jar was removed with a glass condenser. The air then passed through a CO_2 trap (0.4 M KOH), and subsequently through a trap to remove benzene and potential volatile metabolites (2-methoxyethanol). Bell jars stood in a growth chamber kept at $23^0 \pm 1^0$ C, with a 16-hour photoperiod (photosynthetic photon flux = 350 $\mu mol\ m^{-2}s^{-1}$).

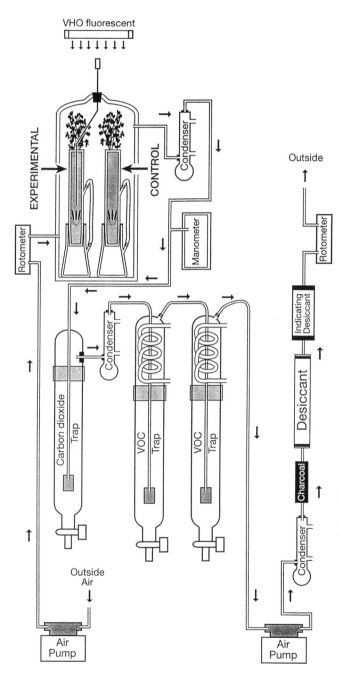

Figure 1. A single high-flow sealed test system. In this diagram, the bell jar (6.3 L) contained two plants: the *experimental* plant (in which the soil was treated with [^{14}C]benzene) and the *control* plant (untreated soil). In some experiments, the bell jars were relatively smaller (1.7 L) and contained only a single planted (or unplanted) soil sample. Flow rates in all experiments were 1 L/min. Condensers and dessicant columns were necessary to remove vapor from the air stream that would otherwise condense in the lines and block the air-flow.

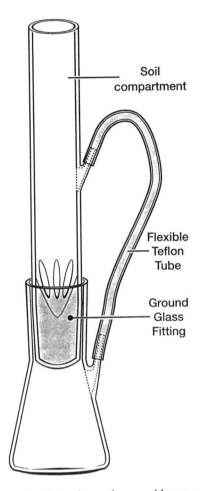

Figure 2. Soils were packed into glass columns with porous bottoms to support the soils. Column capacity ~ 140 g soil. Ground glass fittings were used to connect the columns to flasks. The flasks received leachate. Side arms on the flasks and the soil holders were connected via flexible teflon tubing; this connection released pressure in the flasks which resulted when the soils were irrigated.

Soils. The soil was an artificial loamy sand, made by mixing a native loam soil from the site with coarse washed sand at a ratio of 1:1 (w/w). The loam soil (stored at 4⁰ C before use) was obtained from a former Chevron gas and diesel terminal in Ogden, Utah. The addition of sand to the loam helped maintain adequate aeration, but may have increased [^{14}C]benzene volatilization by increasing the porosity of the soil. Soils at this site have been impacted with petroleum hydrocarbons. The loam soil sample was taken from a zone within the five-acre site where TPH concentrations generally ranged from 300 - 5,000 mg/kg. However, based on GC analysis, the soil sample used in the study had no measurable concentrations of benzene prior to spiking the soil with [^{14}C]benzene (see below). Concentrations of other petroleum hydrocarbons, potentially present in the soil, were not determined. The loamy sand mixture (sand 74%; silt 16%; clay 10%; pH = 7.9; CEC = 77 meq/kg, 1.39% organic carbon; field capacity = 13.8% gravimetric water) was loosely packed into glass columns (Figure 2). The planted and unplanted soils were kept near field capacity by addition of a nutrient solution (0.13 mM KH$_2$PO$_4$), via a long syringe needle, through a septum at the top of the bell jar. Planted systems were watered with 10 ml per day (leachate was less than 0.5 ml/day); this amount of water was sufficient to maintain the plants without creating excessive volumes of leachate. Unplanted systems were watered with ~ 0.8 ml/day. Separate needles were used for the addition of nutrient and [^{14}C]benzene. The root/soil zone was protected from light.

Plants. Alfalfa seeds (*Medicago sativa* Mesa, var. Cimarron VR) were inoculated with *Rhizobium melitoti*. Seeds were sown in the loamy sand, described above, which had been packed into glass columns (Figure 2). Plants were grown for approximately 100 days prior to the start of the experiment, at which time the established plant-soil systems were transferred to the bell jars.

Test Chemical Additions to Soils. [UL-^{14}C]benzene (19.3mCi/mmol) was obtained from Sigma Chemical Company. Ethanolic stock solutions of the [^{14}C]benzene (which in some cases were mixed with solutions of unlabeled reagent grade benzene) were diluted with water prior to the injection into soils. If not used immediately, aqueous solutions of [^{14}C]benzene were stored at 4^0 C in gas-tight glass bottles with no head space. Solutions of [^{14}C]benzene were injected 12 cm below the soils surface via a long syringe needle. As described in Table I, final benzene concentrations in soils were 40 µg/kg ("low" dose experiments) or 622 µg/kg ("high" dose). Leaching of radiolabel added during [^{14}C]benzene addition did not occur because benzene solutions had minimal volumes (however, leachate collection flasks were monitored).
    The following is a list of physicochemical properties of benzene: Molecular weight, 78; vapor pressure, 76 mm Hg (20^0 C); water solubility, 1.8 g/L; density, 0.88 g/ml (20^0 C); log K$_{ow}$ = 2.13; K$_H$ = 0.24 (dimensionless); log K$_{oc}$ = 1.78 (22,23).

Experimental Design. The fate of [^{14}C]benzene in planted systems was determined in trials 1 and 2. One important aspect of these studies was to measure the extent of benzene uptake by the plant roots and translocation to the shoots, and therefore experimental controls were included to measure the extent of foliar uptake (see below). Trial 3 compared the fate of [^{14}C]benzene in planted and unplanted systems, and the "Parafin" experiment specifically examined the potential of alfalfa plants to alter the transfer of benzene from soil to air. Duplicate or triplicate data sets were generated for each trial (Table II) by running two or three flow-through systems

Table I. [^{14}C]Benzene Solutions Added to Planted and Unplanted Soils

Dose	Concentration (mg/L)	Vol. added (ml)	Total addition (μg)	($\mu g/kg$)
Low	1[a]	6[c]	6	40
High	15[b]	6[c]	90	622

[a] Specific radioactivity 110 x 10^3 dpm/μg
[b] Specific radioactivity 28 x 10^3 dpm/μg
[c] 2 ml injected into soils at intervals of 0, 3h, and 6h

simultaneously. Radiolabel was measured in soils, plant tissues, CO_2 traps and volatile organic traps. Subsamples of the [^{14}C]benzene solution that was injected into the soil were analyzed by liquid scintillation counting to determine the amount of [^{14}C]benzene added to each system. The $^{14}CO_2$ and ^{14}C volatile organic traps were monitored at 12h intervals for each system. Means and standard deviations were calculated for each of the duplicate or triplicate sets of data. Extracts of soils and plant tissue were also analyzed for benzene by purge and trap gas chromatography in trial 2 .

Processing of Plant Tissue. At the end of the trial, shoots were cut off at soil level and roots were manually separated from soil. Plant tissues were pulverized in liquid N_2 using a mortar and pestle. The roots were not washed, and therefore some rhizosphere soil was included in the "root fraction" (Table II).

Analysis. Soils and plant tissues from trial 2 were analyzed for benzene using a purge and trap gas chromatographic/flame ionization method (EPA's SW-846 Methods 5030A and 8020, modified by using a flame ionization detector). Samples (Table II) were placed into 40 mL glass vials equipped with teflon lined rubber septa along with sufficient methanol to fill the vials and minimize headspace. The vials containing the soil or plant tissue and methanol were then agitated for 30 minutes in a rotary tumbler. After allowing the soil and methanol phases to separate, the vials were placed in an automated purge and trap system (DynaTrap, DynaTech Precision Sample Corp). The DynaTrap system automatically removed a 1 to 2 mL aliquot of the methanol extract from the sealed vial and introduced it into a purge vessel where it was diluted with 5 mL of deionized water. The water/methanol mixture was purged with helium for 12 minutes and the benzene concentrated on a Tenax sorbent trap. The benzene was thermally desorbed from the trap (275^0 C for 5 minutes) and introduced into a gas chromatograph (Shimadzu GC-14A) equipped with a capillary column (Petrocol 3710; 0.75 mm x 10 m) and flame ionization detector. The retention time for benzene was 5.9 minutes. The detection limits of the chromatographic method were 80 $\mu g/kg$ for soils and root fraction, and 400 $\mu g/kg$ for plant shoots. Methods used for measurement of radiolabel in soils and plant tissues using combustion analysis have been described previously (21). ^{14}C in liquid samples was quantified by scintillation counting.

Table II. Behavior of [^{14}C]Benzene in Planted and Unplanted Soils

Trial	Dose ($\mu g/kg$)	System	Mass Balance[b] (%)	Distribution of Radiolabel[a] (% Recovered)			Plant Tissue	
				VOC	Soil	$^{14}CO_2$	Shoots	Root Fraction
1	40	Planted (n=3)	93 (91-94)	17 $\pm$5	48 $\pm$7	26 $\pm$3	1.7[f] $\pm$1.2	7.5 $\pm$2.0
2[c]	622	Planted (n=2)[e]	119 (109-130)	56	26	13	0.5[f]	4.0
3[d]	40	Planted (n=3)	107 (96-127)	61 $\pm$5	28 $\pm$3	6.3 $\pm$3	1.2 $\pm$0.4	2.0 $\pm$0.9
		Unplant-ed (n=3)	106 (79-123)	70 $\pm$8	21 $\pm$3	8.5 $\pm$5.2	--	--

[a] Data are means $\pm$ 1 standard deviation.
[b] Percent recovery of added radiolabel. Numbers in parentheses are ranges.
[c] Samples of soil (7.0 g), root fraction (7.0 g), and plant shoots (1.2 g) were analyzed for benzene parent compound (Methods). The following are mean values for the total mass of each fraction: soil, 146.6 g; root fraction, 19.0 g; plant shoots 1.4 g.
[d] Because our system accommodated a maximum of 3 modules (Methods), trial 3 was done in two runs, one run with two planted and one unplanted system, and the second run with one planted and two unplanted systems.
[e] In trial 2, a duplicate (rather than triplicate) data set was obtained because one module in the test system (Methods) malfunctioned. Values given in the Table are means. Specific values (%): VOC, 59, 53; Soil, 25, 27; $^{14}CO_2$, 12, 14; Shoots, 0.53, 0.37; Roots, 3.3, 4.6.
[f] Values estimated for foliar uptake have been subtracted from the values shown.

Results

Trial 1; Low Dose of [^{14}C]Benzene Added to Planted Systems. [^{14}C]benzene was added to each of three planted systems at a concentration of 40 $\mu g/kg$ soil (Table II). The volatile organic traps contained measurable levels of ^{14}C, most likely benzene, soon after the application of the [^{14}C]benzene to the subsurface of the soils. The evolution of [^{14}C]VOC continued for approximately 1.5 days reaching a plateau level of 17% (Figure 3). $^{14}CO_2$ was detected after day 1, and evolution continued for about 7 days reaching a plateau level of 26% (Figure 3). Similar time courses for both [^{14}C]VOC and $^{14}CO_2$ evolution were observed for trials 2 and 3 (data not

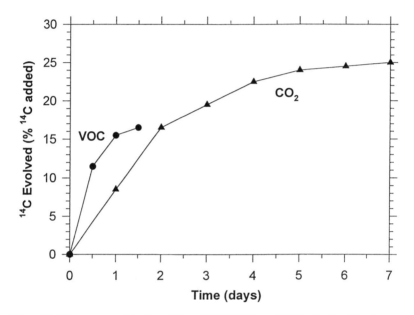

Figure 3. Time course for the efflux of $[^{14}C]VOC$ and $^{14}CO_2$ in Trial 1. A "low" dose of benzene (40 µg/ kg soil) was injected into three planted systems. The dose was applied in three equal portions and into a single place in the soil, at $t = 0$, 3h, and 6h. The efflux of $[^{14}C]VOC$ and $^{14}CO_2$ was monitored at intervals. Efflux of VOC reached a plateau at 1.5 days, after which no further efflux occurred; $^{14}CO_2$ efflux reached a plateau after 7 days. Data points are means (n=3).

shown). At the conclusion of the experiment (at day = 7), the test systems were dismantled and the soils and plant tissues analyzed for ^{14}C. Ninety three percent of the added radiolabel was recovered. Forty eight percent of the recovered radiolabel was found in the soils fraction, 2.9 % in the plant shoots, and 7.5% in the root fraction (Table II).

The ^{14}C recovered in the plant shoots was taken up by the roots and translocated to the shoots and was also the result of foliar uptake (*17, 18, 24*). However, trial 1 (and trial 2, see below) was designed in such a way that we could differentiate between these two sources of radiolabel in the shoots. The flow-through test systems were fitted with bell jars of sufficient size to hold two plants: *Experimental* plants were subirrigated with solutions of [^{14}C]benzene; *control* plants were not subirrigated. The shoots of the control plants were exposed to [^{14}C]benzene only in the air of the bell jar, and served as controls for foliar uptake. For trial 1, experimental plant shoots contained 2.9% of the recovered radiolabel, and the control plant shoots 1.2%. By subtraction, 1.7% of the label in the shoots of the experimental plants was derived from root uptake and translocation (Table III). Thus, for trial 1, our data suggest that 41% of the radiolabel found in the shoots was from foliar uptake and 59% from root uptake and translocation.

Table III. Percent Radiolabel Recovered in Plant Shoots
(Data are means $\pm$ 1 SD).

Trial	Dose (μg/kg)	Experimental Plants (Total[a] uptake)	Control Plants (Foliar uptake only)	Experimental Plants (Root uptake only)
1	40	2.9 ± 1.1	1.2 ± 0.03	1.7 ± 1.2
2	622	0.8 ± 0.3	0.3 ± 0.1	0.5 ± 0.2

[a] Total = foliar uptake + root uptake

Trial 2; High Dose of [^{14}C]Benzene Added to Planted Systems. In trial 2, [^{14}C]benzene was added to yield a soil concentration of 622 μg/kg. Mass balance data, distribution of radiolabel in the various fractions, and data for foliar uptake controls are presented in Tables II and III. Compared to trial 1, a higher percentage of the radiolabel was recovered in the VOC fraction and less in the other fractions (Table II). No detectable levels of benzene were found during the gas chromatographic analysis of soils and plant tissues.

Trial 3; Low Dose of [^{14}C]Benzene Added to Planted and Unplanted Systems. Trial 3 was specifically designed to compare the fate of [^{14}C]benzene between planted and unplanted soils. We were interested in determining if plants could increase the biodegradation of [^{14}C]benzene, or change the extent to which [^{14}C]benzene volatilized from the system. The data summarized in Table II suggested that plants did not cause a statistically significant change in the extent of mineralization of [^{14}C]benzene (6.3 ± 3.0 % of the radiolabel recovered as ^{14}CO$_2$ in the planted systems compared to 8.5 ± 5.2 % in the unplanted systems). Also, the rates of mineralization of [^{14}C]benzene (the time course for ^{14}CO$_2$ efflux) were identical in the planted and unplanted systems (data not shown).

Plants did not significantly change the extent to which [^{14}C]benzene volatilized from the system. The percent of ^{14}C recovered in the volatile traps was 61 ± 5% in the planted systems and 70 ± 8% in the unplanted systems (Table II). However, in comparing the results from various trials, the percentage of ^{14}C recovered in the volatile traps of trial 3 was significantly higher than that observed in trial 1 for a similar loading of [^{14}C]benzene. It is unclear why these large differences were observed, because the mass recoveries were comparable. Therefore, to further examine the potential of plants to alter the transfer of VOCs from soil to air, an additional experiment was performed as described in the next section.

"Parafin" Experiment. In trials 1 to 3, [^{14}C]benzene injected into the root zone could potentially escape from the soil and be collected in the volatile organic traps by direct diffusion through soil pores, as well as by root uptake, translocation, and efflux via the stomates. In order to differentiate between these two possible pathways, an additional experiment was performed in which the surface of the planted soils was sealed with warm liquid parafin, leaving the stems of the alfalfa plants penetrating the seal. This seal should greatly reduce the direct volatilization of benzene. With the seal in place, the plants (in triplicate) were injected with 40 μg/kg dose of benzene. Over the next seven days, the volatile organic and carbon dioxide traps were monitored as previously described for trials 1 to 3. The ^{14}C recovered in the VOC traps was greatly reduced relative to that observed for trials 1 to 3 (Table IV). This result suggested that the majority of the [^{14}C]VOC efflux from the systems was via direct diffusion through soil pores, not through the plants. A small amount of ^{14}CO$_2$ was also recovered (Table IV), suggesting that the parafin seal was slightly leaky: Efflux of ^{14}CO$_2$ probably originated from mineralization of [^{14}C]benzene in soil and escape of ^{14}CO$_2$ from soil pores, rather than originating from plant metabolism.

Discussion

The percent of radiolabel recovered in the VOC traps varied widely in trials 1 to 3 (Table II). If we assume that the majority of this [^{14}C]VOC efflux was due to diffusion through soil pores, as suggested by the "parafin" experiment, then one reason for this variability could be differences in soil moisture content. As discussed by Russell, et al. (15), mathematical models show that transport of VOCs in soil is highly dependent on the soil moisture content. We tried to maintain soil moisture close to field capacity, but this was difficult because the plants were large, relative to the amount of soil, and actively transpiring.

The variability of [^{14}C]VOC efflux complicated the analysis of trial 3, which was a comparison of the fate of [^{14}C]benzene in planted and unplanted soils. In order to facilitate this analysis, we recalculated the results in Table II in terms of "residual" radiolabel: that portion of the recovered radiolabel which was distinct

Table IV. Comparison of [^{14}C]VOC and ^{14}CO$_2$ efflux in planted soils: trials 1 to 3 and in the "Parafin" experiment

Trial	Distribution of Radioactivity (% of Total Added)	
	VOC	^{14}CO$_2$
1 (n=3)	16 ± 4	25 ± 2
2 (n=2)[a]	67	16
3 (Planted; n=3)	65 ± 6	7 ± 5
"Parafin" (n=3)[b]	3.6 ± 3.6	2.1 ± 0.5

[a] Mean values for duplicate measurements are shown in the table.
Specific values (%): VOC, 77, 57; ^{14}CO$_2$, 16, 16
[b] Unplanted soil controls were also performed.
Specific values (%): VOC, 16, 8; ^{14}CO$_2$, 2.6, 2.6

from the [^{14}C]VOC fraction. As shown in Table V, the percent recovery of *residual* radiolabel in the soil fraction in trial 3 was approximately the same in the planted and unplanted systems (~ 72%). Values are also compared for the recovery of residual radiolabel in the ^{14}CO$_2$ fraction (unplanted soil, 25 ± 12 %; planted soil, 16 ± 7 %) and in the plant fraction (8 ± 3 %) (Table V). The analysis in Table V was useful for comparing results for trials 1 and 2 (low dose and high dose experiments, respectively). Note that the distribution of residual radiolabel in the soil and ^{14}CO$_2$ fractions, and in the plant tissues were very similar for trials 1 and 2. Thus, although the [^{14}C]benzene concentration increased 15-fold in trial 2, relative to trial 1, the fate of the benzene apparently remained the same.

The results shown in Table V (trial 3) indicated that alfalfa did not increase the rate or extent of mineralization of [^{14}C]benzene. In a previous study with [^{14}C]phenanthrene (*21*), we observed that crested wheatgrass did not increase the rate or extent of mineralization when planted and unplanted systems were compared (the initial concentration of the compound was 100 mg/kg). However, a "rhizosphere" effect was observed in experiments with [^{14}C]pyrene (*25*). The onset of mineralization occurred sooner in planted systems (mean = 45 days) than in unplanted systems (mean = 75 days). In this latter experiment, again using crested wheatgrass, the extent of [^{14}C]pyrene mineralization was the same in the two types of systems (initial concentrations of pyrene were 100 mg/kg soil). Although plant species-specific effects and compound-specific effects must be further evaluated, one tentative conclusion from these studies is that for readily degradable organics like benzene or phenanthrene, oxygen limitations are more likely to effect the rate and extent of degradation than the presence of plants. In near-surface soils, volatilization may be a more important fate mechanism than degradation for volatile organics. On the other hand, the presence of plant exudates (co-metabolites) may accelerate biodegradation for compounds which might be degraded co-metabolically (e.g., pyrene).

Table V.   Distribution of radiolabel recovered in the residual[a] fraction
for trials 1 to 3

Trial	Dose ($\mu g/Kg$)	System	% Recovered Soil	% Recovered $^{14}CO_2$	% Recovered Plants
1	40	Planted (n=3)	57 ± 6	32 ± 5	11 ± 2
2	622	Planted (n=2)	58 ± 1	30 ± 0.3	10 ± 1
3	40	Planted (n=3)	73 ± 9	16 ± 7	8 ± 3
		Unplanted (n=3)	72 ± 12	25 ± 12	--

[a] That portion of the recovered radiolabel that was distinct from [^{14}C]VOC fraction

Plant uptake of the radiolabel via the roots and translocation to shoots was minimal (trial 1, 1.7% of the recovered radiolabel; trial 2, 0.5%; Table III). These estimations were possible because we carried out controls for foliar uptake. Such controls were essential because the flux of air, even in a high-flow system such as ours, was much lower than would occur outdoors, potentiating the accumulation of high concentrations of VOCs and elevated foliar uptake. Approximately 40% of the radiolabel recovered in the experimental plant shoots was derived from foliar uptake of the radiolabel. However, our gas chromatographic analysis of the experimental shoots did not detect benzene parent compound. The observation that little root-uptake of radiolabel occurred, and no benzene was detected in the shoots, was an important result from a practical standpoint. We are currently using phytoremediation systems at several BTEX-contaminated sites, and the results presented here suggest that the ecotoxicological hazard is minimal.

The extent of plant uptake of organic compounds from soils is dependent upon the physicochemical properties of the compound (14,17,26) as well as its rate of biodegradation in the soil. Rapid biodegradation, of course, greatly limits plant uptake. Radiolabeled benzene was rapidly degraded in the soils used in our experiments. We recovered $^{14}CO_2$ just a few hours after [^{14}C]benzene was injected into the soil (Figure 3), and in trials 1 to 3 a large percentage of the "residual" radiolabel was recovered as $^{14}CO_2$ (Table V). Gas chromatographic analysis of soils at the conclusion of trial 2 did not detect benzene parent compound. Other workers have also observed rapid biodegradation of benzene in soils (16). Therefore, although the physicochemical properties of benzene would predict extensive plant uptake (11,14,17,26), rapid biodegradation evidently prevented this uptake from occurring. These results were similar to those reported for toluene by Narayanan et al. (13).

Several models have been proposed suggesting that VOCs having certain physicochemical properties can be taken up by plant roots, move with the transpirational stream from roots to shoots, and leave the plant via volatilization through the stomatal pathway to air (11-14). This pathway for VOCs in plant systems has recently been observed empirically (19,20). In the "parafin" experiment, the percent of radiolabel recovered in the VOC trap (3.6% evolved in approximately 1.5 days) was close to values for maximum transport flux through plants calculated using a simple model suggested by Russell, et al. (*15*). In this model, J = q C T_{SCF}, where q is the transpiration rate of the plants, C is the concentration of the volatile compound in the soil water in the root zone, and T_{SCF} is the transpiration stream concentration factor proposed by Briggs, et al. (*26*). Using this model for the "parafin" experiment, we calculated that the maximum transport flux through the plants of the added radiolabel equaled 5.6% per day; as with trials 1 to 3, efflux of [^{14}C]VOC was observed for 1.5 days. Of course, this calculation for maximal flux is based on a variety of assumptions, including negligible biodegradation in soil and no plant metabolism of benzene. In fact, the transport flux through the plants was probably less than 3.6%, and some of this [^{14}C]VOC recovered in the "parafin" experiment was probably due to leaks in the parafin seal, as judged by the recovery of $^{14}CO_2$.

Literature Cited

1. Anderson, T. A.; Guthrice, E. A.; Walton, B.T. Bioremediation in the Rhizosphere. *Environ. Sci. Tech.* **1993**, *27*, 2630-2636.

2. Anderson, T. A.; Kruger, E. L.; Coats, J. R. Enhanced Microbial Degradation in the Rhizosphere of Plants from Contaminated Sites. Air and Waste Mgt. Assoc. 86th Annual Meeting 1993, pp 1-11.

3. Shimp, J. F.; Tracy, J. C.; Davis, L. C.; Lee, E.; Huang, W.; Erickson, L. E.; Schnoor, J. L. Beneficial Effects of Plants in the Remediation of Soil and Groundwater Contaminated with Organic Materials. *Crit. Rev. Environ. Sci. Tech.* **1993**, *23*, 41-77.

4. *Bioremediation Through Rhizosphere Technology.* Anderson, T. A.; Coats, J. R., Eds. Am. Chem. Soc.: Washington DC, 1994.

5. Erickson, L. E.; Banks, M. K.; Davis, L. C.; Schwab, A. P.; Muralidharan, N.; Reilley, K.; Tracy, J. C. Using Vegetation to Enhance *In Situ* Bioremediation. *Environ. Progr.* **1994**, *13*, 226-231.

6. Anderson, T. A.; Coats, J. R. An Overview of Microbial Degradation in the Rhizosphere and its Implications for Bioremediation. In: *Bioremediation Science and Applications.* Skipper, H. D.; Turco, R. F., Eds.; Soil Science Society of America Special Publication: Madison, WI, 1995, Vol. 43; pp 135-143.

7. Schnoor, J. L.; Licht, L. A.; McCutcheon, S. C.; Wolfe, N. L.; Carreira, L. H. Phytoremediation of Organic and Nutrient Contaminants. *Environ. Sci. Tech.* **1995**, *29*, 318A-323A.

8. Watkins, J. W.; Sorensen, D. L.; Sims, R. C. Volatilization and Mineralization of Naphthalene in Soil-Grass Microcosms. In: *Bioremediation Through Rhizosphere Technology.* Anderson, T. A.; Coats, J. R., Eds. Am. Chem. Soc.: Washington DC, 1994; pp 123-131.

9. Narayanan, M.; Davis, L. C.; Erickson, L. E. Fate of Volatile Chlorinated Organic Compounds in a Laboratory Chamber with Alfalfa Plants. *Environ. Sci. Technol.* **1995**, *29*, 2437-2444.

10.   Schnabel, W. E.; Dietz, A. C.; Burken, J. G.; Schnoor, J. L.; Alvarez, P. J. Uptake of Trichloroethylene by Edible Garden Plants. *Water Res.* **1996**, (in press).

11.   Trapp, S.; Matthies, M.; Scheunert, I.; Topp, E. M. Modeling the Bioconcentration of Organic Chemicals in Plants. *Environ. Sci. Technol.* **1990**, *24*, 1246-1252.

12.   Lindstrom, F. T.; Boersma, L.; McFarlane, C. Mathematical Model of Plant Uptake and Translocation of Organic Chemicals: Development of the Model. *J. Environ. Qual.* **1991**, *20*, 129-136.

13.   Narayanan, M.; Davis, L. C.; Tracy, J. C.; Erickson, L. E.; Green, R.M. Experimental and Modeling Studies of the Fate of Organic Contaminants in the Presence of Alfalfa Plants. *J. Hazard. Mater.* **1995**, *41*, 229-249.

14.   *Plant Contamination: Modeling and Simulation of Organic Chemical Processes.* Trapp, S.; McFarlane, J. C., Eds.; CRC Press, Inc.: Boca Raton, FL, 1995.

15.   Russell, N. K.; Davis, L. C.; Erickson, L. E. A Review of Contaminant Transport to the Gas Phase above Fields of Vegetation. *Proc. Air and Waste Management Association, 89th Annual Meeting,* Paper No. 96-RP141.01, 1996.

16.   McFarlane, J. C.; Cross, A.; Frank, C.; Rogers, R. D. Atmospheric Benzene Depletion by Soil Microorganisms. *Environ. Monit. and Assess.* **1991**, *1*, 75-81.

17.   Topp, E.; Scheunert, I.; Attar. A.; Korte, F. Factors Affecting the Uptake of ^{14}C-Labeled Organic Chemicals by Plants from Soil. *Ecotoxicol. and Environ. Safety* **1986**, *11*, 219-228.

18.   Schroll, R.; Scheunert, I. A Laboratory System to Determine Separately the Uptake of Organic Chemicals from Soil by Plant Roots and by Leaves after Vaporization. *Chemosphere* **1992**, *24*, 97-108.

19.   Burken, J. G.; Schnoor, J. L. Hybrid Poplar Tree Phytoremediation of Volatile Organic Compounds. American Chemical Society National Meeting, Agrochemicals Division. Orlando, FL, Aug. 25-29, 1996.

20.   Gordon, M.; Choe, N.; Duffy, J.; Ekuan, G.; Heilman, P.; Muiznieks, I.; Newman, L.; Ruszaj, M.; Shurtleff, B. B.; Strand, S.; Wilmoth, J. Phytoremediation of Trichloroethylene with Hybrid Poplars. American Chemical Society National Meeting, Agrochemicals Division. Orlando, FL, Aug. 25-29, 1996.

21.   Ferro, A.M.; Sims, R.; Bugbee, B. Hycrest Crested Wheatgrass Accelerates the Degradation of Pentachlorophenol in Soil. *J. Environ. Qual.* **1994**, *23*, 272-279.

22.   *Chemical Fate and Transport in the Environment.* Hemond, H. F.; Fechner, E. J., Eds.; Academic Press: San Diego, CA, 1994.

23.   Karickhoff, S.W. Semi-Empirical Estimation of Sorption of Hydrophobic Pollutants on Natural Sediments and Soils. *Chemosphere* **1981**, *10*, 833-846.

24.   Kerler, F.; Schönherr, J. Accumulation of Lipophilic Chemicals in Plant Cuticles: Prediction from Octanol/Water Partition Coefficients. *Arch. Environ. Contam. Toxicol.* **1988**, *17*, 1-6.

25. Ferro, A.M.; Stacishin, L.; Doucette, W.; Bugbee, B. Crested Wheatgrass Accelerates the Degradation of Pyrene in Soil. Emerging Technologies in Hazardous Waste Management VI. Industrial and Engineering Chemistry Division, Special Symposium American Chemical Society. September 19-21, 1994.

26. Briggs, G. G.; Bromilow, R. H.; Evans, A. A. Relationships between Lipophilicity and Root Uptake and Translocation of Non-ionized Chemicals by Barley. *Pestic. Sci.* **1982**, *13*, 495-504.

Chapter 17

Metabolism of Chlorinated Phenols by *Lemna gibba*, Duckweed

Harry E. Ensley[1], Hari A. Sharma[2], John T. Barber[2], and Michael A. Polito[2,3]

Departments of [1]Chemistry and [2]Ecology, Evolution and Organismal Biology, Tulane University, New Orleans, LA 70118

The toxicity and metabolism of phenol and a series of chlorinated phenols, 4-chlorophenol to pentachlorophenol, in axenically grown *Lemna gibba* were studied. It was found that the toxicities of the phenols tended to increase with increasing number of chlorine substituents on the phenol ring. Over relatively short incubation periods (< 7 days), the plants metabolized each of the phenols in the same manner, producing compounds that were more polar than the corresponding phenol from which they were derived. The plant-produced metabolites of phenol, 2,4-dichlorophenol and 2,4,5-trichlorophenol were isolated, purified and their structures were identified by high field NMR and chemical ionization MS to be β-glucoside conjugates. It was further shown, by GC/MS, that over longer incubation periods (*ca* 20 days), the plants were able to progressively dechlorinate the phenols. While conversion of the chlorinated phenols to their corresponding phenyl glucosides results in compounds that are more water-soluble and less toxic to the plants than were the parent phenols, the potential for regeneration of the original phenols, as a result of low pH or enzymatic cleavage of the glucoside, remains. In contrast, reductive dechlorination represents a real detoxification since the toxicity of the chlorinated phenols decreases with decreasing number of chlorine substituents. It is possible therefore, that the ability of duckweed to perform reductive dechlorination can be exploited as part of a remediation technology.

Phenol and the 19 chlorinated phenols are used extensively as pesticides, herbicides, fungicides, bactericides, etc. In addition, significant amounts of chlorophenols are generated as a result of the bleaching process in paper mills (*1*), the chlorination of drinking water and the incineration of waste materials (*2*). The magnitude of the problem

[3]Current address: Isotron Corporation, 13152 Chef Menteur Highway, New Orleans, LA 70129

238

is indicated by the production figures for phenol alone; 4.16 billion lbs in 1995, making it 34th on the list of most-produced organics in the U.S. (*3*). The types of usage for phenol and the chlorophenols, the methods used for their application and the casual way in which they were stored or disposed of in the past, ensures that they are common contaminants that almost inevitably end up in aquatic environments. Many Superfund sites are contaminated with significant concentrations of chlorophenols (*4*).

The ways in which organisms deal with phenol and the chlorinated phenols are of considerable importance to human health for a variety of reasons including the potential risks of contamination via food webs, as well as the potential benefits resulting from bioremediation. Higher plants, including many aquatics, have received a certain amount of attention in this regard and Sandermann has termed plants the "green liver" of the earth (*5*). Studies of the metabolism of chlorophenols by terrestrial angiosperms have included 2,4,6-trichlorophenol and tomatoes (*6*); pentachlorophenol (PCP) and rice (*7*); PCP and soybean and wheat (*8*). The metabolism of phenol and chlorophenols by aquatic angiosperms has also been studied, e.g. PCP and *Elodea* (*9*); phenol and *Eichornia* (*10*).

The duckweeds (Lemnaceae) are floating monocotyledenous aquatic angiosperms and are ideal candidates for metabolic studies of xenobiotics. Almost all permanent fresh or brackish surface waters contain members of the duckweed family (*11*) and, as their name implies, they constitute a significant portion of the summer diet of waterfowl as well as several invertebrate and vertebrate species. The duckweeds are, therefore, ecologically important. In addition, the duckweeds can be grown in sterile culture under controlled environmental conditions and are, therefore, experimentally valuable for toxicity and metabolism studies.

Several previous investigations of the toxicity and metabolism of phenol and the chlorophenols have employed members of the Lemnaceae. *Lemna minor* was used to demonstrate the effects of substituted chlorine atoms on the biological activities of chlorinated phenols (*12,13*), as well as to investigate metabolism of PCP (*9, 14*). Most recently, *L. gibba* was used to investigate the toxicity and metabolism of 2,4-dichlorophenol (2,4-DCP) and phenol (*15, 16*).

The conclusions of these and other papers (e.g. *5, 17*) is that transformation, conjugation and compartmentation are common responses of higher plants to challenges by organic contaminants. However, while there is a growing body of information concerning the metabolic capabilities of higher plants with regard to xenobiotics, and while innumerable papers have considered and recommended higher plants as agents for inorganic remediation (e.g. *18*), their potential for organic remediation is less clear. This study presents results that hold promise for the use of a higher plant for this purpose.

Toxicity Toward *L. gibba*

Numerous methods are available to assess the toxicity of xenobiotics to duckweed. These include determination of chlorophyll content, biomass and frond reproduction. We have found the most convenient, reproducible and sensitive method to evaluate toxicity is vegetative frond reproduction.

Stocks of *L. gibba* G3 were maintained in sterile culture in 125-ml Erlenmeyer flasks, containing 50 ml of the medium described by Cleland and Tanaka (*19*), plus sucrose and tryptone to reveal microbial contamination. The cultures were grown in

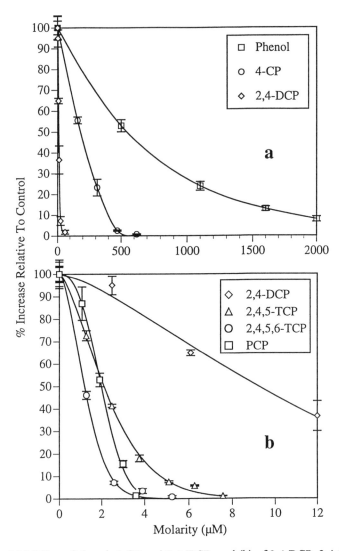

Figure 1.(a) Effect of phenol, 4-CP and 2,4-DCP and (b) of 2,4-DCP, 2,4,5-TCP and PCP on the vegetative reproduction of *L. gibba*. Each data point represents the average of 10 replicates. Absolute numbers of fronds in control cultures at Day 7 were between 375 and 425. Due to the wide range in toxicities of the phenols tested, some data points for certain phenols are off-scale.

chambers under continuous light (2.0×10^{-3} watts/cm^2; 300-1100 nm at plant level) at 82°F. Experimental cultures, ten replicates of each, were started using approximately 15 fronds per culture (i.e. five, three-frond colonies each). The number of fronds in each culture at Day 0 was noted. The phenol under study was added aseptically from a stock solution to give a range of concentrations up to at least EC$_{90}$. After a 7-day growth period, the fronds were counted in each flask and the percent frond increases, relative to controls, were calculated according to the formula:

$$\frac{(\text{Fronds at Day 7 - Fronds at Day 0})_{\text{Exp}} \times (\text{Fronds at Day 0})_{\text{Con}}}{(\text{Fronds at Day 0})_{\text{Exp}} \times (\text{Fronds at Day 7 - Fronds at Day 0})_{\text{Con}}} \times 100 \quad ;$$

where 'Exp' refers to the frond count in the presence of the phenol and 'Con' refers to the frond count in the controls. Since the controls grown in the absence of phenols show frond increases in the range of 2000-3000%, even relatively small toxic effects can be observed.

The results of phenol and chlorinated phenol toxicity are shown in Figure 1a and 1b. As increasing numbers of chlorine substituents are added to the aromatic ring, the toxicity increases until three chlorines are present. At that point a "leveling-off" effect is observed and 2,4,5-tri- (2,4,5-TCP), 2,4,5,6-tetra- (2,4,5,6-TCP), and pentachlorophenol (PCP) have about the same high level of toxicity. Although there are several possible reasons for the similarity in toxicity of 2,4,5-TCP, 2,4,5,6-TCP, and PCP, it has been suggested that chlorophenols interfere with electron transport and/or pH changes that occur during phosphorylation (*20*). Increasing chlorination of the phenol would facilitate electron acceptance. It is possible that for 2,4,5-TCP, 2,4,5,6-TCP, and PCP electron acceptance is no longer rate limiting. Huber and coworkers (*14*) have shown that photosynthesis in *L. minor*, measured in terms of O$_2$ produced, is more sensitive to PCP than any other parameters measured. Schnabl and Youngman (*21*) have presented evidence that PCP blocks the PS II step of photosynthetic electron transport. The EC$_{10}$ and EC$_{50}$ values, with regard to the vegetative frond reproduction are shown in Table I. The values were calculated using the exponential equation derived from the growth data.

Table I. Effective Concentrations of Phenols Studied		
Compound	EC$_{10}$ ($\pm$ SE) in μM	EC$_{50}$ ($\pm$ SE) in μM
Phenol	80.0 (9.2)	540.0 (37.0)
4-Chlorophenol	34.0 (4.4)	183.0 (13.9)
2,4-Dichlorophenol	2.5 (0.2)	9.2 (0.5)
2,4,5-Trichlorophenol	0.6 (0.1)	2.1 (0.2)
2,4,5,6-Tetrachlorophenol	0.4 (0.1)	1.2 (0.2)
Pentachlorophenol	1.0 (0.2)	2.0 (0.3)

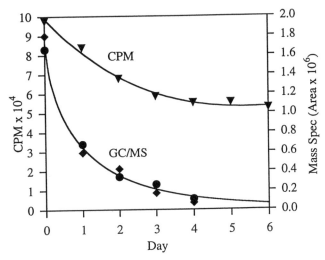

Figure 2. Analyses of *L. gibba* growth media by scintillation spectrometry (upper trace) and GC/MS (lower trace). The data from two replicate GC-MS analyses are shown. SOURCE: Adapted from ref. 15.

Adapted Figures 2, 5, and 6 are reprinted with permission from *Environmental Toxicology and Chemistry,* 1997. "Toxicity and Metabolism of 2,4-Dichlorophenol By The Aquatic Angiosperm *Lemna gibba*" by Harry E. Ensley, John T. Barber, Michael A. Polito, and Ana I. Oliver, 1994, Vol. 13(2). Copyright SETAC 1994.

Metabolism Studies

To investigate the disappearance of the phenols from the growth media of *L. gibba*, plants were grown, as described above, in the presence of sub-lethal concentrations of each phenol. Since the cultures were to be analyzed by GC-MS, the sucrose and tryptone were omitted from the plant growth media and 100 plants were initially present to allow higher concentrations of the phenol to be used. It is important to note that EC_{10} and EC_{50} values reported in Table I depend critically on the experimental conditions. The ratio (number of plants):(quantity of phenols) is equally as important as the concentration of phenol. Since the number of plants used in these large incubations was very high, the concentrations of phenols used were actually below the EC_{50}. The first phenol studied in this way was 2,4-DCP (initial concentration 15 μM). At daily intervals, 100 μl samples of media were aseptically withdrawn from the growth flasks and 2 μl aliquots were injected into a GC/MS operated in the single ion monitoring mode (m/e 162 and 164). The resulting peaks were integrated and the disappearance of 2,4-DCP from the growth media is shown in Figure 2 (lower trace). Surprisingly, the 2,4-DCP concentration was reduced by more than 90% after 4 days. Identical incubations were conducted in which [U-^{14}C] 2,4-DCP (0.5 μCi) was added as a tracer. At daily intervals 100 μl of media were removed and assayed for radioactivity. This study (Figure 2, upper trace) shows that the radioactivity in the media falls to about 50% of the initial activity after 4 days and subsequently remains constant. Apparently, the 2,4-DCP had been taken up by the plants, metabolized and the metabolite(s) had been released to the media.

To confirm the metabolism of the 2,4-DCP, experiments were conducted as described above with sucrose and tryptone omitted from the growth media and 2,4-DCP was added to give a concentration of 15 μM in the flask, along with [U-^{14}C] 2,4-DCP (0.5 μCi) as a tracer. At 24 hour intervals, 1 ml aliquots of growth media were withdrawn aseptically from the cultures and were analyzed by reverse-phase HPLC (RP-HPLC, 250 mm x 4.6 mm i.d.; packing — C18, 5 micron particle size, injection volume 100 μl). A linear elution gradient program was employed using HPLC-grade acetonitrile and deionized-distilled water as the solvent system, with a flow rate of 1 ml min^{-1}. The gradient ran over a 30 minute interval, starting with a mixture of 10% acetonitrile/90% water, and ending with a mixture of 99% acetonitrile/1% water. One ml fractions were collected and assayed for radioactivity by liquid scintillation counting. The results were then plotted as number of counts *vs* fraction number, from which the disappearance of 2,4-DCP and the concomitant appearance of the metabolite(s) were evident, Figure 3a. The incubation time for maximum production of the major metabolite was determined to be 6 days.

Similar metabolism studies were subsequently conducted with phenol (Figure 3b), 2,4,5-TCP, (Figure 4a) and PCP (Figure 4b). In the case of phenol and 2,4,5-TCP, metabolism was very similar to that observed with 2,4-DCP. A major metabolite was formed which was subsequently released back into the media. When *L. gibba* was exposed to PCP under the same conditions, a slightly more polar metabolite was formed initially but this was further converted to a much more polar metabolite. By the end of an eight-day incubation, only the very polar metabolite was present in the growth media. For all of the phenol incubations, harvesting the plants at the end of the incubation period, homogenization and extraction with methanol, followed by reverse-phase HPLC,

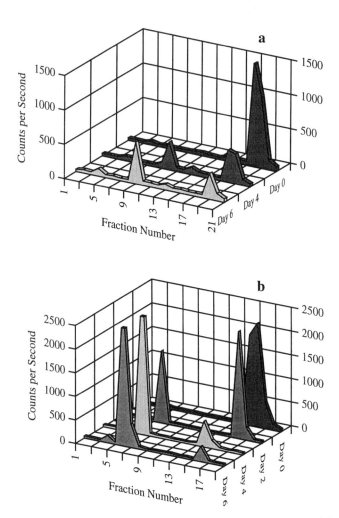

Figure 3.(a) Reversed-phase HPLC of *L. gibba* growth media which initially contained contained 15 μM 2,4-DCP and (b) 540 μM phenol. The growth media also contained 0.5 μCi of the corresponding [UL-^{14}C] labelled compound.

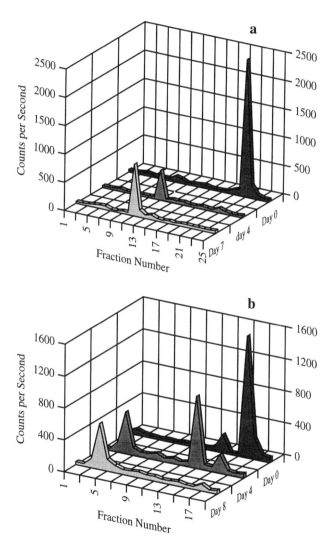

Figure 4.(a) Reversed-phase HPLC of *L. gibba* growth media which initially contained contained 2.1 μM 2,4,5-TCP and (b) 2.0 μM PCP. The growth media also contained 0.5 μCi of the corresponding [UL-¹⁴C] labelled compound.

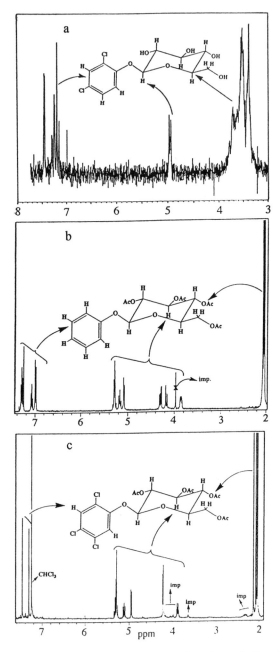

Figure 5. Proton NMR spectrum of the plant-produced metabolite of 2,4-DCP (a, 200 MHz); of the acetylated metabolite of phenol (b, 400 MHz) and of the acetylated metabolite of 2,4,5-TCP (c, 400 MHz). The structures corresponding to the spectra are shown above the spectra. SOURCE: Portions adapted from ref. 15 and 16.

indicated that the same metabolites were also present in the plants, although in smaller quantities.

Isolation and Identification of the Metabolites

Having shown that each of the phenols studied were converted to a major metabolite that was partially released back to the growth media, the next task was to isolate sufficient quantities of the metabolite for identification. The isolation of the major metabolite required large-scale incubations (500 ml of medium in 3-L Fernbach flasks). A sublethal concentration (approximately EC_{50}) of the phenol, spiked with radiolabelled substrate as a tracer, was added to the medium, along with approximately 1,000 fronds. As stated earlier, EC_{10} and EC_{50} values depend on ratio of the number of plants to the quantity of phenols. Since the number of plants used in these large incubations was very high, the concentrations of phenols used were actually below the EC_{50} for that number of plants.

To date we have isolated and identified the metabolites produced from phenol, 2,4-DCP and 2,4,5-TCP. Because of the dramatic differences in toxicities of these three phenols and the resulting concentrations in which the plants will survive, varying numbers of large scale incubations were performed. One large scale incubation of phenol (initial concentration 800 μM) provided sufficient material for the identification of the metabolite. In the case of 2,4-DCP (initial concentration 10 μM) five large scale incubations were required and for 2,4,5-TCP (initial concentration 3 μM) ten were required.

The plants were harvested after 6 days and the growth medium was lyophilized. The plants were mixed with the lyophilized medium and homogenized in methanol. The homogenized mixture was filtered and the filtrate was reduced in volume under vacuum. The resulting methanol extract was passed through a silica gel column (60-200 mesh) and eluted first with ethyl acetate and then with methanol. Radioactivity assays of a sample of the eluate showed that the metabolite was contained in the methanol fraction, which was reduced in volume and subjected to RP-HPLC using a semi-preparatory column (300 mm x 10 mm i.d.; packing — C18, 5 micron particle size). The solvent system and elution gradient were the same as that used for the analysis of the growth medium described above, with the exception that the flow rate was adjusted to 5 ml min^{-1} to compensate for the larger column. The HPLC was equipped with an in-line radioactivity detector and stream splitter. Ten percent of the effluent was assayed for radioactivity and the remaining 90% was collected. The appropriate fractions were combined, reduced in volume using rotary evaporation, and the resulting metabolite analyzed by NMR.

The proton NMR spectrum of the plant-produced metabolite of 2,4-DCP showed that the aromatic nucleus substitution pattern was unchanged but additional signals at 4.95 ppm and in the 3.3-3.9 ppm range were also present (Figure 5a). The doublet at 4.95 ppm is diagnostic of an anomeric proton and the coupling constant of 7.3 Hz shows that it is a β-glucoside. An α-glucoside would have a J-value of 3-4 Hz (22). As further evidence that the metabolite was a β-glucoside, it was treated with β-glucosidase and the products were analyzed using RP-HPLC and scintillation counting. The analysis showed that the metabolite was not present following enzymatic digestion and that free 2,4-DCP, which co-chromatographed with authentic 2,4-DCP, had appeared where there had been very little prior to enzyme treatment (Figure 6). Synthesis of an authentic sample of

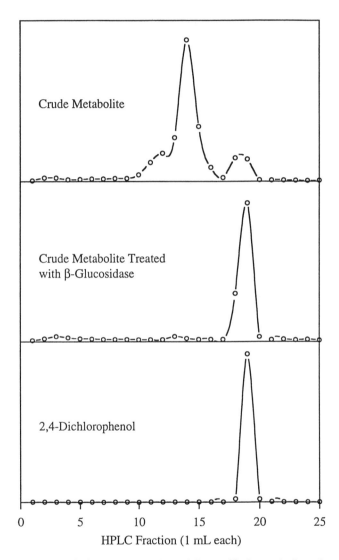

Figure 6.   Reversed-phase HPLC of partially purified metabolite of 2,4-DCP (upper plot), of the same metabolite after incubation with β-glucosidase (center plot), and of authentic 2,4-DCP (lower plot). SOURCE: Adapted from ref. 15.

2,4-dichlorophenyl-β-D- glucopyranoside (*23*) and comparison with the metabolite by NMR and mass spectroscopy confirmed that the two samples were identical. Both the synthetic and natural metabolites were silylated and analyzed by GC-MS. Both derivatives had the same GC retention time and showed m/e values at 614 (M^+), 597 (M^+-CH_3), 507 (M^+-((CH_3)$_3$SiOH + CH_3)) and 451 (M^+-2,4-DCP).

The major metabolites of phenol and 2,4,5-TCP were isolated in a similar fashion but the metabolite could not be completely separated from endogenous material. In both cases the NMR signals attributable to the aromatic ring, the axial anomeric proton and the monosaccharide were observed; however, the presence of numerous inseparable impurities prevented direct identification. In order to identify the metabolites of phenol and 2,4,5-TCP, each metabolite was acetylated (excess acetic anhydride in pyridine) and the corresponding tetra-acetate was further purified by flash chromatography (*24*). The acetylated metabolite of phenol (Figure 5b) was confirmed as 2,3,4,6-tetraacetylphenyl-β-D- glucopyranoside by comparison to an authentic standard (*16*). The acetylated metabolite of 2,4,5-TCP (Figure 5c) was confirmed as 2,3,4,6-tetraacetyl-2,4,5-trichlorophenyl-β-D- glucopyranoside by comparison to an authentic standard (*23*) also. The very polar metabolite formed from PCP has not yet been identified. In the case of phenol it has been shown that the conjugate formed by *L. gibba* (phenyl β-D-glucopyranoside) is half as toxic as phenol itself (Figure 7). The EC_{10} and EC_{50}, with regard to frond reproduction, for the conjugate were 0.15 mM and 1.10 mM, respectively, and for phenol itself were 0.08 mM and 0.54 mM, respectively.

Reductive dechlorination

Incubations of each of the phenols showed that a maximum amount of the major metabolite was produced after 6-8 days. Longer incubations indicated that the major metabolite itself was being further metabolized. A very minor radioactive fraction isolated from the incubation of labeled 2,4-DCP with duckweed after a 10 day growth period showed an NMR spectrum similar to that of 2,4-dichlorophenyl-β-D- glucopyranoside except that four aromatic protons were present and appeared as an AB quartet, indicating the loss of the chlorine atom at the 2-position. Taking this as an indication that duckweed may be able to reductively dechlorinate chlorophenols, long term incubations were conducted to investigate this possibility.

Large scale incubations of 2,4,5-TCP (initial concentration 15 μM, 5 x 500 ml of medium in 3-L Fernbach flasks, 1000 plants at Day 0) were conducted as described above with the exception that the incubations were continued for 20 days. At the end of this period the plants and media were homogenized in a stainless steel homogenizer. The homogenate was then extracted with CH_2Cl_2 (5 x 100 ml). The extract was concentrated and analyzed for phenol, mono-, di- and trichlorophenols by selected ion monitoring GC-MS. Only trace quantities of phenol and 2,4,5-TCP were detectable. Concentrated HCl (100 ml) was then added to the homogenate, and the mixture was refluxed overnight to hydrolyze any glucoside conjugates. The mixture was again extracted with CH_2Cl_2 (5 x 100 ml). Evaporation of the CH_2Cl_2 gave a complex mixture of products which could not be analyzed directly. The phenolic portion was isolated by extraction into 2N NaOH, acidification and extraction with CH_2Cl_2. Analysis of this extract by GC-MS showed very little 2,4,5-TCP remained (3%) but large quantities of dechlorinated phenols had been

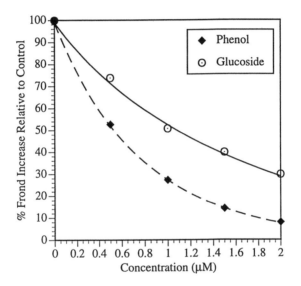

Figure 7. Effect of phenol and phenyl-β-D-glucopyranoside on the vegetative reproduction of *L. gibba*. Each data point represents the average of ten replicates. SOURCE: Adapted from ref. 16.

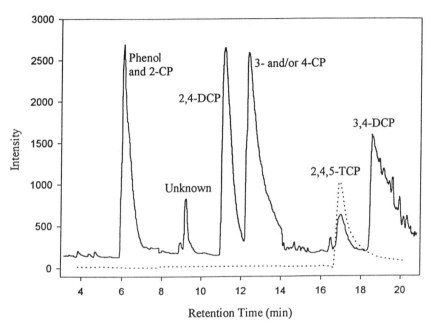

Figure 8. GC-MS of phenolic fraction from a 20-day incubation of 2,4,5-TCP with *L. gibba* (solid line). The 2,4,5-TCP standard is shown for comparison (dotted line).

produced (Figure 8). The peak with retention time 6.0 minutes (14%) is a mixture of both phenol and 2-chlorophenol. The peak at 11.1 minutes (16%) is 2,4-DCP and the peak at 12.3 minutes (24%) is 4-chlorophenol and/or 3-chlorophenol. The peak at 18.4 minutes (19%) is 3,4-DCP. Both the retention times and fragmentation patterns were confirmed by comparison to authentic standards. The dechlorinated phenols were present as the corresponding glucoside conjugates, indicating that conjugation is probably a prerequisite for reductive dechlorination. In addition, dechlorination is non-selective and loss of the *ortho-*, *meta-* and *para-* chlorines appears equally facile.

Dechlorination of chlorophenols is common in certain types of bacteria, fungi and also animals (*25*); however, reductive dechlorination has not previously been reported for higher plants grown under aseptic conditions. Barnett and coworkers (*26*) have studied the metabolism of PCP by soybeans and found the production of 2,3,4,6-tetrachlorophenol, methoxytetrachlorophenol, 2,3,4,6-tetrachloroanisole and pentachloroanisole. Haque and coworkers (*27*) have observed the formation of 2,3,4,6-tetrachlorophenol on treatment of rice plants with PCP. However, these studies were not conducted under sterile conditions and the possibility that the dechlorination was a result of bacterial action was noted by both authors.

While conjugation of phenol or the chlorinated phenols to form β-D-glucopyranosides results in decreases in their toxicities (Figure 7), it is not a permanent detoxification. We have shown that 2,4-dichlorophenyl-β-D-glucopyranoside contained in duckweed fed to red swamp crayfish (*Procambarus clarkii*) is rapidly cleaved in the crayfish stomach back to glucose and 2,4-DCP (unpublished results). In contrast, progressive dechlorination of a chlorophenol by duckweed results in progressive decreases in the toxicity (Figures 1a and 1b) which are irreversible. While bacterial dechlorination of chlorophenols has been demonstrated and while it has been suggested for higher plants, this paper shows that higher plants could find use in constructed wetlands for the remediation of these types of organic contaminants.

ACKNOWLEDGEMENTS

This work was supported in part by the Department of the Interior, U.S. Geological Survey through the Louisiana Water Resources Research Institute, the Defense Nuclear Agency (Grant #2, Order # 89, 116, 88-150) and the Department of Energy (Grant # DE-FG01-93EW53023).

LITERATURE CITED

1. Parker, W.J.; Farquhar, G.J.; Hall, E.R. *Environ. Sci, Technol.* **1993**, *27*, 1783 - 1789.
2. WHO, 1993. Chlorophenols other than pentachlorophenol. World Health Organization Publication, Geneva, Switzerland.
3. Chemical and Engineering News. 1996. Facts and figures for the chemical industry, June 24, p 41, American Chemical Society, Washington, D.C.
4. U.S. Environmental Protection Agency, 1993. EPA updates CERCLA priority list of hazardous substances. Haz. Waste Consult. May/June, 2.26 - 2.30.
5. Sandermann, H., Jr. *Trends in Biochem.Sci.* **1992**, *17*, 82-84
6. Fragiadakis, A.; Sotiriou, N.; Korte, F. *Chemosphere* **1981**, *10*, 1315-1320.

7. Weiss, U.M.; Moza, P.; Scheunert, I.; Haque, A.; Korte, F *J. Agric. Food Chem.* **1982**, *30*, 1186-1190.

8. Langebartels, C.; Harms, H. *Z. Pflanzenphysiol.* **1984**, *113*.S. 201 - 211.

9. Hedtke, S.F.; West, C.W.; Allen, K.N.; Norberg-King, T.J.; Mount, D.I. *Environ. Toxicol. Chem.* **1986**, *5*, 531-542.

10. O'Keefe, D.H.; Wiese, T.E. ; Brummet, S.R.; Miller, T.W. *Recent Adv. Phytochem.* **1987**, *21*, 101-129.

11. Hillman, W.S. *Bot. Rev.* **1961**, *27*, 221-289.

12. Blackman, G.E.; Parke, M.H.; Garton, G. *Arch. Biochem. Biophys.* **1955**, *54*, 45 - 54.

13. Blackman, G.E.; Parke, M.H.; Garton, G. *Arch. Biochem. Biophys.* **1955**, *54*, 55 - 71.

14. Huber, W.; Schubert, V.; Sauter, C. *Environ. Pollut. Ser. A Ecol. Biol.* **1982**, *29*, 215 - 223.

15. Ensley, H.E.; Barber, J.T.; Polito, M.A.; Oliver, A.I. *Environ. Toxicol. Chem.* **1994**, *13*, 325 - 331.

16. Barber, J. T.; Sharma, H. A.; Ensley, H. E.; Polito, M. A.; Thomas, D. A. *Chemosphere* **1995**, *31*, 3567-3574.

17. Lamoureux, G.L. and Rusness, D.G. In *Xenobiotic Conjugation Chemistry.*, Editors, G.D. Paulsen, J.J. Caldwell, D.H. Huston and J.J. Menn; ACS Symposium Series 299; American Chemical Society: Washington, D.C., **1986**; pp 62-105.

18. Wang, T.C.; Weissman, J.C.; Ramesh, G.; Varadarajan, R.; Benemann, J.R. *In* Bioremediation of Inorganics; Batelle Press: Columbus, OH, **1995**, pp 65-69.

19. Cleland, C. F.; Tanaka, O. *Plant Physiol.* **1979**, *64*, 421-424.

20. Dedonder, A.; Van Sumere, C. F. *Z. Pflanzenphysiol.* **1971**, *65*, 70-80.

21. Schnabl, H.; Youngman, R. H. *Angew. Botanik* **1987**, *61*, 493-504.

22. Casu, B.; Reggiani, M.; Gallo, G. G.; Vigevani, A. *Tetrahedron* **1966**, *22*, 3061-3083.

23. Sinnott, M. L.;Souchard, I. J. L. *Biochem. J.* **1973**, *133*, 89-97.

24. Still, W.C., Kahn, M., Mitra, M. *J. Org. Chem.* **1978**, *43*, 2923.

25. Ahlborg, U. G. In *Pentachlorophenol: Chemistry, Pharmacology and Environmental Toxicology*, Editor, K. R. Rao; Plenum Press: New York, N. Y., **1978**; pp 115-130.

26. Casterline, J. L.; Barnett, N. M.; Ku, Y. *Environ. Res.* **1985**, *37*, 101-118.

27. Haque, A.; Scheunert, I.; Korte, F. *Chemosphere*, **1978**, *1*, 65-69.

Chapter 18

Rhizosphere Effects on the Degradation of Pyrene and Anthracene in Soil

S. C. Wetzel[1], M. K. Banks[1], and A. P. Schwab[2,3]

Departments of [1]Civil Engineering and [2]Agronomy, Kansas State University, Manhattan, KS 66506

Polycyclic aromatic hydrocarbons (PAHs) are among the more resistant compounds found in petroleum contaminated soils and persist even after extensive bioremediation. Phytoremediation has been demonstrated to enhance the degradation of PAHs, but the mechanisms of dissipation have not been identified. The degradation of pyrene and anthracene was investigated in a laboratory study in which soil was removed from the rhizosphere of a long-term stand of alfalfa and compared to degradation in non-rhizosphere and sterile soil. Low molecular weight organic acids typically found in the rhizosphere were added to the soils to determine if exudation of simple organic compounds may be part of the rhizosphere effect. Dissipation in non-sterile soils was found to be much greater than in sterile soil, but there was no rhizosphere effect and the addition of organic acids did not enhance degradation. The effect of the rhizosphere on PAH degradation seems to be short-lived and requires the continued presence of roots.

Polycyclic aromatic hydrocarbons (PAHs) are a group of hydrophobic organic compounds composed of varying numbers of condensed aromatic (benzene) rings and arranged in different configurations. These lipophilic chemicals are ubiquitous in the environment (usually in minute concentrations as contaminants) and are generally formed during the combustion, pyrolysis, and pyrosynthesis of organic matter. PAHs contain only hydrogen and carbon atoms with two or more benzene rings. The arrangement and number of the rings contained within the molecule results in a wide range of physical and chemical characteristics. The concern over PAHs stems from evidence of mutagenic effect in bacterial and animal cells and the carcinogenic effects in animals (1, 2). Therefore, the bioavailability of PAH contaminants is considered to be the major concern for polluted soils.

 The fate and transport of PAHs in soil are affected by those factors that determine the partitioning of the compounds between the solid, aqueous, and vapor

[3]Corresponding author

phases. Some PAHs are volatile enough that atmospheric transport can be an important environmental consideration (*3*), but Park (*4*) observed negligible volatilization from soil for PAH molecules with three or more rings. This observation appears to be the combined result of very low aqueous solubility (1 µg/L or less) and high adsorption to soil (log K_{oc} > 4). Leaching of PAHs also is limited by low water solubility and strong partitioning to the soil surfaces, but this can be modified by cosolvent effects and dissolution in dense nonaqueous phase liquids (5, 6). Transport of these insoluble compounds in soils also may be facilitated to a small degree by adsorption to and moving with mobile organic and inorganic colloids (*7*). Overall, PAHs are immobile and persistent in soil.

Biodegradation is a potentially important means of removing PAHs from contaminated soil. Properly stimulated, indigenous soil microorganisms can degrade PAHs through complete mineralization (*8*), cometabolic degradation (*9*), and non-specific radical oxidation (*10*). The availability of PAHs for biodegradation is limited by the combination of their very low aqueous solubilities and high degree of adsorption. Thus, these compounds tend to remain in the soil at relatively high concentrations after the successful bioremediation of other compounds.

The soil region under the immediate influence of plant roots and in which there is proliferation of microorganisms is called the rhizosphere (*11*). This zone has properties that have the potential to enhance bioremediation of recalcitrant compounds because of elevated concentrations of naturally-occurring organic materials and high microbial activities. Organic materials include low molecular weight exudates, metabolic secretions, plant mucilages, and gelatinous mucigel, and these compounds are involved in many processes including providing microbial substrates; decreased adsorption of otherwise strongly sorbed contaminants through surfactant activity; and, enhanced physical contact between roots, microorganisms, and water. The rhizosphere influence can project as much as 20 mm from the root (11).

There is persuasive evidence that the rhizosphere community of plant roots and elevated microbial populations offer a potentially significant means by which to remediate chemically contaminated sites *in situ*. Laboratory and greenhouse studies have demonstrated degradation in the rhizosphere of herbicides (*12, 13, 14*), insecticides (15, 16, 17), surfactants (*18*), and petroleum products (*19, 20*).

Research Objectives

The objectives of this laboratory research study were to:
(1) Quantify the potentially beneficial effects of rhizosphere soil on the degradation of pyrene and anthracene. In this study, soil from the rhizosphere of alfalfa was used.
(2) Assess the importance of low molecular weight organic acids typically found in root exudates on the degradation of pyrene and anthracene.
(3) Apply a rigorous statistical design to the experiment to allow complete comparisons.

The results of these experiments will provide important information with respect to the biodegradation of PAHs in the rhizosphere.

Materials and Methods

Soils and Treatments. The soil used in this study was an Ivan silt loam (fine-silty, mixed, mesic, Cumulic Hapludolls) obtained from the Department of Agronomy agricultural farm, Manhattan, KS. The soil was analyzed for a variety of properties and background PAH concentrations (Table I) and sieved.

Table I. Chemical and physical properties of the soil used in this study

Parameter	Rhizosphere Soil	Non-rhizosphere soil
P (mg/kg)	31	50
K (mg/kg)	195	160
NH_4^+-N (mg/kg)	5.2	3.5
NO_3-N (mg/kg)	4.9	7.6
pH	7.03	6.9
CEC (cmol/kg)	16	18
% Organic matter	3.1	2.5
Field Capacity (%)	21	21
PAH Background		
Anthracene (mg/kg)	<0.2	<0.2
Pyrene (mg/kg)	<0.3	<0.3

Three kinds of soil were used: Control sterilized soil, non-rhizosphere soil, and rhizosphere soil. Non-rhizosphere soil was autoclaved three times at 120°C and 20 psi for 30 minutes to obtain control sterilized soil. The three soil types received either of two amendments: an aqueous solution of 0.01 M calcium chloride or an organic acid solution. The organic acid solution contained 40 µM acetic acid, 15 µM succinic acid, and 10 µM formic acid and was used to simulate organic acid production in the rhizosphere. These acids were by far the dominant organic compounds in the solution phase of this soil, and the concentrations are representative of those determined experimentally (21). All solutions were mixed in deionized, distilled water which was autoclaved and filtered through a 0.22 µm filter prior to application to the soil.

The three soils were amended with two PAH compounds, anthracene (Aldrich Chemical Company, Milwaukee, WI) and pyrene (Sigma Chemical Company, St. Louis, MO). The target compounds were dissolved in Optima Grade acetone (Fisher Scientific, St. Louis, MO) and applied to the soil in three separate stages to obtain a concentration of 100 mg/kg.

Experimental Design. Rhizosphere soil from alfalfa (*Meticago sativa*) was collected by shaking loose soil adjacent to the root. Non-rhizosphere soil was collected from an unvegetated area near the alfalfa field. The alfalfa had been growing in this site for 3

years, and the unvegetated site had been devoid of vegetation for at least 2 years. The soil was amended and placed in 125-mL Erlenmeyer flasks with foam stoppers. Eight time periods were selected for analyses of PAHs in the soil: 0, 3, 7, 14, 25, 35, 45, and 56 days. All treatments were applied in quadruplicate. Six treatments were imposed: sterile soil with water, sterile soil with organic acids, non-rhizosphere soil with water, non-rhizosphere soil with organic acids, rhizosphere soil with water, and rhizosphere soil with organic acids.

Six grams of moist soil were placed in each flask. The total weight of the flask with and without soil was recorded, as was the initial moisture content for each of the three soil types. The soil contained in each flask was maintained at the optimum water content of 21% moisture by weight by daily additions of the appropriate solution (water or organic acids). Abiotic samples stored at 4°C for each soil treatment were analyzed by the same methods and at identical intervals as previously described.

At each sampling time, one block of 24 flasks was analyzed. The soil in each flask was homogenized and a subsample of 2.5 g was placed in a 20-mL glass scintillation vial. The remaining soil was used to determine soil moisture content. The PAH compounds were extracted using 10 mL of acetone which was added to the vials and shaken for 60 minutes. The slurry was filtered through Whatman #42 paper and the clear filtrate was collected. One mL of filtrate was transferred to a 1.8-mL glass gas chromatography (GC) vial for GC analysis. Recoveries of PAHs in excess of 90% were found in a preliminary study.

Analytical Methods. Analyses of samples were performed using a Hewlett-Packard 5890A Gas Chromatograph equipped with an HP 3396 Integrator and HP7673A Autosampler. A J&W DB-5 capillary column (J&W Scientific, Folson, CA) and flame ionization detector (FID) were used. The column had an inside diameter of 0.25 mm and length of 30 m, with a film thickness of 0.25 μm. The carrier gas and the fuel source for the FID was H_2. The fuel gas was delivered at 45.0 mL/minute and the carrier gas at 5.0 mL/minute. The make-up gas consisted of N_2 delivered at a flow rate of 30.0 mL/minute, and zero grade air was delivered at a rate of 420 mL/minute. An extended initial isothermal period (35°C for 6 minutes) was implemented to ensure the FID signal returned to near baseline after the solvent peak. The oven temperature increased at an initial rate of 70°C/minute to 110°C, followed by a secondary heating rate of 20°C/minute to a temperature of 300°C. The volume of the injected sample was 2 μL, and the temperatures of the injection port and the detector were 250°C and 310°C, respectively.

Statistical Analyses. The statistical design was completely randomized with two factors (time and soil treatment) and four replications. Analysis of variance was performed using Cohort software (Cohort, Berkeley, CA) with Duncan's multiple range test (P<0.05) for mean separation.

Safety Considerations. None of the chemicals used in this experiment were considered to be unusually hazardous; nevertheless, certain precautions were taken to ensure the

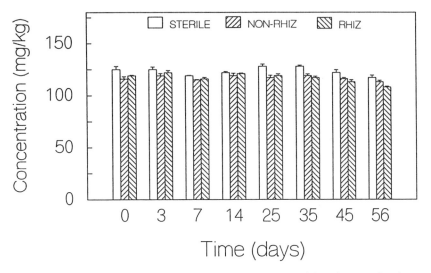

Figure 1. Concentrations of anthracene in soil as a function of time for samples that were kept at 4 °C to minimize biological degradation. Error bars represent one standard deviation. NON-RHIZ refers to non-rhizosphere soil, and RHIZ refers to rhizosphere soil.

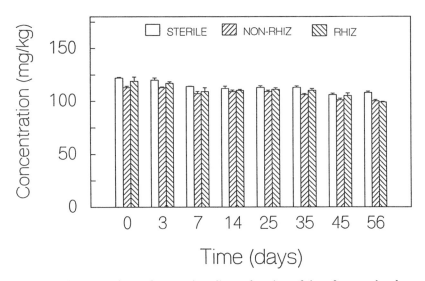

Figure 2. Concentrations of pyrene in soil as a function of time for samples that were kept at 4 °C to minimize biological degradation. Error bars represent one standard deviation. NON-RHIZ refers to non-rhizosphere soil, and RHIZ refers to rhizosphere soil.

health and safety of the laboratory personnel. All solvents were transferred under a hood. Anthracene and pyrene have not been demonstrated to be carcinogenic or mutagenic, but complete protective clothing was used whenever these materials were used. All waste materials were disposed through the Kansas State University Office of Campus Safety.

Results and Discussion

Abiotic Dissipation. Samples of rhizosphere, non-rhizosphere, and sterile contaminated soil were stored at 4 °C to restrict microbial degradation and allow us to evaluate non-biological dissipation including irreversible adsorption, abiotic degradation, and photolysis. Because the bags of soil were sealed, volatilization was limited. For anthracene in the abiotic samples (Fig. 1), the initial sampling indicated average concentrations of 125 mg/kg for the sterile soils, 116 mg/kg for non-rhizosphere, and 119 mg/kg for rhizosphere soil. On day 56, the concentrations had decreased to 117 mg/kg for the sterile soil, 113 mg/kg for the non-rhizosphere soil, and 108 mg/kg for the rhizosphere soil. For all soil samples, abiotic dissipation of anthracene was less than 10%.

For pyrene in the abiotic samples (Fig. 2), the initial concentrations were 122 mg/kg for the sterile soil, 113 mg/kg for non-rhizosphere soil, and 118 mg/kg for the rhizosphere soil. After 56 days, the concentrations had decreased to 108, 100, and 99 mg/kg, respectively, for an average dissipation of 13%. Thus, when the temperature of incubation is higher than 4 °C, dissipation must be greater than 10% to be considered more than simple non-biological dissipation.

Degradation as Affected by Soil Type and Amendment. An analysis of variance was performed to determine statistical differences in pyrene and anthracene concentrations as a function of time, source of soil, and treatment. For both PAHs, there was a significant interaction between source of soil and sampling date, indicating that differences between soil types changed with time. Both pyrene and anthracene were degraded to significantly lower concentrations in the non-rhizosphere and rhizosphere soils than those of the sterilized soil (Figs. 3 and 4). Differences in concentrations between sterile and non-sterile soils for anthracene and pyrene were significant at 35 days and beyond. This is despite significant biological dissipation (>10%) of these PAHs in the sterile controls, probably the result of microbiological contamination or regrowth in the sterile soils.

For both pyrene and anthracene, there was not a statistically significant difference in dissipation between rhizosphere and non-rhizosphere soils. Initially, anthracene was degraded more slowly in the rhizosphere soil than in the non-rhizosphere soil, but by day 56 both soils had performed the same. Pyrene degradation followed the same trends; however, by day 56 the non-rhizosphere soil actually had slightly greater degradation than the rhizosphere soil. This is in contrast to observations of other researchers (12,19,22). In this experiment, the rhizosphere soil came from a stand of alfalfa that had been in place for approximately 3 years; the non-rhizosphere soil had

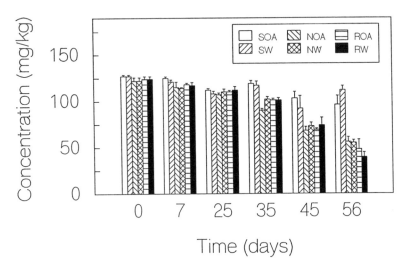

Figure 3. Concentrations of anthracene in experimental soils as a function of time. Error bars represent one standard deviation. S - sterile soil; N - non-rhizosphere soil; R- rhizosphere soil; W - irrigated with water only; OA - irrigated with a solution of organic acids.

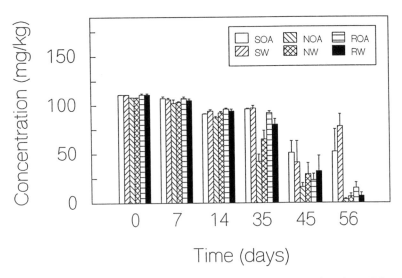

Figure 4. Concentrations of pyrene in experimental soils as a function of time. Error bars represent one standard deviation. S - sterile soil; N - non-rhizosphere soil; R- rhizosphere soil; W - irrigated with water only; OA - irrigated with a solution of organic acids.

been bare for at least two years. It appears that any rhizosphere effect that was responsible for accelerated PAH degradation in the presence of alfalfa (*22*) quickly disappeared when the soil was removed from the root zone.

Low molecular weight organic acids are commonly exuded in the root zone, and these exudates may be responsible for previously observed rhizosphere effects. Therefore, one of the soil treatments was to irrigate the soil with a solution of typical organic acids to help stimulate degradation. There were no significant differences in final pyrene and anthracene concentrations between those soils with organic acids added and those without, regardless of the source of the soil. As a result, we would conclude that a) rhizosphere effects are transient and disappear quickly in the absence of plants, and b) this effect is more than the simple exudation of these organic acids.

The degradation of anthracene was less than that for pyrene, an observation that contrasts with previous studies. Typically, the recalcitrance of PAHs is a direct function of the molecular weight of the compound (*8*). However, the reverse trend in degradation between anthracene and pyrene could possibly be the result of the soil moisture content, solubilities, and bioavailability of the compounds, as was shown by Verschueren (*23*). Our observations are in partial agreement with those of Stetzenbach et al (*24*) who reported that pyrene degradation exceeded that of anthracene in a mineral salts medium, used to simulate ground water. Only after eight weeks did anthracene degradation finally exceed that of pyrene.

Previous studies in our laboratory clearly demonstrated that anthracene and pyrene degraded more rapidly in soil in which alfalfa was growing than in unvegetated soil (22). However, in this study, the rhizosphere effect was no longer apparent in the absence of the plant, and the addition of organic acids neither simulated nor restored this effect. Therefore, until the mechanisms of phytoremediation have been determined, studies concerning the degradation or mineralization of recalictrant organic contaminants should be conducted in the presence of growing plants.

Literature Cited

1. Dipple, A., Cheng, S.C., and Bigger, C.A.H. In *Mutagens and Carcinogens in the Diet*; Pariza, M.W., Aeschbacher, H.U., Felton, J.S., and Sato, S., Eds. Wiley-Liss, New York, NY, 1990.
2. Thakker, D.R., Yagi, H., Levin, W., Wood, A.W., Connery, A.H., and Jerina, D.M.

   In *Bioactiviation of Foreign Compounds*; Anders, M.W., Ed. Academic Press, Orlando, FL, 1985.
3. Charbeneau, R.J., and Weaver, J.W. *J. Hazard. Mater.* 1992, *32*, 293-311.
4. Park, K.S., Sims, R.C., Dupont, R.R., Doucette, W.J., and Mathews, J.E. *Environ. Toxicol. Chem.* 1990, *9*, pp. 187-195.
5. McCarthy, J., and Zachara, J.. Environ. Sci. Technol. 1989, *23*, 496-502.
6. Hall, C.W. *J. Hazard. Mater.* 1992, *32*, pp. 215-223.
7. McCarthy, J.F., Williams, T.M., Liang, L., Jardine, P.M., Jolley, L.W., Taylor, D.L., Palumbo, A.V., and Cooper, L.W. *Environ. Sci. Technol.* 1993, *27*, pp. 667-676.

8. Cerniglia, C.E. In *Organic Substances and Sediments in Water: Biological;* Baker, R.A., Ed. Lewis Publishers, Chelsea, MI, vol 3, 1991.

9. Mahro, B., Schaefer, G., and Kastner, M. In *Bioremediatioin of Chlorinated and Polycyclic Aromatic Hydrocarbon Compounds;* Hinchee, R.E., Leeson, A., Semprini, L., and Ong, S.K., Eds. Lewis Publishers, Boca Raton, FL, 1994.

10. Bumpus, J.A., Tien, M., Wright, D., and Aust, S.D. *Science,* 1985, *228,* pp. 1434-1436.

11. Paul, E.A., and Clark, F.E. *Soil Microbiology and Biochemistry,* Academic Press, Inc, San Diego, CA, 1989.

12. Sandmann, E.R.I.C., and Loos, M.A., *Chemosphere,* 1984, *13,* pp. 1073-1084.

13. Lappin, H.M., Greaves, M.P., and Slater, J.H. *Appl. Environ. Microbiol.,* 1985, *49,* pp. 429-433

14. Liu, C.M., McClean, P.A., Sookedeo, C.C., and Cannon, F.C. 1991, *Appl. Environ. Microbiol.,* *57,* pp. 1799-1804.

15. Sato, K. In *Interrelationships Between Microorganisms and Plants in Soil;* Vancura, V., and Kunc, F. Eds. Publishing House of the Czechoslovak Academy of Sciences, Prague, 1989.

16. Reddy, B.R., and Sethunathan, N., *Appl. Environ. Microbiol.,* 1983, *45,* pp. 826-829.

17. Hsu, T.S., and Bartha, R., *Appl. Environ. Microbiol.,* 1979, *37,* pp. 36-41.

18. Federle, T.W., and Schwab, B.S., *Appl. Environ. Microbiol.,* 1989, *55,* pp. 2092-2094.

19. Walton, B.T., and Anderson, T.A. *Appl. Environ. Microbiol.,* 1990, *56,* pp. 1012-1016.

20. Aprill, W., and Sims, R.C. Chemosphere, 1990, *20,* pp. 253-265.

21. Jones, R.D. 1991. Soil chemistry in a brome meadow under long-term fietilization. Ph.D. Dissert., Kans. St. Univ., Manhattan. (Diss. Abstr. AAC 9201009).

22. Reilley, K, Banks, M.K., and Schwab, A.P. *J. Environ. Qual.* 1996, *25,* pp. 212-219.

23. Verschueren, K., and Visschers, M.J., *Toxicol. Environ. Chem.,* 1988, *16,* pp. 245-258.

24. Stetzenbach, L.D., Kelley, L.M., and Sinclair, N.A. In *Groundwater Contamination and Reclamation.* Amer. Water Resour. Assoc., Bethesda, MD.

PHYTOREMEDIATION OF METALS

Chapter 19

Arabidopsis thaliana as a Model System for Studying Lead Accumulation and Tolerance in Plants

J. Chen[1], J. W. Huang[1,3], T. Caspar[2], and S. D. Cunningham[1]

[1]Dupont Central Research and Development, Environmental Biotechnology, GBC–301, P.O. Box 6101, Newark, DE 19714–6101
[2]Dupont Central Research and Development, Experimental Station 402–4220, Wilmington, DE 19880–0402

In addition to the often-cited advantages of using *Arabidopsis thaliana* as a model system in plant biological research (*1*), *Arabidopsis* has many additional characteristics that make it an attractive experimental organism for studying lead (Pb) accumulation and tolerance in plants. These include its fortuitous familial relationship to many known metal hyperaccumulators (Brassicaceae), as well as similar Pb-accumulation patterns to most other plants. Using nutrient-agar plates, hydroponic culture, and Pb-contaminated soils as growth media, we found significant variation in *Arabidopsis thaliana* ecotypes in accumulation and tolerance of Pb. In addition, we have found that Pb accumulation is not obligatorily linked with Pb tolerance, suggesting that different genetic factors control these two processes. We also screened ethyl methanesulfonate-mutagenized M_2 populations and identified several Pb-accumulating mutants. Current characterization of these mutants indicates that their phenotypes are likely due to alteration of general metal ion uptake or translocation processes since these mutants also accumulate many other metals in shoots. We expect that further characterization of the ecotypes and mutants will shed light on the basic genetic and physiological underpinnings of plant-based Pb remediation.

Recently, the use of plants to remove toxic heavy metals from contaminated land and water has had increasing attention (*2*). Among the many metals which are being studied, Pb is often considered as a primary target due to its wide distribution and potentially serious effects on children who ingest it. It is for this reason that we are currently seeking plants that take-up, translocate, and tolerate large amounts of Pb. Unfortunately, unlike other heavy metal contaminants, few naturally occurred

[3]Current address: Phytotech Inc., 1 Deer Park Drive, Suite I, Monmouth Junction, NJ 08852

"hyperaccumulators" of Pb have been reported (*3*). The scarcity of confirmed Pb-hyperaccumulating plants, combined with little basic physiological and genetic information, suggests that a model plant system should be studied. *Arabidopsis thaliana*, a widely used weed in plant biological research (*4*), has many characteristics to be a useful model system for the study of Pb accumulation and tolerance in plants.

The genus *Arabidopsis* belongs to the mustard or crucifer family (Brassicaceae or Cruciferae), a widely distributed family of approximately 340 genera and 3350 species, with greatest abundance of species and genera in the temperate zone of the northern hemisphere (*5*). There are about 27 species assigned to the genus *Arabidopsis*, 22 of which occur within central Asia and adjacent areas of the Himalayas, which could be the center of greatest diversity of the genus in its current circumscription. This region is also very rich in apparently indigenous populations of *Arabidopsis thaliana*, which now has become a cosmopolitan weed with diverse ecotypes around world (*6*). *Arabidopsis thaliana* is a self-pollinated diploid plant with five pairs of chromosomes (2n = 10). *Arabidopsis* is particularly useful as a plant molecular biology tool as it has the smallest genome among the higher plants (the haploid nuclear DNA content is considered to be 100 Mb) with remarkably little dispersed repetitive DNA (*7*). In addition, *Arabidopsis* has a small plant size, and can be readily grown in confined laboratory environments with only six to eight weeks to fulfill a life cycle and to bear prolific seeds (*4*).

Beyond these general advantages for genetics and molecular biology, *Arabidopsis* also has a rich history of being used in studying plant responses to nutrient deficiency and metal toxicity. The earliest attempt at using *Arabidopsis* to isolate nutritional mutants dates to the 1960s, when first thiamine auxotrophs (*8*), and then mutants with lesions in nitrate utilization were isolated (*9, 10*). Recently, classical genetics, combined with molecular techniques, has greatly accelerated mutant isolation and characterization from *Arabidopsis*. For example, mutants showing increased uptake of iron (Fe) were isolated and the phenotype was attributed to increased activity of Fe(III) chelator reductase regardless of iron deficiency or sufficiency in growth media (*11*). Other examples include two phosphorus (P) translocation mutants. One, *pho1*, with low phosphate (Pi) in leaves was found due to a defect in translocating P from roots to shoots (*12*), whereas another, *pho2*, with high P in shoots had abnormal regulation of Pi translocation (*13*). More recently, *Arabidopsis* has been used for isolating mutants in metal tolerance and/or sensitivity. For example, an *Arabidopsis* mutant with increased resistance to aluminum (Al) was shown to release citrate into the rhizosphere, and citrate subsequently acted as a chelator which prevented uptake of Al into the root apex (*14*). Howden et al (*15, 16, 17*) screened mutagenized *Arabidopsis* M2 populations and isolated several mutants which were defective in phytochelatin biosynthesis, and thus sensitive to cadmium (Cd).

We recently initiated a research program using *Arabidopsis thaliana* to study Pb accumulation and tolerance in plants. In this chapter we report on: (A) the overall feasibility of using *Arabidopsis* as a model system for studying plant responses to Pb, (B) development of culture media for studying Pb accumulation and tolerance in *Arabidopsis*, (C) the natural variation of *Arabidopsis* ecotypes in accumulation and tolerance of Pb, (D) screening of ethyl methanesulfonate (EMS) mutagenized *Arabidopsis* M2 populations to identify mutants with increasing Pb accumulation and tolerance, (E) results of some partially characterized mutants.

Arabidopsis as a model system for studying plant responses to Pb

Although *Arabidopsis* has been used in almost all aspects of plant biological research (*1, 4*), the utility of this plant as a model system for studying plant accumulation and tolerance of Pb remains unproved. We selected *Arabidopsis* not only because of those often-cited advantages, but also its familial relationship with many known metal hyperaccumulators which are also members of the Brassicaceae. For example, *Thlaspi caerulescens* which is reported to be a Zn hyperaccumulator, and *Thlaspi rotundifolium*, a Pb hyperaccumulator, belong to this family (*3*). Recently, *Brassica juncea*, another member of the Brassicaceae, was identified as a Pb accumulator (*18*) and is being used in field trials for Pb phytoextraction by Phytotech, Inc. (Raskin, I., Rutgers University, Personal communication, 1996). In order to determine whether *Arabidopsis* is a good system for modeling plants in accumulation and tolerance of Pb, we compared Pb-accumulation patterns of *Arabidopsis* with those of some "normal" and reported Pb-hyperaccumulating plants using a Pb-contaminated soil, as well as observed phenotypic changes of *Arabidopsis* after exposure to Pb.

Seeds of *Ambrosia artemissifolia*, *Arabidopsis thaliana* (Columbia), *Brassica juncea*, *Thlaspi caerulescens*, *Thlaspi rotundifolium*, and *Zea may* (cv. Fiesta) were germinated in potting mixture supplemented with nutrient solution. Seedlings were transplanted into pots containing a Pb-contaminated soil 10 days after germination. The soil properties were described previously (*19*) with pH 5.1 and total Pb 2,500 mg kg^{-1}. Lead in the soil solution, recovered through centrifugation, was at 3.5 mg L^{-1}. Plants were harvested three weeks after transplanting, and Pb in shoots and roots was measured by inductively coupled argon-plasma emission spectrometry (ICP) (*19*). The pattern of Pb accumulation in *Arabidopsis* was similar to that of the other plant species, including the reported hyperaccumulators (Table I), i.e. Pb accumulated primarily in roots with only small amounts translocated to shoots.

The appearance of *Arabidopsis* after exposure to Pb resembled that of the other plants as well. The first visible symptom was the inhibition of root elongation, and subsequently, reduction of the root size. Effects on shoot growth were evident five days after germination. Purple coloration and eventual chlorosis occurred in lower leaves, and subsequently, shoot growth was reduced. The purple coloration of leaves could be reversed by foliar application of phosphorus (P), suggesting that P deficiency was implicated in Pb toxicity. The phenotypic appearance, together with the patterns of Pb accumulation after exposure to Pb, indicated that responses of *Arabidopsis* to elevated Pb were similar to the responses of other plant species. *Arabidopsis thaliana*, therefore, should be useful as a model system for studying Pb accumulation and tolerance in plants, and potentially isolating genes which might be of use in converting "normal" high biomass plants to Pb hyper-accumulators.

Culture media for studying Pb accumulation and tolerance in *Arabidopsis*

Lead, compared with other heavy metals, has some unique properties that make it a difficult metal to study (*20*). First, Pb is a soft Lewis acid which means it can form strong covalent bonds with other soil ions. As a result, Pb is reported to be the least mobile among the heavy metals, and the concentration of Pb in soil solution is generally low even in the soils with large amounts of Pb (*21*). Secondly, unlike

Table I. Pb in roots and shoots of selected plants grown in a Pb-contaminated soil[a]

Plant species	Root	Shoot
	(mg kg^{-1})	
Ambrosia artemissifolia	2100	75
Arabidopsis thaliana (Columbia)	2500	308
Brassica juncea	2400	129
Thlaspi caerulescens	5000	58
Thlaspi rotundifolium	6400	79
Zea may (cv. Fiesta)	1300	225

[a]Total Pb in the soil was 2,500 mg kg^{-1}. Lead in the soil solution was 3.5 mg L^{-1}. Plants were harvested three weeks after transplanting, and Pb in the plants was measured by ICP.

some other target elements, Pb is a nonessential but toxic metal to plants. Plants do not normally accumulate Pb. Thirdly, Pb often interacts with other nutrient elements, such as Ca, P, and Zn, during the processes of plant absorption, translocation, and use of the nutrients (*21*). These characteristics suggest that all culture solutions and other culture media used for the study of Pb should have certain basic parameters, such as pH and concentrations of P and Pb, similar to natural soil conditions. Furthermore, these parameters should be standardized between culture media and maintained over the course of the experiment to ensure that results from one culture medium are reproducible in another.

We have collected and characterized many Pb-contaminated soils and found that total Pb in the soils ranged from 2,000 to 8,000 mg kg^{-1}, but Pb in the soil solution was only about 4.0 mg L^{-1}. The soil pH was around 5.0. We then simulated the effects of pH and P concentrations on Pb activity in nutrient solution using the GEOCHEM-PC program (*22*). The simulation suggested that with pH at 4.5-5.0 and P concentration less than 0.3 mg L^{-1}, most of the additional Pb in a range of 2-21 mg L^{-1} was in a soluble form. Based on the information obtained from Pb-contaminated soils as well as from the simulation, one Pb contaminated soil collected from an industrial site in northern New Jersey (19) was used as culture medium for *Arabidopsis* growth. Total Pb in the soil was 2,500 mg kg^{-1}, while soluble Pb, recovered by centrifugation, was 3.5 mg L^{-1}. The soil pH (1:1 soil to water ratio) was 5.1.

In addition, a nutrient solution, representing soil solution characteristics of the Pb-contaminated soil, was made. The composition of the nutrient solution was 78 mg L^{-1} K (KNO_3, KH_2PO_4, and KCl), 20 mg L^{-1} Ca ($Ca(NO)_2$), 4.9 mg L^{-1} Mg and 6.4 mg L^{-1} (MgSO$_4$), 1.8 mg L^{-1} NH$_4$ (NH$_4$NO$_3$), 192 mg L^{-1} NO$_3$ (NH$_4$NO$_3$, Ca(NO$_3$)$_2$, and KNO$_3$), 0.3 mg L^{-1} P (KH$_2$PO$_4$), 1.8 mg L^{-1} Cl (KCl), 0.1 mg L^{-1} B (H$_3$BO$_4$), 0.1 mg L^{-1} Mn (MnSO$_4$), 33 μg L^{-1} Zn (ZnSO$_4$), 13 μg L^{-1} Cu (CuSO$_4$), 9.6 μg L^{-1} Mo (Na$_2$MoO$_4$), 5.9 μg L^{-1} Ni (NiSO$_4$), 4.1 mg L^{-1} Pb (PbNO$_4$), and 1.1 mg L^{-1} Fe as Fe-HEDTA. The solution pH was maintained between 4.5 and 5.0. Using this nutrient solution, a hydroponic culture system was set up with a continuous flow of the nutrient solution (100 mL h^{-1}) from reservoir controlled by multi-channel cartridge pump (Cole-Parmer Instruments). We also designed a nutrient agar medium for growing *Arabidopsis* seedlings. The nutrient agar medium was prepared using the same nutrient solution with the addition of 0.8% agar and 0.5% sucrose. The pH in the nutrient agar medium was 5.0.

The nutrient solution, Pb-contaminated soil, and nutrient agar medium were then used as culture media for evaluation of *Arabidopsis*. Table II presents Pb concentration in roots and shoots of *Arabidopsis* grown in the three culture media. The Pb concentration was measured from seedlings 10 days after sowing seeds onto the agar medium, and from plants three weeks after transplanting to solution culture or Pb-contaminated soil. Lead in roots varied depending on the growth medium, while Pb in shoots was comparable among the three media . These three culture media were therefore used for further experimentation.

Table II. Lead in roots and shoots of *Arabidopsis thaliana* (Columbia) grown in three culture media

Culture medium	Root	Shoot
	(mg kg^{-1})	
Nutrient agar[a] (4.1 mg L^{-1} Pb)	2600	305
Pb-contaminated soil[b] (2,500 mg kg^{-1})	2500	325
Solution culture (4.1 mg L^{-1} Pb)	54300	358

[a]Lead in seedlings was measured 10 days after sowing seeds onto agar medium.
[b]Lead in plants was measured three weeks after transplanting to solution culture or Pb-contaminated soil, Pb in the soil solution was 3.5 mg L^{-1}.

Variation of *Arabidopsis* ecotypes in Pb accumulation and tolerance

There is a long history of soil contamination by Pb (*23*). It is generally believed that plants that survive in heavily contaminated soils with high available levels of

Pb, must evolve specific tolerance mechanisms. Tolerance strategies have been documented as either metal exclusion or accumulation (*24*), but few Pb-hyperaccumulating plants have been reported (*3, 18, 24*). Those plants that have been identified as hyperaccumulators are either non-cultivated weeds or open-pollinated plants, invariably with large genome size and are not often suitable for genetic analysis. Thus far, information on plant intraspecific variation in accumulation and tolerance of Pb is limited (*18*). Moreover, few have addressed the relationship between Pb accumulation and tolerance. The goals of the initial phase of this study was to use *Arabidopsis* ecotypes to address: (i) the existence of intraspecific variation in Pb accumulation and tolerance, and (ii) the relationship between Pb accumulation and tolerance.

Seventy-four *Arabidopsis* ecotypes, representing collections from widely diverse geographic areas, were evaluated in nutrient agar medium with 15.5 mg L^{-1} Pb. Root length was measured 10 days after sowing seeds. A tolerance index (TI), defined as a ratio of root length on agar medium containing 15.5 mg L^{-1} Pb to root length on control medium (no added Pb), was calculated. Based on the root response, i.e. the TI, the ecotypes were categorized into tolerant (TI > 0.5), intermediate (0.2 < TI < 0.5), and sensitive (TI < 0.2) types. The 74 ecotypes were also grown in the Pb-contaminated soil and Pb concentration in shoots was determined three weeks after transplanting. Based on the shoot Pb concentration, the ecotypes were classified as accumulator (Pb > 400 mg kg^{-1}), indicator (200 mg kg^{-1} < Pb < 400 mg kg^{-1}), and nonaccumulator (Pb < 200 mg kg^{-1}) types.

The TI ranged from 0.17 to 1.2, and Pb in shoots varied between 80 to 800 mg kg^{-1}(Fig. 1), suggesting significant intraspecific variation occurred in *Arabidopsis* for accumulation and tolerance of Pb. In addition, the three shoot-response types (accumulator, indicator, and nonaccumulator) were found in each root-response group (sensitive, intermediate, and tolerant). This result indicated that there is no direct relationship between accumulation and tolerance of Pb in *Arabidopsis,* suggesting that separate mechanisms are involved in Pb accumulation and tolerance. Genetic analysis of the Pb accumulation and tolerance are underway using crosses made among the identified tolerant, sensitive, accumulating, and nonaccumulating ecotypes in all combinations. Initial evaluation of F$_1$ progenies indicated that Pb accumulation is dominant to nonaccumulation and tolerance is dominant to sensitivity, (data not shown).

Identifying mutants in Pb accumulation from mutagenized M$_2$ populations

Since Pb tolerance and accumulation are under genetic control, the next step was to screen mutagenized populations to identify Pb-accumulating mutants. Our selection was based on root growth on nutrient agar medium containing 21 mg L^{-1} Pb, i.e. individuals that grew much longer roots than control were selected as putative Pb-tolerant plants. The notion of using root length as an index to identify Pb accumulators was based on our results from the ecotype survey which demonstrated that Pb accumulators can be found as a sub-group of Pb-tolerant plants. We screened 500,000 EMS-mutagenized *Arabidopsis* M$_2$ plants and identified 250 putative Pb-tolerant mutants. Thirty out of the 250 putative mutants have been evaluated to date in hydroponic culture with 4.1 mg L^{-1} Pb. Three Pb accumulating mutants were identified (Table III). Mutants APb2, 7, and 8 were

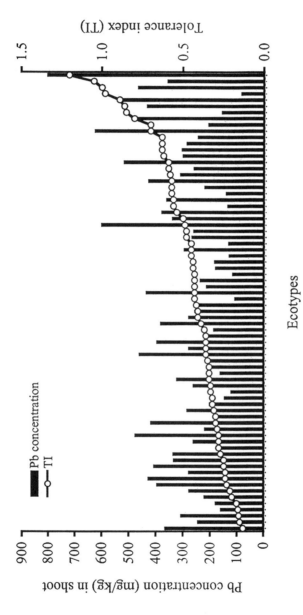

Figure 1. Tolerance index (TI) and shoot Pb concentration of 74 *Arabidopsis* ecotypes. Bars represent the Pb concentration in the shoot of a single ecotype, and open circles represent the TI of the ecotype.

able to accumulate more than twice the Pb in shoots than the wild type. We are in the process of completing the evaluation of the remaining putative mutants.

Table III. Shoot dry weight and Pb concentration of *Arabidopsis* mutants (APb) grown in nutrient solution culture containing 4.1 mg L^{-1} Pb[a]

Mutant	Dry weight (mg plant^{-1})	Pb concentration (mg kg^{-1})
Wild type	43	373
APb2	20	932
APb7	23	834
APb8	52	990

[a]Plants were harvested three weeks after adding Pb into nutrient solution, and Pb concentration in shoots was measured by ICP.

Comparing the concentrations of mineral nutrients in shoots of the three mutants APb2, 7, and 8 with those of the wild type, we found that not only Pb concentration, but also the concentration of many other nutrient elements increased (Table IV). This result is quite similar to a recent report (25) that an *Arabidopsis* mutant, *man1*, was not only able to accumulate Mn, but also Cu, Mg, Zn, and S. The author concluded that the *man1* mutation disrupted the regulation of metal-ion uptake or homeostasis in *Arabidopsis* This may be possible for mutants APb2, 7,

Table IV. Concentrations of mineral nutrients and Pb in shoots of *Arabidopsis* mutants grown in nutrient solution culture containing 4.1 mg L^{-1} Pb[a]

Mutant	Ca	Fe	Mg	Mn	Na	Pb	Zn
				mg kg^{-1}			
Wild type	6513	106	1957	29	3252	373	33
APb2	12597	320	3583	62	4718	932	54
APb7	11965	713	3923	52	9238	834	73
Apb8	10764	205	3056	33	4746	990	57

[a]Plants were harvested three weeks after transplanting, and concentrations of the elements were measured by ICP.

and 8; however, an alternative interpretation that cannot yet be ruled out is that the mutations in Apb2, 7, and 8 alter either normal uptake or translocation systems, resulting in plant shoots which over-accumulate many metal ions. We have yet to identify other mutations that may occur which specifically affect Pb uptake or translocation, and thus cause hyperaccumulation of Pb only.

Summary

Arabidopsis thaliana is not a plant of choice for practical phytoextraction of Pb because of its small size. However, it is an attractive model system for studying Pb accumulation and tolerance in plants. *Arabidopsis* is a member of the Brassicaceae family from which many metal hyperaccumulating plants have been identified. Thousands of *Arabidopsis* ecotypes collected around the world and populations mutagenized with a variety of agents are now available. *Arabidopsis* mimics many of the same characteristics as other plants in response to elevated Pb. We have found a wide array of intraspecific variation in Pb accumulation and tolerance, and demonstrated that Pb tolerance is not obligatorily linked with Pb accumulation. Lead accumulators, however, can be found amongst tolerant plants. Genetic analysis is underway to elucidate the genetic mechanisms controlling Pb accumulation and tolerance. Using simple screening and selection methods, we have identified several mutants affecting Pb accumulation. Further characterization of the ecotypes and mutants should reveal underlying genetic, biochemical, and physiological mechanisms controlling Pb accumulation and tolerance. Since *Arabidopsis* is closely related to other metal accumulating plants, an understanding of the mechanisms of uptake, translocation, tolerance, and accumulation of Pb and other metals in *Arabidopsis* should be an aid in the analysis of metal metabolism in these other species. Moreover, the eventual characterization of *Arabidopsis* genes controlling Pb accumulation and tolerance should provide tools for engineering plants which have the larger biomass necessary for Pb phytoextraction.

Acknowledgments

We would like to thank Robert Cox, Marty Holmes, Stacey Pepe, and Steve Germani of DuPont CR &D for their technical assistance in plant culture and sample analysis. We also want to thank the staff at the *Arabidopsis* Biological Resource Center at Ohio State University for the their quick responses in providing many of the *Arabidopsis* ecotype seeds used in this study, Dr. Alan Baker at University of Sheffield, UK for providing *Thlaspi caerulescens* and *T. rotundifolium*, and Dr. Jinglan Wu of DuPont CR&D for critically reading this manuscript.

Literature Cited

(1)   Meyerowitz, E. M. *Cell*, **1989**, *56*, 263-269.
(2)   Cunningham, S. D.; Ow, D. W. *Plant Physiol*. **1996**, *110*, 715-719.
(3)   Baker, A. J. M.; Brooks, R. R. *Biorecovery*, **1989**, *1*, 81-126.

(4)   *Arabidopsis*; Meyerowitz, E. M.; Somerville, C. R., Eds.; Cold Spring Harbor Laboratory Press: New York, NY, 1994.
(5)   Al-Shehbaz, I. A. *Contrib. Gray Herb.* **1973**, *204*, 3-148.
(6)   Price, R. A.; Palmer, J. D.; Al-Shehbaz, I. A. In *Arabidopsis*, Meyerowitz, E. M.; Somerville, C. R., Eds.; Cold Spring Harbor Laboratory Press: New York, NY, 1994, pp 7-19.
(7)   Pruitt, R. E.; Meyerowitz, E. M. *J. Mol. Biol.* **1986**, *187*, 169-183.
(8)   Feenstra, W. J. *Genetica*, **1964**, *35*, 259-269.
(9)   Jacobs, M. *Arabidopsis Inform. Serv.*, **1964**, *1*, 22-23
(10)   Oostindier-Braaksma, F. J.; Feenstra, W. J. *Mutation Research*, **1973**, *19*, 175-185.
(11)   Guerinot, M. L.; Yi, Y.; Friday, E.; Parry, A. P.; Fett, J. *Abstract of Sixth International Conference on Arabidopsis Research*, Madison, WI, 1995, pp 216.
(12)   Poirier, Y.; Thoma, S.; Somerville, C.; Schiefelbein, J. *Plant Physiol.* **1991**, *97*, 1087-1093.
(13)   Delhaize, E.; Randall, P. *Plant Physiol.* **1995**, *107*, 207-213.
(14)   Larsen, P. B.; Tai, C.; Kochian, L. V.; Howell, S. H. *Plant Physiol.* **1996**, *110*, 743-751.
(15)   Howden, R.; Cobalt, C. S. *Plant Physiol.* **1992**, *99*, 100-107.
(16)   Howden, R.; Goldsbrough, P. B.; Andersen, C. R.; Cobalt., C. S. *Plant Physiol.* **1995**, *107*, 1059-1066.
(17)   Howden, R.; Andersen, C. R.; Goldsbrough, P. B.; Cobalt., C. S. *Plant Physiol.* **1995**, *107*, 1067-1073.
(18)   Kumar, N. P. B. A.; Dushenkov, V.; Motto, H.; Raskin, I. *Enviorn. Sci. and Technol.* **1995**, *29*, 1232-1238.
(19)   Huang, J. W., Cunningham, S. D. *New Phytol.* **1996**, *134*, 75-84.
(20)   Woolhouse, H. W. In *Physiological Plant Ecology III: Responses to Chemical and Biological Environment*; Lange, O. L.; Nobel, P. S.; Osmond, C. B.; Ziegler, H., Eds.; Springer-Verlag: New York, NY, 1983, pp 283-285.
(21)   Kabata-Pendias, A.; Pendias, H. *Trace Elements in Soils and Plants*; CRC Press: Boca Raton, FL, 1992, pp 187-198.
(22)   Parker, D. R.; Norvel, W. A.; Chaney, R. L. In *Chemical Equilibrium and Reaction Models*; Loeppert, R. H.; Schwab, A. P.; Goldberg, S. Eds.; Soil Science of American: Madison, WI, 1995, pp 163-200.
(23)   *Lead in Soil: Recommended Guidelines*; Wixson, B. G; Davies, B. E., Eds, Science Reviews: Northwood, 1993, pp25-38.
(24)   Baker, A. J. M; Walker, P. L. In *Heavy Metal Tolerance in Plants: Evolutionary Aspect*, Shaw, A. J. Ed, CRC Press: Boca Raton, FL, 1990, pp 155-177.
(25)   Delhaize, E. *Plant Physiol.* **1996**, *111*, 849-855.

Chapter 20

Bioremediation of Chromium from Water and Soil by Vascular Aquatic Plants

P. Chandra, S. Sinha, and U. N. Rai

Aquatic Botany Laboratory, National Botanical Research Institute,
Lucknow–226 001, India

The ability of aquatic plants to absorb, translocate and concentrate metals has led to the development of various plant-based treatment systems. The potential to accumulate chromium by *Scirpus lacustris*, *Phragmites karka* and *Bacopa monnieri* was assessed by subjecting them to different chromium concentrations under laboratory conditions. Plants showed the ability to accumulate substantial amounts of chromium during a short span of one week. When the plants were grown in tannery effluent and sludge containing 2.31 μg ml^{-1} and 214 mg kg^{-1} Cr, respectively, they caused significant reduction in chromium concentrations. While there was an increase in biomass, no visible phytotoxic symptoms were shown by treated plants. The plants can then be harvested easily and utilized for biogas production.

Chromium is one of the toxic metals widely distributed in nature. Of the two forms found in the environment, trivalent and hexavalent, hexavalent chromium is the form considered to be the greatest threat because of its high solubility, its ability to penetrate cell membranes, and its strong oxidizing ability *(1)*. The large-scale uses of chromium in metallurgical, pigment and dye, and in textile, and electroplating makes these industries potential sources of chromium pollution. The tanning industry is also a major contributor of chromium pollution of water resources. In India, according to a recent estimate, ca 2,000-3,200 tonnes of elemental chromium escape into the environment annually from the tanning industries alone *(2)*. Chromium concentrations in effluents usually range between 2,000-5,000 μg ml^{-1} compared to the recommended permissible limit of 2 μg ml^{-1}. The management of large amounts of effluent discharged by the tanneries has become a formidable task in developing countries. Contamination of water resources by these effluents is posing serious health hazards and is a threat to aquatic ecosystems.

The treatment of enormous volumes of tannery wastewater by conventional methods prior to its release is energy intensive and involves huge expenditures. It will also result in vast quantities of chrome sludge causing a problem of solid waste disposal. In India, if all chromium used in the tanning industry is treated by conventional methods, as much as 75,000-100,000 tonnes of chrome sludge will be generated every year (2). To overcome this problem, cost effective biological systems seem to be a feasible alternative.

In recent years, considerable efforts have been made to develop aquatic macrophyte-based wastewater treatment systems (3-5). Constructed wetlands planted with *Phragmites*, *Typha* and *Scirpus* spp. have also been found effective in treatment of municipal and industrial wastewaters (6-7). During the course of this study, a large number of promising metal-accumulating species have been identified (8-14). However, large-scale exploitation of these plants in metal abatement programs has yet to be undertaken.

While working on a National Mission Project on drinking water, several promising chromium-accumulating macrophytes were identified from chromium-polluted water bodies adjacent to tanneries (15). Plants of the same species were collected from unpolluted ponds and lakes located at different sites, and a series of experiments were carried out with tannery effluent and sludge to assess their chromium uptake potential. The study includes the results of experiments carried out on *Bacopa monnieri*, *Scirpus lacustris*, *Phragmites karka*, and *Nymphaea alba* under the laboratory and field conditions.

Material and Methods

For the study, plants of *Bacopa monnieri*, *Scirpus lacustris*, *Phragmites karka* and *Nymphaea alba* were collected from unpolluted water bodies and grown in large hydroponic tubs for 8 weeks. Young plants of each species were detached from mother plants and two sets of plants of each species were prepared in 10% Hoagland's solution. One set of plants (*N. alba*) was acclimatized under laboratory conditions and the other set in a natural unpolluted environment for conducting experiments with effluent. The plants under the laboratory conditions were provided with 16 h of illumination using fluorescent tube light (Philips), 115 μmol m^{-2} s^{-1} and 8 h dark photoperiods at 25 $\pm$ 2 °C. For conducting experiments with tannery sludge, young plants of *B. monnieri*, *P. karka* and *S. lacustris* were acclimatized in an uncontaminated garden soil for 8 weeks.

Laboratory Studies

The acclimatized plants of *N. alba* were subjected to five different concentrations (0.1, 0.5, 1.0, 2.0 and 5.0 μg ml^{-1}) of Cr prepared using K_2CrO_4 supplied at zero time without further addition during the treatment. Plants in 10% Hoagland's solution without metal served as controls. Three sets of plants (each set consisting of four beakers of 2 litres capacity containing two plants of *N. alba* in 1.5 litre treatment solution) were used for each concentration. One set of plants each was harvested after 48, 72 and 168 h and washed thoroughly with distilled water. Root, leaf and rhizome were separated manually, dried in an oven at 105°C and digested in a mixture of HNO_3: $HClO_4$ (5:1, v/v). Chromium was estimated with a Perkin Elmer 2380 Atomic Absorption Spectrophotometer.

Studies under Field Conditions

Experiment with Tannery Effluent: Effluent was collected manually in acid washed plastic containers (N = 5) of 20 litres capacity each from the different outlets of the Northern tannery (Unnao, U.P.) India, brought to the field laboratory, and allowed to settle for one week. Effluent was filtered using muslin cloth and physicochemical properties were determined using standard methods of APHA *(16)*. The effluent was digested in HNO_3: $HClO_4$ (5:2, v/v) and chromium was estimated by Atomic Absorbtion Spectrophotometry. The mean value (N = 10, two replicates from each container) of Cr was 2.31 ± 0.40 µg ml^{-1}. The effluents from all the containers were mixed and used in this study.

Five litres of filtered effluent containing 2.31 µg ml^{-1} Cr were treated with plants (50-g fresh weight each of *N. alba, S. lacustris, B. monnieri* and *P. karka*) in round plastic troughs (12" diameter, 8" depth). Though the biomass weight of each species was equal in each treatment, numbers of plants varied in different species, it was 3, 5, 15 and 20 in *N. alba, S. lacustris, B. monnieri* and *P. karka*, respectively. Each experiment consisted of two sets (each having four replicates), one was harvested after one week and others after 2 weeks. The experimental sets were kept in a net house to prevent natural contamination. The plants in tap water containing no chromium served as controls. Harvested plants were dried and processed for metal estimation following the method described earlier.

Experiment with Tannery Sludge: The sludge samples collected manually in plastic bags from the Northern tannery (Unnao, U. P.) India were brought to the field laboratory, dried, powdered and sieved. Sludge samples (N = 20) were digested in HNO_3: $HClO_4$ (5:2, v/v) and Cr content was estimated by Atomic Absorption Spectrophotometry. The mean value of Cr was 2.14 ± 18.05 mg kg^{-1}.

All the above plants (30-g fresh weight) were planted in round plastic troughs (12" diameter, 8" depth) containing powdered, sieved sludge (5 kg). The experiment for each test plant (four replicates each separately) was set up for different harvesting periods. A thin layer of water (2-4 cm) was maintained over the sludge throughout the experiment. Plants were harvested after 4, 8 and 12 weeks. The biomass was determined on fresh weight basis from each trough. For metal estimation, plants were oven dried and digested with a mixture of HNO_3: $HClO_4$ (5:1, v/v). Chromium was determined by Atomic Absorption Spectrophotometry.

Plants were analyzed for their Cr content at time zero and values of Cr in control plants were deducted from the accumulation values of treated plants.

Statistical Analysis

Analysis of variance of the data was performed for calculating the statistical significance using completely randomized block design. Student 't' test (two tailed) was applied to see the significance level as compared to controls *(17)*.

Quality Control and Quality Assurance

Analytical data for quality of Cr were ensured through repeated analysis (N = 5) of EPA quality control samples (Lot TMA 989) in water, and results were found within 5.41% of the certified values. For plants, recoveries of Cr from the plant tissues were found to be 98.2 ± 4.65 % as determined by digesting five samples each from untreated plants with known volumes of Cr. The blanks were run all the time, and duplicate analysis was carried out to check the precision of the method.

Some of the data on *B. monnieri* and *S. lacustris* published earlier *(18,19)* have also been used for comparison in this study.

Results

Physicochemical analysis of effluent (N = 5) showed high values for pH (8.84 ± 0.21), biological oxygen demand (BOD: 790 ± 52.52 µg ml⁻¹), chemical oxygen demand (COD: 1870 ± 98.5 µg ml⁻¹) and total dissolved solid (TDS: 850 ± 36.8 µg ml⁻¹). The value for dissolved oxygen (DO) was quite low (2.3 ± 0.015 µg ml⁻¹). Effluent used in the experiments contained 2.31 ± 0.40 µg ml⁻¹ Cr. Tannery sludge had high pH values of 9.2 ± 0.2 and contained 214 ± 18.05 mg kg⁻¹ Cr.

Figure 1 (a-c) shows chromium accumulation in the root, leaf and rhizome of *N. alba* at different ambient chromium concentrations (0.1-5.0 µg ml⁻¹) and treatment durations. The maximum accumulation of chromium was in the root followed by rhizome, and leaf. During the first 48 h, the accumulation in the roots was significant, and it reached steady state conditions at 168 h (Figure 1a). The maximum accumulation of chromium (1207 µg g⁻¹ dw) was shown by the roots at 5 µg ml⁻¹ ambient chromium concentration after 168 h. Accumulation of chromium in the leaf of *N. alba* was comparatively low, maximum being 515 µg g⁻¹ dw after 168 h at 5.0 µg ml⁻¹ background Cr concentration (Figure 1b). The accumulation of chromium in the rhizome of *N. alba* was lowest (Figure 1c). At the highest background concentration of chromium, maximum accumulation of 80.06 µg g⁻¹ dw was shown by the rhizome of *N. alba* after 168 h.

The young plants of *B. monnieri*, *N. alba*, *S. lacustris* and *P. karka* were treated with 100% tannery effluent for 1 and 2 weeks (Figure 2). The plants showed different levels of chromium accumulation during this period. A maximum accumulation of chromium was shown by the plants of *P. karka* (615 µg g⁻¹ dw) followed by S *lacustris* (515 µg g⁻¹ dw), *N. alba* (465 µg g⁻¹ dw) and *B. monnieri* (340 µg g⁻¹ dw) after 2 weeks.

Table 1 shows the results of experiments carried out with high chromium containing sludge collected from the tanneries. All of the three plants (*P. karka*, *S lacustris* and *B. monnieri*) showed duration dependent accumulation of chromium. The maximum accumulation was in the roots of *P. karka* (816 µg g⁻¹ dw) followed by *S. lacustris* (460 µg g⁻¹ dw) and *B. monnieri* (310 µg g⁻¹ dw) after 12 weeks. In comparison to *B. monnieri* and *S. lacustris* chromium concentrations were higher in the above ground parts of *P. karka* (110 µg g⁻¹ dw).

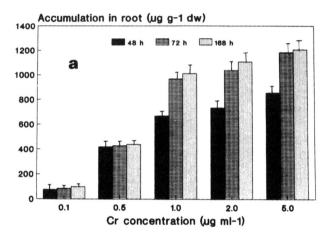

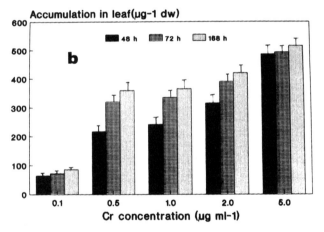

Figure 1. (a-c). Chromium accumulation by *N. alba* in different plant parts (a = root, b = leaf, c = rhizome) as a function of Cr concentration and exposure. All values are means of four replicates ± SD. Root, F-value (concentration) = 49.91a; F-value (exposure) = 6.92b. For leaf, F-value (concentration) = 101.4a; F-value (exposure) = 13.5. For rhizome, F-value (concentration) = 132.23a; F-value (exposure) = 8.50b; a = p < 0.01; b = p < 0.05.

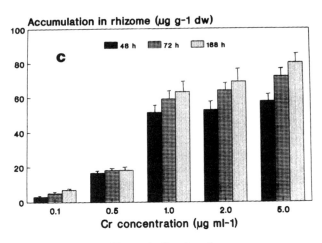

Figure 1. *Continued*

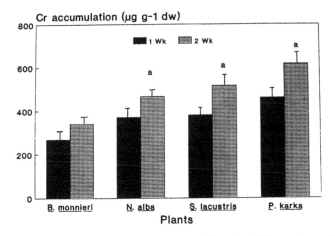

Figure 2. Accumulation of chromium from tannery effluent by different plants after 1 and 2 weeks. All values are means of four replicates $\pm$ SD. T-test (two tailed): a = p < 0.025 with comparisons made at 1 week.

Table 1. Chromium concentrations (μg g^{-1} dw) in the plants treated with tannery sludge

Plants	Root + Rhizome		
	4 Weeks	8 Weeks	12 Weeks
S. lacustris	203 ± 19	363 ± 26b	460 ± 47
P. Karka	645 ± 75	752 ± 64	816 ± 69
B. monnieri	133 ± 12	207 ± 14a	310 ± 25c
	Shoot		
S. lacustris	36.6 ± 4.2	60.3 ± 7.5	75.2 ± 8.8
P. Karka	69.4 ± 6.8	82.4 ± 9.2	110.3 ± 12.5
B. monnieri	53.4 ± 3.9	74.6 ± 7.8	94.6 ± 8.2

All values are means of four replicates ± SD. T-tests (two tailed): a = p < 0.02 with comparisons made at 4 weeks; b = p < 0.01 with comparisons made at 4 weeks; c = p < 0.02 with comparisons made at 8 weeks.

Plants grown in tannery sludge showed an increase in biomass (Figure 3) at 12 weeks, and it was in order of *P. karka* < *S. lacustris* < *B. monnieri*. The percent increases in biomass were 633%, 620% and 433% in *P. karka, S. lacustris* and *B. monnieri,* respectively.

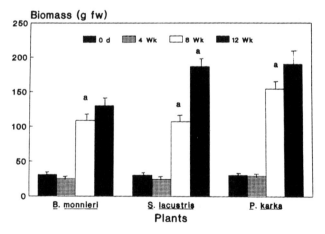

Figure 3. The effect of chromium accumulation on biomass (g dw) in different plants treated with tannery sludge. All values are means of four replicates ± SD. T-test (two tailed): a = p < 0.05 with comparisons made at day 0, b = p < 0.05 with comparisons made at 8 weeks.

Discussion

The efficiency of aquatic macrophyte-based phytoremediation systems depends mainly on their design, and the nature of wastewater and soil. A plant species may be tolerant to one metal but may not be so with another one. Plant forms (i.e., free-floating, submerged and emergent) also play a significant role in the removal of metals. Submerged species accumulate more metals than the floating and emergent ones *(20, 21)*. A rootless submerged species *(Ceratophyllum demersum)* has been found to possess the potential to accumulate substantial amounts of chromium *(10)* conforming with the earlier reports.

Rooted aquatics with floating foliage are important constituents of aquatic ecosystems. The plants of *Nymphaea* are reported to be capable of accumulating metals from polluted waters *(22)*. During the present study, *N. alba* also showed the potential to accumulate substantial amounts of chromium under laboratory conditions as well as in field experiments with tannery effluent. The accumulation of chromium was higher in the root in comparision to leaf and rhizome. The lower concentration of chromium in the leaf is probably due to the slower mobility of Cr transport from root to the shoot. Trivalent chromium forms complex compounds with COOH groups that inhibit the translocation of metal from the root to the shoot *(23)*. The ability of fine roots to accumulate high quantities of metal has also been reported *(24)*. The tuft of fine roots in *B. monnieri* and *S. lacustris* has also shown substantial accumulation of chromium and copper *(19)*.

Chromium accumulation from soil sludge depends upon the nature of plants and conversion of Cr (III) into Cr (VI), the ions of which easily penetrate the cell membrane. To make chromium ions freely available to roots requires oxygen from the air for a few weeks to facilitate the oxidation of Cr (III) into Cr (VI) *(25)*. The selection of aquatic macrophytes is, therefore, of utmost importance as they are the only such species which are capable of transporting atmospheric oxygen into the rhizosphere zone through their leaves. This is important in reducing chromium concentrations in sludge soil. In the present study, roots of *P. karka* showed the potential to accumulate chromium from tannery sludge. This might have been caused by a large supply of oxygen to the rhizosphere zone through the huge foliage of the plants. Similarly, higher accumulation of chromium was shown by the plants of *S. lacustris* which possess long, cylindrical, hollow, tube-like leaves capable of transporting atmospheric oxygen to the rhizosphere zone.

This study shows that chromium uptake performance of *B. monnieri, S. lacustris, P. karka* and *N. alba* plants from tannery effluent is promising. The chromium accumulation was significant in *P. karka* (615 μg g^{-1} dw) followed by *S. lacustris* (515 μg g^{-1} dw), and *N. alba* (465 μg g^{-1} dw) after 14 days. The accumulation for this period was lowest (340 μg g^{-1} dw) in *B. monnieri*. In our earlier study *(18)*, tannery effluent (containing 0.7 μg ml^{-1} Cr) treated with a combination of plants *(Spirodela polyrrhiza, Hydrilla verticillata)* showed maximum reduction in chromium concentration at 25% dilution of the effluent.

The results of the present study are quite encouraging. Overall performance of plants in the removal of chromium from tannery sludge and effluent was quite satisfactory.

Plants showed high levels of tolerance to chromium. Though, there was a slight decrease in chlorophyll content in *B. monnieri*, and *S. lacustris (19)*, plants showed no visible phytotoxic symptoms. The results indicate the suitability of these plants in phytoremediation of chromium in water and soil.

Acknowledgement

We thank Dr. P.V. Sane, Director, National Botanical Research Institute, Lucknow, India for his help and encouragement in this work. NBRI Publication No. (452) NS.

References

1. Riedel, G.F., In *Aquatic Toxicology and Environmental Fate;* Eds. Suter,G.W. II; Lewis, 1 M.A. ASTM, STP, **1989**, Vol.11, *pp. 537-548.*
2. Thyagarajan, G. *Hindu.* **1992**,143-145.
3. Athie, D.; Cerri, C.C. *Water Sci. Tech.* **1987**, 14(10), 1-177.
4. Reddy, KR; Smith W.H. *Aquatic Plants for Water Treatment and Resource Recovery.* Mangolia Publishing Inc. Orlando, Florida, **1987**, p.1032.
5. Brix, H.; Scheirup, H.H. *Ambio.* **1989**, 18, 100-107.
6. Gersberg, M.; Elkins, B.V.; Goldman, C.R. *Water Research.* **1986,** 20(3), 363-368.
7. Hammer, D.A. *Constructed Wetlands for Wastewater Treatment. Municpal, Industrial and Agricultural.* Lewis Publishers Inc. Michigan, USA, **1989**, p. 831.
8. Wolverton, B.C.; Mc Donald, R.C. *NASA Tech. Memorandum, TM-X72727,* **1975a.**
9. Wolverton, B.C.; Mc Donald, R.C. *NASA Tech. Memorandum, TM-X- 72727,* **1975b.**
10. Garg, P.; Chandra, P. *Bull. Environ. Contam. Toxicol.* **1990**, 44, 473-478.
11. Sinha, S.; Chandra, P. *Water, Air, Soil Pollut.* **1990**, 51, 271-276.
12. Nir, R; Gasith, A.; Rerry, A.S. *Bull. Environ. Contam. Toxicol.* **1990**, 44, 149-157.
13. Deepak, D.; Gupta, A.K. *Indian J. Environ. Health.* **1991**, 33 (2), 297-305.
14. Rai, U.N.; Chandra, P. *Sci. Total Environ.* **1992**, 116, 203-211.
15. Chandra, P. *Identification of Suitable Plant Species for Biological Water Purification.* NBRI, Lucknow, India, **1988**, p. 44.
16. American Public Health Association. **1989.**
17. Gomez, K A.; Gomez, A. A. *Statistical Procedures for Agricultural Research.* John Wiley & Sons. New York, U.S.A., **1984.**
18. Vajpayee, P.; Rai, U.N.; Sinha, S.; Tripathi, R D.; Chandra, P. *Bull. Environ. Contam. Toxicol.* **1995**, 55, 546-553.
19. Gupta, M.; Sinha, S., Chandra, P. *J. Environ. Sci. Health.* **1994**, 29 (10), 2185-2202.
20. Baudo, R.G.; Galanti, P.; Guilizzoni, P.; Varini P. *Mem. 1st Ital. Idrobiol.* **1981**, 203-225.
21. Gulizzoni, P. *Aquatic Botany.* **1991**, 41, 87-109.
22. Aulio, K. *Bull. Environ. Contam. Toxicol.* **1980**, 25, 713-717.
23. Sigel, H. *Metal Ions in Biological Systems.* Marcel Dekkar Inc. **1973**, Vol. 2.
24. Sincorpe, T.R.; Langis, R.M.; Gergberg, M.J.; Busnardo, M.J.; Zedler, J.B. *Ecol. Engr.* **1992**, 1, 309-322.
25. Cary, E.E.; Allaway, W.H.; Olson, O.E. J. *Agric. Food. Chem.* **197**

Chapter 21

Phytoextraction of Lead from Contaminated Soils

J. W. Huang[1], J. Chen[2], and S. D. Cunningham[2]

[1]Phytotech Inc., 1 Deer Park Drive, Suite I, Monmouth Junction, NJ 08852
[2]Dupont Central Research and Development, Environmental Biotechnology,
GBC–301, P.O. Box 6101, Newark, DE 19714–6101

Lead phytoextraction, the use of plants to extract Pb from
contaminated soils, is an emerging technology. To develop this
technology, we have conducted extensive research to study
physiological and cellular mechanisms of Pb uptake, translocation,
and accumulation in crops, weeds, and known metal
hyperaccumulators. This paper will review current progress from our
laboratories and several other laboratories in the development of Pb
phytoextraction technology. We will focus on the following subjects:
(1) physiological and cellular aspects of Pb transport in plants, (2)
plant species/cultivars variation in Pb uptake and translocation, (3)
role of plant genetic engineering in Pb phytoextraction, and (4) role
of synthetic chelates in enchancing Pb phytoextraction from
contaminated soils. With the addition of selected chelates to soils
collected from Pb-contaminated sites, we are now able to increase
shoot Pb concentration from less than 500 mg kg^{-1} to more than
10000 mg kg^{-1}, which is the value targeted for commercial Pb
phytoextraction. Field testing of Pb phytoextraction technology is
currently underway in a dozen sites in the United States. Current
results indicate that this technology may provide an environmentally
sound and cost-effective strategy for the clean-up of Pb-
contaminated soils.

CONTENTS

Role of Plant Genetic Engineering in Pb Phytoextraction
Lead Phytoextraction: from Greenhouse to the Field
Exploring Potential Methods to Enhance Pb Phytoextraction
Phosphorus Nutrition and Pb Phytoextraction
Effects of Electroosmosis on Pb Accumulation in Plants
Role of Synthetic Chelates in Enhancing Pb Accumulation in Plants
Future Research Needs

Heavy metal contamination of the environment poses serious problems to human and animal health as well as agricultural production. Lead is one of the most frequently encountered heavy metals of environmental concern and is the subject of extensive remediation research (1-2). Lead contamination of surface soils has resulted from industrial activities such as mining and smelting, production and disposal of lead-acid batteries, the use of paints, the disposal of municipal wastes and sewage enriched in Pb, and the use of certain pesticides (3-5). Excess Pb in the human and animal body may cause a variety of health disorders (6-7). The finding of Pb related human health problems and the recognition of widespread Pb contamination of urban soils have made Pb contamination an important public issue (5). The remediation of Pb-contaminated soils represents a significant expense to many industries and governmental agencies. Currently, soils with severe Pb contamination are remediated through a wide variety of engineering-based technologies (8-10). This process is expensive and frequently requires additional site restoration.

In a search of cost-effective alternatives for the clean-up of contaminated soils, Chaney (11) proposed the use of plants to remediate heavy metal contaminated soils. Over the last 10 years, there have been increasing interests in developing this plant-based technology (phytoremediation) to remediate Pb and other heavy metal contaminated soils (1-2, 11-12). Phytoremediation of Pb-contaminated soils has two different strategies: phytostabilization and phytoextraction (8, 10). Phytostabilization is the use of plants and soil amendments to reduce Pb bioavailability in the contaminated soils, thus reducing the intrinsic hazard of the contaminated soil (8). Phytoextraction is the use of plants to extract Pb from contaminated soils. By continued cultivation of selected plant species on the contaminated sites, the soil could eventually be decontaminated.

As a remediation technology, Pb phytoextraction is still in the developmental stage. The success of using plants to extract Pb from contaminated soils requires a better understanding of the mechanisms of Pb uptake, translocation and accumulation by plants. Lead is not a known essential nutrient for higher plants, however, there is circumstantial evidence indicating that various plant species have the ability to absorb Pb by roots and translocate Pb from roots to shoots (13-15). There is a small number of plant species endemic to metalliferous soils that can tolerate and accumulate high levels of Pb and other heavy metals. These plants are called metal hyperaccumulators and can accumulate more than 0.1% of Pb, Co, Cr, or more than 1% of Mn, Ni or Zn in any above-ground tissue when growing in their natural habitat (16). There are 400 known metal hyperaccumulators in the world. Among these, only five species are Pb hyperaccumulators (Table 1). However, all

Table 1. Lead concentrations (on a dry weight basis) in plant shoots of reported Pb hyperaccumulators growing on Pb-contaminated sites

Plant Species	Shoot Pb Concentration	References
	mg kg^{-1}	
Polycarpaea synandra	1040	(33)
Minuartia verna	1400	(34)
Armeria maritima	1600	(14)
Thlaspi alpestre	2740	(34)
Thlaspi rotundifolium	130-8200	(13)

these reported Pb hyperaccumulators have a very slow growth rate and are not practically suited for phytoextraction of Pb from contaminated soils. For the last few years, our laboratories and several other laboratories have carried out extensive research to screen for Pb accumulating plants from high biomass species (*15, 17*). Results from these studies demonstrate that plant species differ significantly in Pb uptake and translocation, and that shoot Pb concentrations among plant species vary by more than 10 to 20 fold. However, even the best shoot Pb accumulating plants could hardly accumulate a shoot Pb concentration greater than 0.1% when they were grown on Pb contaminated soils without soil amendments. Recently, we have explored various potential methods to enhance Pb accumulation (15, Huang et al., *Environ Sci. Technol.*, in press). We found that certain synthetic chelates could dramatically enhance Pb desorption from soil to soil solution and thus significantly increase Pb accumulation in plants. This paper will review recent research progress in the development of Pb phytoextraction technology. We will focus on the understanding of physiological and cellular mechanisms of Pb transport in plants, screening of Pb accumulating plant species with high biomass production, and exploring various potential methods to enhance both Pb desorption from soil to soil solution and Pb accumulation in plants.

Physiological Aspects of Lead Transport in Plants

Hydroponic Studies of Pb Transport in Plants. One of the difficulties in studying Pb transport in plants is the complexity of Pb solution chemistry. Lead is classified as soft Lewis acid which implies a strong covalent character to many ionic bonds which it forms in soils and plants. For Pb-contaminated soils, there are many factors (such as soil fertility, soil texture, pH, and possible other heavy metal contamination) affecting Pb uptake and translocation in plants. In order to simplify the system and to study Pb transport in plants under precisely controlled conditions, we used a hydroponic system to study the basic physiology of Pb uptake, translocation, and accumulation in plants (*15*). In traditional solution culture with a pH of 6.0-7.0 and a P level of 0.1-2.0 mM, most of the Pb added to the solution is precipitated as Pb hydroxide and/or Pb phosphate. Researchers have attempted to avoid the Pb precipitation problems by using simple salt solutions or complete nutrient solution without P. However, in order to mimic soil conditions, the use of a complete nutrient solution with similar composition as to the soil solution from Pb contaminated sites is preferred. In our hydroponic studies, we used complete nutrient solution with low P concentrations (15). We used GEOCHEM-PC programs (18) to simulate the effects of pH and P level on free Pb activity in the nutrient solution. Through the simulation, we found that at pH values of 4.5-5.0 and P levels less than 10 μM, most of the added Pb in the range of 10-100 μM was in a soluble form. According to the calculation, we used P concentrations ranging from 2-10 μM depending on the experiments, and we maintained the solution pH at 4.5-5.0. In order to maintain the selected P levels and pH, we used a continuous flow of nutrient solution which was accomplished by using a microprocessor-controlled multichannel cartridge pump (15). The advantage of a hydroponic study is that the plant-growth media can be controlled precisely. This technique is useful in studying the mechanisms of Pb uptake and translocation as well as dissecting the factors affecting Pb uptake, translocation and accumulation. Such basic understanding of Pb transport is important for the development of Pb phytoextraction technology.

Lead Uptake by Roots. Using the hydroponic system, we found that Pb is rapidly accumulated in the root-cell apoplasm and symplasm. For example, in short-term (60 min) Pb transport experiments with 20 μM Pb in nutrient solution, intact corn roots accumulated 500 mg Pb kg^{-1}, while ragweed and sunflower roots accumulated 1500 mg Pb kg^{-1} (15). The Pb concentrations in the roots were more than 100-fold higher than that in the root growth medium. A 30-min desorption period following the 60-min Pb uptake removed 60% of the absorbed Pb from the roots. The difference in Pb uptake between corn and sunflower was independent of the desorption solution used. We also found that 40% of short-term absorbed Pb was not exchangeable, probably representing the symplasm-associated Pb pool. These results suggest that Pb is rapidly transported into the root cells. To further examine the cellular distribution of Pb in sunflower roots, we used Scanning Electron Microscopy with an ion microprobe to examine the Pb distribution within root cells via back-scattered electrons. Lead was localized in the cytoplasm of cortical cells and associated with cell walls in the developing vascular tissue. Using x-ray probe analysis, Qureshi et al. (19) reported that Pb was accumulated in both cell walls and cytoplasm of root tips excised from Pb-sensitive clone of *Anthoxanthum odoratum*.

There is little information in the literature concerning the mechanisms of Pb transport into plant cells. Lead transported into root cells would have crossed the root-cell plasma membrane (PM). One possible transport pathway could be through PM cation channels, such as Ca^{2+}-channels. Voltage-gated Ca^{2+} channels in the PM of root cells have been identified in both corn and wheat roots (20-21). Using isolated root-cell plasma membrane vesicles, we have recently found that Pb can significantly inhibit voltage-gated Ca^{2+} channels (unpublished results). The inhibition of the Ca^{2+} channel activity could result from Pb blockage of the channel, or Pb competing with Ca^{2+} for the transport pathway (Ca^{2+} channel). These results parallel those found in animal systems where there is experimental evidence showing that Pb^{2+} is transported into cells via the Ca^{2+} channels. For example, Tomsig and

Suszkiw (*22*) detected permeation of Pb^{2+} into isolated bovine chromatin cells through the Ca^{2+} channels. These authors also found that voltage-gated Pb^{2+} transport was blocked by Ca^{2+} channel blocker (nifedipine) and enhanced by Ca^{2+} channel agonist (BAY K8644). Audesirk (*23*) also reported that Pb^{2+} was highly permeable through voltage-sensitive Ca^{2+} channels in animal cells. It is very likely that Pb^{2+} is transported into plant cells via divalent cation channels.

Lead Translocation from Roots to Shoots. In contrast to root Pb accumulation, shoot Pb accumulation is much lower, and a major limiting step is Pb translocation from roots to shoots. For both monocots and dicots we have studied, only a small proportion of absorbed Pb is translocated from roots to shoots (*15, 17, 24*). We have used a shoot to root Pb concentration ratio to estimate Pb translocation. Using this method, we found that there are 50-fold differences in Pb translocation among plant species tested (*8*). Plants with higher Pb translocation will yield a higher shoot Pb concentration. Therefore, it is important to select plant species with higher Pb translocation from roots to shoots because only the shoots are harvested in Pb phytoextraction.

Because of the poor translocation of Pb from roots to shoots, roots have been considered to be the main barrier for Pb translocation. To test whether root removal would alter Pb translocation, we investigated Pb accumulation in excised shoots of corn, goldenrod, ragweed, and sunflower. In this experiment, plants were grown in nutrient solution in the absence of Pb, and the shoots were excised under the solution surface at the root-shoot junction. The excised shoots (30 cm in height) were transferred into nutrient solution with 5 cm of the shoots immersed in the aerated nutrient solution containing 20 μM Pb. Lead concentration in the portion of excised shoots immersed in the solution was several orders of magnitude higher than the portion of the shoots not in contact with the solution (Figure 1). For the portion of excised shoots immersed in solution, Pb concentration reached a value similar to that observed in intact roots (*15*). For intact plants, Pb translocation of this corn cultivar was the highest among the species/cultivars tested (*8, 15*). However, in excised shoots, Pb translocation was lower in corn than in ragweed and other plant species tested (Figure 1). For the same Pb level and Pb exposure period, Pb concentration in excised corn shoots was 35% of that in intact corn shoots. The results indicate that for intact plants, the higher Pb translocation in corn and low Pb translocation in ragweed are physiologically controlled by roots, and this control is lost when roots are removed. The physiological mechanisms involved are not known.

Species Variation in Pb Accumulation. The success of using plants to extract Pb from contaminated soils requires the identification of Pb-accumulating high-biomass plant species. The existence of certain plants with remarkable metal accumulation capacities is well documented. For example, certain trees are able to accumulate Ni in their sap as high as 25% by dry weight (*25*). *Thlaspi caerulescens* can accumulate 3-5% of Zn in its shoots (*26-27*). Unlike other heavy metals, there are only a few Pb hyperaccumulators available (Table 1). However, none of these Pb hyperaccumulators are practically suited for Pb phytoextraction because of their slow growth rates.

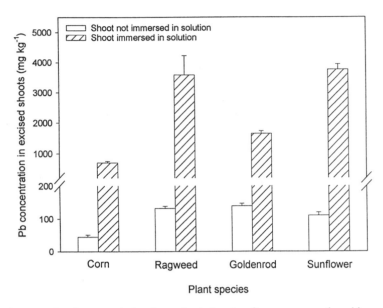

Figure 1. Lead accumulation in excised shoots of corn, ragweed, goldenrod, and sunflower plants exposed for 14 days to nutrient solution containing 0 or 20 μM Pb. Error bars represent ± SE (n=3).

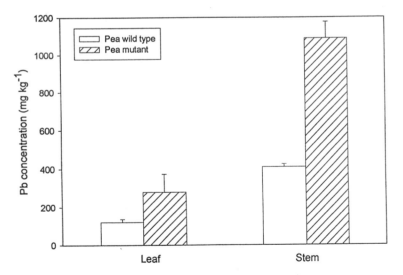

Figure 2. Lead accumulation in the leaf and stem of a single-gene pea mutant (E107) and wild type pea (Sparkle) grown in complete nutrient solution (pH 4.5) containing 20 μM Pb. The plants were grown in nutrient solutions in the absence of Pb for two weeks before exposure to solution containing 20 μM Pb. Plants were harvested after 14 days of the Pb exposure.

Using both Pb-contaminated soils and nutrient solutions set to mimic soil solutions from Pb-contaminated soils, we have screened crops, weeds, and known metal hyperaccumulators for Pb-accumulating plants. Our results demonstrate that plant species differ significantly in Pb uptake and translocation (*15*). For 50 species/cultivars we have tested so far, shoot Pb concentrations differed among plant species by more than 10 fold when grown in nutrient solution containing 20 μM Pb, and by more than 20 fold when grown in Pb contaminated soils with a soil solution Pb of 17 μM. Using hydroponics and sand culture, Kumar et al. (*17*) carried out an extensive screening for Pb accumulating plants from *Brassicaceae* family. These authors found that some cultivars of *Brassica juncea* showed a strong ability to accumulate Pb in shoots. We compared the Pb-phytoextraction potential of the three best shoot Pb accumulating cultivars (21100, 426308, 531268) and the three worst shoot Pb accumulating cultivars (175607, 180269, 184290) of *Brasscia juncea* reported by Kumar et al (*17*) with that of corn (cultivar Fiesta) using Pb-contaminated soils and nutrient solutions. In solution culture, Pb concentration in shoots of corn was similar to that in shoots of the two best Pb accumulating cultivars (211000 and 426308) of *Brassica juncea* (Table 2). In the Pb-contaminated soil, shoot Pb concentration in corn was significantly higher than that of cultivars 211000 and 531268. Cultivar 426308 was extremely sensitive to severely Pb contaminated soil (total soil Pb, 2500 mg kg^{-1}), and the plants died a few days after transplanting to the soil. The soil used in the experiment was severely contaminated with Pb during the manufacture of explosives over a period of 50 years. For contaminated soils with a total soil Pb content less than 1000 mg kg^{-1}, the *B. juncea* cultivar 426308 accumulated significantly higher shoot Pb concentration than did Fiesta (unpublished results).

Role of Plant Genetic Engineering in Pb Phytoextraction. It is generally believed that metal tolerance and accumulation are genetically controlled (*28*). Molecular biology has provided a useful tool for identifying genes, and mutation techniques have traditionally provided the means for identifying and isolating specific genes (*29*). Isolation and characterization of mutations with specific mutant phenotypes provides good insight into specific gene functions. Using mutation techniques, researchers have succeeded in identifying a number of metal-accumulating mutants. For example, a Fe hyperaccumulating pea mutant (E107) has been identified from the ethyl methanesulfonate (EMS) mutagenized pea (Sparkle) seeds (*30*). This pea mutant, resulted from a single gene mutation, accumulated 50-100 fold higher shoot Fe concentration than the wild type pea, and the mutant also accumulated higher concentrations of Mn, Mg, and Ca (*31*). We investigated Pb accumulation by this pea mutant along with the pea wild type. In solution culture, this pea mutant accumulated significantly higher Pb concentration in both leaves and stems (Figure 2). However, root Pb concentrations were not significantly different between the mutant (1.26%) and wild type (1.25%). These results suggest that this single-gene pea mutant has a higher efficiency in Pb translocation than wild type pea. When transplanted to a Pb-contaminated soil (total soil Pb, 2500 mg kg^{-1}), this pea mutant accumulated 20% more Pb in shoots than the wild type. The results suggest that Pb translocation from roots to shoots may be genetically controlled. The advantage of screening Pb-accumulating mutants from fast-growing plants is

Table 2. Lead concentrations[1] in selected plant species grown in nutrient solution containing 20 µM Pb and in a Pb-contaminated soil (total soil Pb 2500 mg kg^{-1})

Plant Species	Solution experiment		Soil experiment	
	Shoots	Roots	Shoots	Roots
	---------------------- mg kg^{-1} --------------------			
Zea mays (cv. Fiesta)	375	2280	225	1250
Brassica juncea (211000)	347	14500	129	2390
Brassica juncea (426308)	329	6650	ND[2]	ND
Brassica juncea (531268)	241	19500	97	3460
Thlaspi rotundifolium	226	28700	79	6350
Brassica juncea (175607)	176	18200	ND	ND
Triticum aestivum (cv. Scout 66)	139	5330	120	1890
Ambrosia artemissifolia	95	4670	75	2050
Brassica juncea Czern	65	9580	45	3580
Thlaspi caerulescens	64	26200	58	5010
Brassica juncea (180269)	59	4840	ND	ND
Brassica juncea (184290)	32	5260	30	2310

[1]Adapted from Huang and Cunningham (15). [2]ND denotes no data because these cultivars did not survive in the Pb-contaminated soil.

that the created mutants can be used for Pb phytoextraction immediately. By using this approach, however, it is difficult to identify gene(s) involved because of the large genome and long life cycle. Therefore, we initiated a program to identify Pb accumulating mutants from *Arabidopsis* (Chen et al., in this book). In our preliminary studies with 74 *Arabidopsis* ecotypes, we found a significant intraspecific variation in Pb accumulation and tolerance. We also found that higher Pb-accumulating ecotypes could be found in both Pb-sensitive and Pb-tolerant ecotype groups. Although Pb hyperaccumulation does not necessary require Pb tolerance, it is desirable if Pb hyperaccumulating plants are also Pb tolerant because Pb-tolerant plants will easily establish the vegetation on Pb-contaminated sites. Based on this information, we decided to screen for Pb-accumulating mutants by first selecting putative Pb tolerant mutants. After screening one-half million EMS mutagenized *Arabidopsis* seedlings, we identified 250 putative Pb-tolerant mutants. Three Pb-accumulating mutants identified from these putative mutants accumulated 2-3 fold higher shoot Pb concentration than their wild type. These results suggest that molecular biology could play an important role in creating Pb-hyperaccumulating plants.

Lead Phytoextraction: from Greenhouse to the Field. In our greenhouse studies, we found that the rank of shoot-Pb concentration observed in solution culture was similar to that in the pot experiments (Table 2). Now, the question was whether the results from pot experiments were positively correlated to those of field trials. Table 3 shows data from 1995 field trial and the data from pot experiments in the greenhouse at the same season. For the plant species tested, the rank of shoot-

Table 3. Shoot Pb concentrations of plants grown on a Pb-contaminated soil in field experiment and greenhouse pot experiments[1]

Plant species	Shoot Pb concentration	
	Field experiment	Pot experiment
	-------------------- mg kg^{-1} ----------------------	
Corn	340-490	280-320
Sunflower	270-390	180-210
Ragweed	190-210	140-160
Goldenrod	130-190	130-150
Bermudagrass	80-100	90-100
Hemp dogbane	70-100	60-90

[1]The soil used in the pot experiments was collected from the same site where the field experiment was conducted. The soil was a sandy loam with total soil Pb of 2500 mg kg^{-1} .

Pb concentration in the pot experiments was the same as that in the field experiment. This study suggests that under similar conditions, the results obtained in greenhouse pot experiments can be reproduced in the field. Field experiments are expensive, therefore, it is important to get enough information from pot experiments before the initiation of field trials for Pb contaminated sites.

Exploring Potential Methods to Enhance Lead Phytoextraction

Phosphorus Nutrition and Pb Phytoextraction. For most Pb-contaminated soils, P bioavailability is very low due to Pb phosphate precipitation in the soil. Plants grown in Pb-contaminated soils with total soil Pb greater than 1000 mg kg^{-1} showed severe P-deficiency symptoms after 1 to 2 weeks from planting or transplanting, and further plant biomass increase was either very little or completely ceased if the P deficiency symptom was not corrected (unpublished results). To correct P deficiency problem, we tested foliar P application, and we found that spraying 10 mM P (KH$_2$PO$_4$, pH 6.0) solution on the shoots could correct P deficiency within a few days for a number of plant species. In a pot experiment, foliar P increased plant dry weight by more than 4 fold for both shoots and roots of goldenrod within a month after the foliar P application (Table 4). Foliar P decreased shoot-Pb concentration by 55% and root-Pb concentration by 20%, however, the total Pb accumulation was increased by 115% in shoots and 300% in the roots because of the huge biomass increase which resulted from the foliar P treatment (Table 4). We also investigated the effects of soil-applied P with a range of P levels (0, 50, 150, 300, 500 and 1000 mg P kg^{-1} soil as KH$_2$PO$_4$) on total Pb phytoextraction by corn plants. Shoot Pb concentration was significantly decreased at all P levels tested. However, the shoot biomass increased dramatically with increasing P levels up to 500 mg P kg^{-1} soil, and then it became constant with further increases in P levels. Total Pb phytoextraction is a function of a number of factors, and the two main parameters are shoot Pb

Table 4. Effects of foliar P on biomass production and Pb phytoextraction by goldenrod plants[1] grown on a Pb-contaminated soil (total soil Pb 3500 mg kg^{-1})

Parameters	Shoot		Root	
	Control	+Foliar P	Control	+Foliar P
Dry weight (g plant^{-1})	3.2	14.8	2.1	8.8
Pb concentration (mg kg^{-1})	340	155	2260	1830
Total Pb content (mg plant^{-1})	1.1	2.3	5.2	16.7

[1]Data are means of the treatment (n=3, unpublished results).

concentration and total biomass. The addition of 500 mg kg^{-1} P increased total shoot-Pb accumulation by more than 2 fold compared to the control (unpublished results). This result indicates that managing soil P levels are necessary to produce maximum Pb removal by plant shoots.

Effects of Electroosmosis on Pb Accumulation in Plants. Electroosmosis is the transport of charged particles in an electrical field. Reed et al. (*32*) evaluated the potential application of electroosmotic technologies for *in situ* remediation of Pb contaminated soils. These authors found that a significant amount of soil Pb could be transported through the soil, however, Pb was precipitated or reabsorbed onto the soil adjacent to the cathode (*32*). As a major limiting factor for Pb uptake by plant roots is the low Pb availability in soil solution, we have made an attempt to couple Pb phytoextraction to electroosmosis. The hypothesis was that electroosmosis could increase Pb mobility in the soil, thus increase Pb accumulation in plants. In this experiment, we applied direct electric currents (DC) across the soil by vertically inserting cathodes and anodes into the soil (unpublished results). Using corn and ragweed, we found that applied DC currents doubled shoot-Pb concentrations in both corn and ragweed, however, root-Pb concentration was not significantly affected by the treatment. The increased shoot-Pb concentration could be the result of increased Pb mobility in the soil and/or increased Pb translocation in the plants.

Role of Synthetic Chelates in Enhancing Pb Phytoextraction. The goal of Pb phytoextraction is to reduce Pb levels in the contaminated soils to acceptable levels within a reasonable time frame (3 to 20 years). To achieve this goal, we need to use plants which are able to accumulate greater than 1% Pb in shoots and produce more than 20 metric tons of shoot biomass ha^{-1} year^{-1}. From both pot experiments and field experiments, it is clear that shoot-Pb concentrations are far below the level targeted for Pb phytoextraction. From hydroponic studies, we know that shoot-Pb concentrations in plants increased dramatically as Pb levels in the nutrient solution increased (*15*). This leads us to speculate that a significant increase in Pb level in soil solution could yield a significantly higher shoot-Pb concentration. To test this hypothesis, we investigated the effects of adding synthetic chelates to Pb-contaminated soils on Pb desorption from soil to soil solution and Pb accumulation in plants. Within 24 h after the application of a chelate (HEDTA) to the soil, the Pb concentration in soil solution increased from 17 μM for the control (without added HEDTA) to19000 μM for the HEDTA (2 g kg^{-1}) treatment (*15*). The surge of Pb concentration in soil solution was closely related to the surge of Pb concentrations in plants (Figure 3). One week after transplanting the plants into the Pb-contaminated soil, the shoot Pb concentrations increased from 40 mg kg^{-1} for plants grown on the Pb-contaminated soil without HEDTA to 10600 mg kg^{-1} for plants grown on the HEDTA treated Pb-contaminated soil (Figure 3).

To search for an ideal chelate to enhance Pb phytoextraction from contaminated soils, we investigated the relative efficiency of five synthetic chelates (HEDTA, EDTA, EGTA, EDDHA, and DTPA) in enhancing both Pb desorption from soil to soil solution and Pb accumulation in plants. For the chelates tested, EDTA was the most efficient chelate in increasing soil-Pb desorption and shoot-Pb accumulations in plants (Huang et. al., *Environ. Sci. Technol.*, in press). For the soil tested, the order of the effectiveness in increasing Pb desorption from soil to soil solution was EDTA > HEDTA > DTPA > EGTA > EDDHA. The effects of these chelates on enhancing Pb accumulation in plants paralleled the effects of these chelates on the soil-Pb desorption (Figure 4). We also examined the correlation between Pb concentrations in plants and Pb levels in soil solution. Lead concentrations in plant shoots increased linearly with increasing Pb levels in the soil solution, accomplished by applying chelates to the contaminated soils. Our results demonstrate that a key to Pb phytoextraction is to increase soil Pb desorption, and that synthetic chelates can play an important role in this process.

In order to understand the mechanisms involved in chelate-triggered Pb hyperaccumulation in plants, we further investigated the effects of EDTA on Pb translocation from roots to shoots. Applying EDTA to Pb-contaminated soils significantly increased Pb transport into the xylem of plant shoots (Huang et al., *Environ. Sci. Technol.*, in press). For example, 24 h after applying EDTA to corn plants grown on the contaminated soil, Pb concentration in the shoot xylem sap increased more than 100 fold for the treatment of 1.0 g EDTA kg^{-1} soil (Table 5).

Table 5. Effects of adding EDTA to a Pb-contaminated soil[1] (total soil Pb 2500 mg kg^{-1}) on Pb transport in 21-d-old corn plants grown in the contaminated soil

EDTA added to soil	Pb concentration in xylem sap	Pb translocation to shoots
g kg^{-1}	mg L^{-1}	μg plant^{-1} day^{-1}
0.0	0.2	0.8
0.5	6.9	28.0
1.0	21.3	99.8
ANOVA P>F	0.001	0.001

[1]Adapted from Huang et al. (*Environ. Sci. Technol.*, in press). EDTA treatment was initiated by adding the appropriate EDTA solution to the soil surface for each pots. Shoots were cut 1 cm above the root-shoot junction 24 h after applying the EDTA. Immediately following the harvest, xylem sap was collected for 8 h for each treatment.

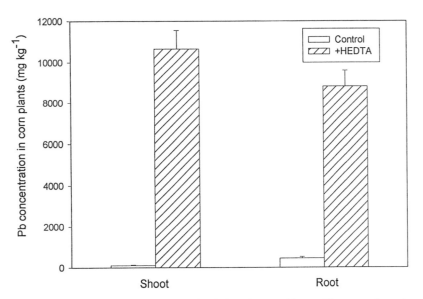

Figure 3. Effects of adding a synthetic chelate (HEDTA) to a Pb-contaminated soil (total soil Pb 2500 mg kg^{-1}) on Pb accumulation in corn plants. Corn seedlings (10-d-old) were transplanted to the Pb contaminated soil treated with HEDTA (0 or 2 g HEDTA kg^{-1} soil), and the plants were harvested 7 days after transplanting to the contaminated soil. Error bars represent ± SE (n=3). (Redrawn from Huang and Cunningham, 1996).

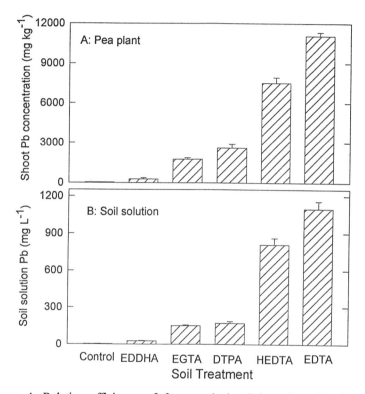

Figure 4. Relative efficiency of four synthetic chelates in enhancing Pb accumulation in shoots of pea (A) grown on the chelate-treated Pb-contaminated soil, and Pb desorption from soil to soil solution (B). Control denotes the Pb contaminated soil (total soil Pb 2500 mg kg^{-1}) without added chelates. The level of each chelate used was 0.5 g kg^{-1} soil. Error bars represent ± SE (n=3). (Adapted from Huang et al., *Environ. Sci. Technol.*, in press).

The increase of Pb concentration in xylem sap is positively correlated with the increase in Pb translocation from roots to shoots. Compared to the control, added EDTA (1.0 g kg^{-1}) increased Pb translocation from roots to shoots by 120 fold within 24 h (Table 5). Our results suggest that chelates enhanced Pb desorption from soil to soil solution, facilitated Pb transport into the xylem, and increased Pb translocation from roots to shoots.

Synthetic chelates can play a significant role in enchancing soil-Pb desorption, and Pb accumulation in plants. It is important to point out that the addition of chelates to a contaminated site has to be done in a carefully controlled manner so that there is no risk of contaminant movement from soil to groundwater. We are currently investigating the best management practices to accomplish this. An important factor which could affect Pb movement in the soil profile is soil water, from rainfall, irrigation, or groundwater. Certain irrigation techniques are under active consideration to control quantitatively water flux in plant root zones. With careful management, chelate-assisted Pb phytoextraction may provide a cost-effective means for the clean-up of Pb-contaminated soils.

Future Research Needs

Significant progress has been made in the development of Pb phytoextraction technology. The success of Pb phytoextraction requires skills and techniques from soil science, agronomy, engineering, and plant biology. Additional research and development is needed in a number of areas. We need a better understanding of the physiology of Pb uptake, translocation, and accumulation in plants. This includes identifying the barrier of Pb translocation from roots to shoots, as well as identifying the form and the species of Pb translocated from cell to cell. We also need to understand the cellular mechanism of Pb transport in plants. For example, how is Pb transported across the root-cell plasma membrane and then distributed between different cell compartments? Further understanding of the mechanisms of chelate-enchanced Pb accumulation in plants is needed. Is the chelate-Pb complex transported into plant cell and translocated from roots to shoots? It is also needed to search for Pb hyperaccumulators that produce more biomass than known Pb hyperaccumulators. Eventually, we want plants to do the entire job, and to limit the use of chelates or other soil amendments. Finally, we should take advantage of technologies in plant molecular biology. Screening Pb-accumulating mutants from Pb-tolerant and high-biomass plant species may yield ideal plants for Pb phytoextraction. To understand the molecular and genetic mechanisms of Pb accumulation, we need to screen Pb-accumulating mutants from *Arabidopsis*. Results from our *Arabidopsis* work (Chen et al., in this book) indicate that we may be able to identify Pb-hyperaccumulating mutants from mutagenized *Arabidopsis* populations. If this objective can be accomplished, we then have a greater chance to create transgenic Pb hyperaccumulators which are practically suited for phytoextraction of Pb from contaminated soils.

Literature Cited

(1) Cunningham, S. D.; Berti, W. R. *In Vitro Cell. Dev. Biol.* **1993**, *29P*, 207-212.
(2) Raskin, I.; Kumar, N.; Dushenkov, S.; Salt, D. E. *Current Opinion Biotech.* **1994**, *5*, 285-290.
(3) Jackson, D. R.; Watson, A. P. *J. Environ. Qual.* **1977**, *6*, 331-338.
(4) Levine, M. B.; Stall, A. T.; Barrett, G. W.; Taylor, D. H. **1989**, *J. Environ. Qual. 18*, 411-418.
(5) Chaney, R. L.; Ryan, J. A. *Risk based standards for arsenic, lead and cadmium in urban soils*; Frankfurt: Dechema, 1994.
(6) Forbes, R. M.; Sanderson, G. C. In *The biogeochemistry of lead in the environment*; Nriagu, J. O. Ed.; Elsevier/North-Hollanbd Inc., NY, 1978; pp. 225-278.
(7) Cheremisinoff, N. P. *Lead: a guidebook to hazard detection, remediation, and control;* PTR Prentice-Hall, Inc., Englewood Cliffs, NJ, 1993.
(8) Cunningham, S. D.; Berti, W. R.; Huang, J. W. In *Bioremediation of Inorganics*; Hinchee, R. E., Means, J.L., Burris, D.R., Eds.; Battelle Press, Columbus, OH, 1995, pp.33-54.
(9) Cunningham, S. D.; Berti, W. R.; Huang, J. W. *Trends Biotech.* **1995**, *13*, 393-397.
(10) Salt, D. E.; Blaylock, M.; Kumar, N.; Dushenkov, V.; Ensley, B.D.; Chet, I.; Raskin, I. *Biotechnol.* **1995**, *13*, 468-474.
(11) Chaney, R. L. In *Land Treatment of Hazardous Wastes*; Parr, J. F., Marsh, P. B., Kla, J. M., Eds.; Noyes Data Corp., Park Ridge, NJ, 1983. pp.50-76.
(12) Baker, A. J. M.; McGrath, S. P.; Sidoli, C. M. D.; Reeves, R. D. *Res. Conserv. Recycl.* **1994**, *11*, 41-49.
(13) Reeves, R. D.; Brooks, R. R. *Environ. Pollution Ser. A*, **1983**, *31*, 277-285.
(14) Baker, A. J. M; Brooks, R. R. *Biorecovery*, **1989**, *1*, 81-126.
(15) Huang, J. W.; Cunningham, S. D. *New Phytol.* **1996**, *134*, 75-84.
(16) Brooks, R. R; Lee, J.; Reeves, E. D.; Jaffre, T. *J. Geochem. Expor.* **1977**, *7*, 49-58.
(17) Kumar, N.; Dushenkov, V.; Motto, H.; Raskin, I. *Environ. Sci. Technol.* **1995**, *29*, 1232-1238.
(18) Parker, D. R.; Norvell, W. A.; Chaney, R. L. In *Chemical Equilibrium and Reaction Models*; Loeppert, R.H.; Schwab, A.P.; Goldberg, S. Eds.; Soil Sci. Soc. Am., Madison, WI, 1995, pp.163-200.
(19) Qureshi, J.A.; Hardwick, K.; Collin, H.A.. *J. Plant Physiol.* **1986**, *122*, 357-364.
(20) Huang, J. W.; Grunes, D. L.; Kochian, L.V. *Proc. Natl. Acad. Sci. USA*, **1994**, *91*, 3473-3477.
(21) Marshall, J.; Corzo, A.; Leigh, R. A.; Sanders, D. *Plant J.* **1994**, *5*, 683-694.
(22) Tomsig, J. L.; Suszkiw, J. B. *Bioch. Biophy. Acta*. **1991**, *1069*, 197-200.
(23) Audesirk, G. *Neurotoxicol.* **1993**, *14*, 137-148.
(24) Jones, L. H. P.; Clement, C. R.; Hopper, M. J. *Plant Soil*, **1973**, *38*, 403-414.
(25) Jaffre, T.; Brooks, R. R.; Lee, J.; Reeves, R. D. *Science*, **1976**, *193*, 579-580.

(26) Baker, A. J. M.; Reeves, R. D.; Hajar, A. S. M. *New Phytol.* **1994,** *127,* 61-68
(27) Brown, S. L.; Chaney, R. L.; Angle, J. S.; Baker, A. J. M. *J. Environ. Qual.* **1994,** *23,* 1151-1157.
(28) Macnair, M. R. In *Heavy Metal Tolerance in Plants: Evolutionary Aspects;* Shaw, A.J. Ed.; CRC Press Inc., Florida, 1989, pp.235-253.
(29) Chrispeels, M. J.; Sadava, D. E. *Plants, Genes, and Agriculture.* Jones and Bartlett Publishers, Boston, MA. 1994.
(30) Kneen, B. E.; LaRue, T. A.; Welch, R. M.; Weeden, N. R. *Plant Physiol.* **1989,** *93,* 717-722.
(31) Welch R. M.; LaRue, T. A. *Plant Physiol.* **1990,** *93,* 723-729.
(32) Reed, B. E.; Berg, M. T.; Thompson, J. C.; Hatfield, J. H. *J. Environ. Eng.* **1995,** *121,* 805-815.
(33) Cole, M. M., Provan, D. M. J., Tooms, J. S. *Trans. Inst. Mining Metal. Section B.* **1968,** *77,* 81-104.
(34) Baker, A. J. M. *J. Plant Nutri.* **1981,** *3,* 643-653.

Chapter 22

Phytoremediation and Reclamation of Soils Contaminated with Radionuclides

James A. Entry[1], Lidia S. Watrud[2], Robin S. Manasse[3], and Nan C. Vance[4]

[1]Department of Agronomy and Soils, College of Agriculture, 202 Funchess Hall, Auburn University, Auburn, AL 36849–5412
[2]Terrestrial Plant Ecology Branch and [3]National Research Council, National Health and Environmental Effects Laboratory, U.S. Environmental Protection Agency, 200 Southwest 35th Street, Corvallis, OR 97333
[4]Pacific Northwest Research Laboratory, Forest Service, U.S. Department of Agriculture, 3200 Jefferson Way, Corvallis, OR 97331

As a result of nuclear testing and nuclear reactor accidents, large areas of land have become contaminated with low concentrations of radionuclides. Removal, transport and treatment of large volumes of soil may be logistically difficult and prohibitively costly. Using plants to remove low concentrations of radionuclides from soil *in situ* is expected to be less expensive than mechanical, physical or chemical methods, particularly for treatment of large areas. Phytoremediation is applicable to a wide range of terrestrial environments and plants can be selected for given soil and climatic conditions. Phytoremediation of contaminated sites should also leave treated sites amenable to subsequent reclamation efforts. The points to consider for initial phytoremediation and subsequent reclamation of contaminated soils include:enhancement of plant accumulation of radionuclides by addition of mycorrhizal or bacterial inocula, chelating agents or organic amendments, periodic harvests to recover or dispose of radionuclides in the ashed plant materials minimization of potential environmental effects on non-target organisms and replacement or augmentation of the initial remediating species with a complex plant community.

Large areas of land have been contaminated by fission by-products resulting from nuclear bombs, (1) above ground nuclear testing (2-4), nuclear reactor operations (5, 6) and nuclear accidents (7). Unlike organic pollutants, radionuclides typically are elements which cannot be degraded. Radionuclides can be distributed to soil and plants in the contaminated area by physically and biologically mediated nutrient cycling processes (8 -11). Radionuclides, especially ^{137}Cs and ^{90}Sr, can accumulate as they move up the food chain (12). Significant concentrations of radionuclides have been found in crops (4,13,14), livestock (14), fish (5, 15) and wildlife (16, 17). Human exposure to harmful radionuclides can occur from inhalation following atmospheric releases, or by ingestion of food contaminated by atmospheric fallout or by accumulation through the food chain (18). Exposure may result in detrimental health effects, such as cancers and genetic mutations (9, 19, 20). This chapter provides an overview of the current progress in phytoremediation of radionuclides from soils and will propose a strategy for initial remediation, (ie. reduction of contaminant levels and subsequent reclamation or revegetation of impacted areas).

Benefits of Phytoremediation

Remediation of soil contaminated with low concentrations of radionuclides using present technology requires that soil be removed from the contaminated site and treated with various dispersing and chelating chemicals. Transport of soil requires heavy equipment, is time consuming and expensive; it may also result in additional dispersal of pollutants through possible spill and/or leaks. Therefore, few attempts have been made to remediate large areas of land contaminated with low concentrations of radionuclides. Furthermore, the cost to dispose of large liquid volumes of chemicals used to decontaminate soils polluted with radionuclides may be prohibitive; estimates of $200 to $300 billion for radionuclide cleanup in the U. S. alone, are considered conservative (21).

In addition to the logistical and cost limitations for treating soils with present technology, physical and chemical alteration of soils may inhibit reclamation of the site. The heavy equipment needed to remove or transport soil compacts the remaining soil, adversely affecting the porosity, bulk density and water holding capacity of soil. Changes in aeration and water availability can in turn negatively impact plant growth and nutrition. If the soil is returned to the site, use of dispersing compounds such as detergents and surfactants and chelating agents during soil washing procedures to extract pollutants from soil can promote the loss of soil nutrients and cofactors needed by plants, microbes and other soil biota. In large quantities, chelating compounds may also adversely affect soil physical characteristics such as cation exchange capacity. Surfactants and detergents may adversely affect the viability of prokaryotic and eukaryotic soil biota by causing membrane damage. The establishment of plants on physically, chemically and microbiologically compromised soils can therefore become problematic.

In contrast, phytoremediation-based approaches, particularly those designed with planned, successive in situ harvests and simultaneous or sequential perimeter plantings of other species, may not only remediate a site, but may eventually reclaim it, by fostering the establishment of a plant community. The ensuing sections will (a) review plant species that have been evaluated for in situ phytoremediation of radionuclides in contaminated soils, (b) propose criteria for selecting and developing plant species to be used for phytoremediation, (c) suggest points to consider for minimizing non-target effects of plants and (d) highlight basic and applied research needed to help ensure environmental and human safety, as well as efficacy of phytoremediation.

Survey of Plants that Accumulate ^{137}Cs and [90]Sr

Although the ability to accumulate radionuclides varies among a wide array of plant species occupying different habitats, many plants growing on contaminated soils have been shown to accumulate large amounts of radionuclides. Numerous reports have described plant accumulation of radionuclides, especially ^{137}Cs and [90]Sr (14, 22-25). Laboratory experiments indicate that certain plants may be able to remove radionuclides, especially ^{137}Cs and ^{90}Sr from soil over a time period of 5 to 20 years. Nifontova *et al.* (6) found that plants accumulated between 530 and 1500 Bq of ^{137}Cs and between 300 and 1100 Bq of ^{90}Sr over a 10 year period in 12 forest and 5 meadow plant communities in the vicinity of the Beloyarsk atomic power station in the Urals pine mountain region of Russia. Wallace and Romney (22) found that a large number of plant species in the desert area near the Nevada Test Site, USA, accumulated substantial quantities of radionuclides from soils contaminated by above-ground nuclear testing.

Trees can also accumulate substantial quantities of radionuclides. Pinder *et al.* (23) reported that *Acer rubrum, Liquidambar stryaciflua* and *Liriodendron tulipifera* accumulated significant quantities of [244]Cm, ^{137}Cs, [238]Pu, [226]Ra and [90]Sr. Robison and Stone (4) found that *Cocos nucifera* accumulated substantial amounts of ^{137}Cs from soils contaminated by nuclear weapons testing on Bikini Atoll. They also reported that additions of K and P to the soil decreased the amount of ^{137}Cs taken up by the trees. Entry *et al.* (26) found that *Pinus radiata* and *Pinus ponderosa* seedlings accumulated substantial quantities of ^{137}Cs and [90]Sr. Entry and Emmingham (27) found that potted *Eucalyptus tereticornis* seedlings removed 31.0 % of the ^{137}Cs and 11.3 % of the ^{90}Sr in sphagnum peat soil after one month of exposure.

Accumulation of ^{137}Cs and ^{90}Sr in grasses and other herbaceous plants has also been widely documented. Dahlman *et al.* (28) reported that *Festuca arundinacea* accumulated 42,143 kBq of [137]Cs m[-2] in 8 months, in an area where the total amount of ^{137}Cs above-ground runoff and sediment was less than 444 kBq of [137]Cs. Salt *et al.* (14) reported that *Lolium perenne, Festuca rubra, Trifolium repens* and *Cerastium fontanum* accumulated from 28 to 1040 Bq [137]Cs g[-1] of plant tissue in a re-seeded pasture in Scotland. Coughtery *et al.* (24) found that a *Festuca / Agrostis* plant community in the United Kingdom accumulated 4-19% of the ^{137}Cs deposited by Chernobyl fallout. Accumulation of ^{137}Cs was higher in *Carex* spp than in 9 species of grasses in an upland area in Great Britain (24).

Radionuclides such as ^{90}Sr and ^{137}Cs often accumulate as they move up the food chain. Radionuclides are accumulated by zooplankton, aquatic plants, fungi and invertebrates such as earthworms. Penntilla *et al.*(15) reported that ^{137}Cs and[90] Sr bioaccumulation in aquatic and terrestrial animals that consume plants eventually leads to incorporation into many foods consumed by humans. Haselwandter and Berreck (29) have recently reviewed accumulation of radionuclides by arbuscular mycorrhizal and ectomycorrhizal fungi and their bioaccumulation in the food chain.

Criteria and Approach for Selecting Candidate Phytoremediation Species

Candidate plant species to remove radionuclides from contaminated soils can be selected and evaluated for efficacy and for environmental safety by using a multi-staged screening approach (30). The first step is to identify plant species or cultivars from the literature or from contaminated sites for their potential abilities to take up given radionuclides. Subsequent points to consider for revegetation of damaged terrestrial ecosystems include assessment of potential ecological effects, agronomic requirements and reproductive characteristics (31, 32). Specific examples of points to consider in selecting or developing plant species for remediation, reclamation or restoration purposes are summarized in Table I. For example, major ecological considerations should address (a) the risks and benefits of using native or exotic species (b) ability of the introduced plants to hybridize with other species, especially weeds, (c) whether the plants are insect or wind pollinated and (d) whether the species are annuals or perennials and (e) effects of concentrated radaionuclides on potential herbivores and pollinators. Agronomic considerations should take into account biomass production, water, nutrients and cultural requirements; ability to form beneficial associations with mycorrhizal fungi, nitrogen fixing or other plant growth promoting soil or rhizobacteria; tolerance to disease, insects and temperature; and to salt or pH extremes. Reproductive considerations should include seed and pollen production. Morphology of the root system and its ability to penetrate and spread in different soil types could be important both in maximizing uptake of radionuclides or other pollutants, and in stabilizing the soil to minimize aerial dissemination of contaminated soils. Some of these properties can be tested in the greenhouse using soils representative of contaminated sites. The effects of organic amendments and mycorrhizal or other microbial inoculants on efficacy can be evaluated initially in greenhouse tests and later in field tests. Similarly, effects on non-target organisms can be tested initially in greenhouse systems and later under field conditions.

Minimizing Non-Target Ecological Effects of Phytoremediation Practices

Ecological risk from weedy or invasive species can be minimized with precautions. For example, the potential for hybridization with weeds and for pollen and seed dissemination can be reduced by harvesting before flowering or seed set. In the future, using infertile hybrids or sterile males of selected species may be a viable option. Avoidance of insect pollinated plants can reduce exposure of accumulated toxics to insect pollinators such as bees, butterflies or other insects. Selecting plants that are less palatable to grazing by vertebrate or invertebrate herbivores, and wind-pollinated rather than insect-pollinated plants, reduces the potential for food-chain accumulation. Periodic harvests of the above-ground portion of plants may maximize continued uptake, reduce potential phytotoxicity associated with bioaccumulation and reduce exposure to potential pollinators, and to vertebrate and invertebrate herbivores. By reducing the potential for phytotoxicity to the accumulating species, the useful life of a given planting may be increased, thereby reducing the need for re-seeding or replanting.

Strategy for *In Situ* Phytoremediation of Radionuclide Contaminated Soils

Soil Amendments The most desirable soil conditions are those that enhance plant uptake of radionuclides without increasing radionuclide mobility in the soil. Achieving this condition in soils *in situ* may be approached in two ways: 1) by improving the ability of the plant to take up radionuclides, (*eg.*, via mycorrhizal associations) (29, 33) or by inoculation with plant-growth promoting rhizobacteria (34, 35), and /or 2) by altering the chemical form of the radionuclide in the soil to increase its availability to plants.

Table I. Points to Consider in Selecting Plants for Phytoremediation Purposes

A. Ecological Aspects	• Native or exotic species • Invasiveness • Ability to hybridize with weed species • Wind or insect dissemination of pollen • Potential impacts on herbivores • Longevity
B. Agronomic Considerations	• Availability of seeds or transplants • Biomass production • Fertilizer requirements • Water requirements • Availability of harvesting methods • Suitability for multiple harvests • Disease and insect tolerance • Potential for symbiotic associations • Types of suitable soils and climates • Tolerance to environmental stresses
C. Reproductive Characteristics	• Seed production and viability • Pollen production
D. Efficacy and Economics	• Effective limits of remediation • Time required for adequate remediation • Harvest, recovery and disposal costs • Potential for recovery of pollutants • Costs for re-seeding or re-planting

Inoculation of plants or soil with specific mycorrhizal fungi or other root associated microflora may additionally maximize plant uptake and accumulation of radionuclides (30). The availability of radionuclides to plants can be enhanced, while decreasing the mobility of these radionuclides in soil, by *in situ* addition of organic amendments, and carefully managed concentrations of chelating agents and fertilizer (30). Chelators, such as diethylenetriamine-pentaacetic acid (DTPA), alter the radionuclide form so it is more biologically available to the plant but not increasing its mobility to the point where it is easily leach from the soil. Organic matter will complex with ^{137}Cs and ^{90}Sr to remove them from absorption sites on mineral solids and reduce soil pH and base saturation, thereby increasing radionuclide availability to plants. Many naturally occurring soil organic compounds as well as synthetic chelators, are bound by soil clays, oxides and mineral surfaces, preventing their downward movement in the soil. Mechanisms proposed for the binding include oxygen bonds, cation bridges and chelated metal bridges (36). If oxygen bonds or other cations are providing bonding to clay sites, then chelated radionuclides should be expected to be accessible for uptake into plants. Harvested plant materials would then be subjected to high temperature combustion or smelting to oxidize and concentrate radionuclides in ash for disposal or recovery.

Strategy for Reclamation of the Phytoremediated Area   In the long term, plants native to an area would be the most desirable from an ecological viewpoint. In non-agricultural lands, if appropriate, the area surrounding the targeted remediation site can be planted with a mixture of seeds from native species. Perimeter plants serve several purposes. They (a) reduce erosion and dissemination of contaminated soil and (b) produce a source of propagules (roots, rhizomes, stolons, seeds, etc.), that can grow in the phytoremediated area and as suggested by Chambers and Mac Mahon (37), facilitate the growth and spread of indigenous or introduced mycorrhizal inocula. With appropriate species selection, one might anticipate that over a period of several years, the species in the phytoremediation area would gradually become replaced by native species. Planting of the perimeter areas could be achieved by mulching with plant canopies from adjacent areas (38). Conventional low cost methods such as tilling and broadcasting or drilling seed, or higher cost transplant methods (of seedlings or of plugs of soil with plants from adjacent plant communities) can also be considered, depending on the size of the area to be remediated.

Research Needs

Numerous research needs exist, ranging from maximization of efficacy in the field to ecological and health effects risk assessment studies. Genetic manipulation of candidate plant species and of associated rhizosphere microflora may be necessary to maximize efficacy while minimizing ecological risks. For example, molecular techniques are available (39), which permit the isolation and introduction of genes to enhance uptake, sequestration, bioaccumulation or biotransformation of given inorganic or organic pollutants (40-43). Plant breeding techniques may be developed that will localize the times or places of radionuclide accumulation in plants to specific plant parts so that environmental exposure to non-target species can be minimized. Cultural and harvest practices aimed at reducing pollen or seed spread, and selection or development of lines with reduced fertility, seed production or seed bank viability may also be useful in minimizing dissemination, persistence and invasion of introduced species. Data are also needed on non-target effects to plant symbionts such as mycorrhizal fungi and nitrogen fixing bacteria and on soil foodweb components such as bacteria, fungi, nematodes and protozoans. Research is also needed on toxicity to invertebrate and vertebrate herbivores of above ground and below ground plant parts. Depending on the specific radiological characteristics of given radionuclides, effects on the mutagenicity of the concentrated radionuclides to target and non-target plants, microbes, invertebrates and vertebrates may also need to be addressed.

Literature Cited

1. Mahara,Y. *J. Environ. Qual.* **1993,** 22, 722-730.
2. Paasikallo, A. *Ann. Agric. Fenn.* **1984,** 23, 109-120.
3. Eisenbud, M. *Experimental Radioactivity.* **1987,** Academic Press, Orlando, FL, pp 475.
4. Robison, W.L.; Stone, E.L. *Health Physics* **1992,** 62, 496-511.
5. Whicker, F.W.; Pinder, J.E.; Bowling, J.W.; Alberts, J.J.; Brisbin, Jr., L. *Ecol. Monogr.* **1990,** 60, 471- 496.
6. Nifontova, M.G.; Kulikov, G.I.; Tarshis , G.I.; Yachenko, D. *Ekologiya* **1989,** 3, 40-45.
7. Clark, M.J.; Smith, F.B. *Nature* **1988,** 332, 245-249.
8. Abbott, M.L.; Rood, A.S. *Health Physics* **1994,** 66,17-29.
9. Breshears, D.D.; Kirchner,T.B; Whicker, F.W. *Ecol. Appl.* **1992,** 2, 285-297.
10. Howard, B.J.; Bresford, N.A.; Hove, K. *Health Physics* **1991,** 61, 715-722.
11. Berg, M. T.; Shuman, L. *Ecological Modeling* **1995,** 83, 387-404.
12. Hoffman, F.O.; Bergstrom, U.; Gyllander, C. A.; Wilikins, A.B. *Nuclear Safety* **1984,** 25, 533-546.
13. Sanzharova, D.I.; Aleksakhin, R.M. *Pochvovedeniye* **1982,** 9, 59-64.
14. Salt, C.A.; Mayes, D.; Elston, A. *J. Appl. Ecol.* **1992,** 29, 378-387.
15. Pennttila, S.; Kairesalo, T.; Uusi-Rauva, A. *Environ. Pollut.* **1993,** 82, 47-55.
16. Lowe, V.P.W.; Horrill, A.D. *Environ. Pollut.* **1991,** 70, 93-107.
17. Rickard, W.H.; Ebrhard, L.E. *Northwest Sci.* **1993,** 67, 25-31.
18. Church, B.W.; Wheeler, D.L.; Campbell, C.M.; Nutley, R.V.; Ansphaugh, L.R. *Health Physics* **1990,** 59, 503-510.
19. Ansphaugh, L.R.; Catlin, R.J.; Goldman, M. *Science* **1988,** 242, 1513-1519.
20. Lange, R.; Dickerson, M.H.; Gudiksen, P.H. *Nuclear Technology* **1988,** 82, 311-223.
21. Watson, R.; Glick, D.; Horsenball, M.; McCormick, J.; Begley, S.; Miller, S.; Carroll, G.; Keene-Osborn, S. *Newsweek,* **December 27, 1993,** pp 14-18.
22. Wallace, A.; Romney, E.M. *Radioecology and Ecophysiology of Desert Plants at the Nevada Test Site.* **1972,** Environmental Radiation Division, Laboratory of Nuclear Medicine University of California, Riverside. pp 432.
23. Pinder, J.E. III; McLeod, K.W.; Alberts, J.J.; Adriano, D.C.; Corey, J.C. *Health Physics* **1984,** 47, 375-384.
24. Coughtry, P.J.; Kirton, J.A.; Mitchell, R.B. *Environ. Pollut.* **1989,** 62, 281-315.
25. Murphy, C.E.; Johnson, T.L. *J. Environ. Qual.* **1993,** 22, 793-799.
26. Entry, J.A.; Rygiewicz, P.T.; Emmingham, W.H. *J. Environ. Qual.* **1993,** 22, 742-746.

27. Entry, J.A.; Emmingham,W.H. *Can. J. For. Res.* **1994**, 25, 1044-1047.
28. Dahlman, R.C., Auerbach, S.I.; Dunaway, P. B. *Environmental Contamination by Radioactive Materials*, **1969**, International Atomic Energy Agency and World Health Organization, Vienna Austria. pp 153-165.
29. Haselwandter, K.; Berreck, M. *Metal Ions in Fungi,* In Winkelmann, G; Winge,D.R. Eds., Marcel Dekker, Inc. New York, **1988**, p 259-277.
30. Entry, J.A.; Vance, N.C.; Hamilton, M.A.; Zabowski, D.; Watrud, L.S.; Adriano, D.C. *Water, Air and Soil Pollution* **1996**, 88, 167-176.
31. Brown, D.; Hallman, R.G.; Lee, C.R.; Skogerboe, J.G.; Eskew, K.; Price, R.A.; Page, N. R.; Clar, M.; Kort, R.; Hopkins, H. *Reclamation and Vegetative Restoration of Problem Soils and Disturbed Lands.* **1986**, Noyes Data Corp., Park Ridge, N. J. pp 560.
32. Manasse, R.S.; Watrud, L.S. *Principles and Methods of Terrestrial Revegetation. United States Environmental Protection Agency 600/R-996/073.* **1996**, Office of Research and Development, Washington, D. C. pp 66.
33. Entry, J.A; Rygiewicz, P.T. *Environ. Pollut.* **1994**, 86, 201-206.
34. Beauchamp, C.J.; Dion, P.; Kloepper, J W.; Antoun, H. *Plant and Soil* **1991**, 132, 273- 279.
35. Kloepper, J.W;, Zablotowicz, R.M.; Tipping, E.M.; Lifshitz, R. In *The Rhizosphere and Plant Growth.* Keister D.L.; Cregan, P.B. Eds. **1994**, Plant and Soil Press, Wageningen, The Netherlands. pp 315-326.
36. Wallace, G.A.; Wallace, A. *J. of Plant Nutrit.* **1983**, 6, 439-446.
37. Chambers, J.C.; MacMahon, J.A. *Annu. Rev. Ecol. Syst.* **1994**, 25, 263-292.
38. Bell, D.T.; Plumer, J.A.; Taylor, S.K. *Bot. Rev.* **1993**, 59, 24-73.
39. Watrud, L. S.; Metz, S. G.; Fischhoff, D.A. In *Engineered Organisms in Environmental Settings.* Levin, M.A.; Israeli, E. Eds. **1996**, CRC Press, Boca Raton, FL pp 165-189.
40. Misra, S.; Gedamu, L. *Theor. Appl. Genet.* **1989**, 78, 161-168.
41. Ortiz, D.F.; Kreppel, L.; Speiser, D.M.; Scheel, G.; McDonald, G.; Ow. D.W. *EMBO J.* **1992**, 11, 3491-3499.
42. Cunningham, S.C. ; Ow, D. W. *Plant Physiol.* **1996**, 110, 715-719.
43. Rugh, C.L.; Wilde, H.D.; Stack, N.M.; Thompson, D.M.; Summers, A.O.; Meagher, R.B. *Proc. Natl. Acad. Sci. USA* **1996**, 93, 3182-3187.

INDEXES

Author Index

Affiliation Index

Subject Index

A